W9-CZO-080

IN DARWIN'S SHADOW

IN DARWIN'S SHADOW

THE LIFE AND SCIENCE OF ALFRED RUSSEL WALLACE

A Biographical Study on the Psychology of History

Michael Shermer

OXFORD
UNIVERSITY PRESS
2002

OXFORD
UNIVERSITY PRESS

Oxford New York
Auckland Bangkok Buenos Aires Cape Town Chennai
Dar es Salaam Delhi Hong Kong Istanbul Karachi Kolkata
Kuala Lumpur Madrid Melbourne Mexico City Mumbai Nairobi
São Paulo Shanghai Singapore Taipei Tokyo Toronto
and an associated company in Berlin

Author contact:
Skeptic Magazine
P.O. Box 338
Altadena, California 91001
626/794-3119
skepticmag@aol.com
www.skeptic.com

Published by Oxford University Press, Inc.
198 Madison Avenue, New York, New York 10016

www.oup.com

Library of Congress Cataloging-in-Publication Data
Shermer, Michael.
In Darwin's shadow : the life and science of
Alfred Russel Wallace / Michael Shermer.
p. cm. Includes bibliographical references.
ISBN 0-19-514830-4
1. Wallace, Alfred Russel, 1823–1913.
2. Naturalist—England—Biography.
3. Natural selection. I. Title.
QH31.W2 S44 2002 508'.092—dc21 [B] 2001055721

All illustrations, unless otherwise noted, were prepared by Pat Linse.

1 3 5 7 9 8 6 4 2

Printed in the United States of America
on acid-free paper

To

Kimberly

With unencumbered love and unmitigated gratitude for fifteen years (plus one honeymoon) spent in fifteen archives reading through 1,500 letters, 750 articles, and 22 books; her deep insight into the complex psychology of the human condition provided both historical and personal enlightenment. She has made me whole.

CONTENTS

ILLUSTRATIONS

PREFACE

GENESIS AND REVELATION

The assignment sounded forbidding. The final project for the colloquial seminar in philosophy appeared in my hands three weeks before the end of the spring semester, 1973. Though it was straightforward enough, selecting the dozen most influential individuals in history and defending the choices was a task this college sophomore found at once both onerous and intriguing. The professor did not *really* expect college students to come up with twelve names and defend them, did he? He did.

This pedagogical tool was to cajole us into thinking about who really mattered in history. Only one name appeared on every student's list—Charles Darwin. The name was on the professor's list as well, who published a book based on the seminar entitled *Upon the Shoulders of Giants,* which opened with these words: "The builders of the world may be divided into two classes, those who construct with stone and mortar and those who build with ideas. This book is concerned with the latter, a small group of giants . . . upon whose shoulders we stand, and it is their concepts that have produced the major intellectual revolutions of history."[1] In the chapter on Darwin, there was an ever-so-brief mention of the "co-discoverer of evolution," the man who "forced Darwin's hand," and the naturalist whom "Darwin offered to help with publication"—Alfred Russel Wallace. It was to be the first exposure to the individual who would later occupy my full-time attention.

On one level this book began with that seminar, if in hindsight we look back to the origins of an event in the contingencies of the past that constructed later necessities—the conjuncture of past events that compelled a certain course of action. As one of the themes of this book deals with the interplay of contingency and necessity in the development of Wallace's thought within his culture, the same analysis might be made in constructing the past of the historical work itself. Beginnings, of course, do have a subjective element to them when reconstructed by later observers (since history is contiguous), but certain events and people stand out above most others, and Richard Hardison and his seminar must be considered as the genesis of this work. He instilled a sense of intellectual curiosity that would later be manifested in a driving pursuit to better understand who lurked within the shadow of Darwin. Other contingencies abound. I took a course in evolution in my first stint of graduate training from Professor Bayard Brattstrom, whose passion for overarching theory coupled to attention to detail taught me a healthy balance between the general and the specific that is reflected (I hope) in my analysis of Wallace. I was hit with the importance of evolutionary theory when, following Brattstrom's course, I came across the geneticist Theodosius Dobzhansky's observation that "Nothing in biology makes sense except in the light of evolution." Professor Meg White helped me understand the nuances and intricacies of animal behavior and its evolution. At Glendale College, biologists Tom Rike, Greg Forbes, and Ron Harlan trusted that my knowledge would catch up to my passion for the field, and for nearly a decade I taught the course in evolutionary theory, which then mutated into a course in the history of evolutionary thought at Occidental College. Professor Earl Livingood, the finest storyteller I have ever had the pleasure of hearing, made history come alive and helped me understand that psychological insights also pertain to those who lived before. To Jay Stuart Snelson I owe my gratitude for demonstrating the importance of semantic precision in the construction of a scientific analysis. And to Richard Milner I am indebted for his contribution of a number of important photographs and illustrations that appear within this biography, as well as for the primary documents on the Slade spiritualism trial in which Wallace was involved, and, finally, for so many interesting and important insights into Wallace, Darwin, and their contemporaries.

As to this work specifically, I owe allegiance to my mentors at Claremont Graduate School: James Rogers, who helped me get my mind around the ever-expanding Darwin industry; Richard Olson, who introduced me to and then shaped my thinking about the interface of science and culture; Michael Roth, who showed me the proper balance between theory and practice; Harry Liebersohn, who convinced me there is history outside the history of science; and Mario DiGregorio, whose historical vision is sharper than most. All of them made important contributions to this work, both structurally and semantically, such that whatever usefulness it may have is owed a good deal to their patience in carefully reading the original manuscript. Since that time—ten years ago to the month that I graduated with my Ph.D.—Wallace archivist Charles Smith has been exceptionally receptive to my numerous queries about Wallace, and was good enough to read parts of the finished manuscript. I acknowledge as well the historians of science who served as expert raters for my assessment of Wallace's personality: Janet Browne, Gina Douglas, Michael Ghiselin, David Hull, John Marsden, Richard Milner, James Moore, Charles Smith, and Frank Sulloway. To the many archivists at the various sources of Wallace material in England I acknowledge their contributions in Appendix I: Wallace Archival Sources.

As always, I thank *Skeptic* magazine Art Director Pat Linse for her important contributions in preparing the illustrations, graphs, and charts for this and my other works, as well as for her insights into the nature of science. Special thanks go to my agents Katinka Matson and John Brockman, and to my editor Kirk Jensen, who helped me find the right balance between biographical narrative and analysis.

As I have done in my previous books, I wish to acknowledge the debt of gratitude owed to *Skeptic* magazine's board members: Richard Abanes, David Alexander, the late Steve Allen, Arthur Benjamin, Roger Bingham, Napoleon Chagnon, K. C. Cole, Jared Diamond, Clayton J. Drees, Mark Edward, George Fischbeck, Greg Forbes, Stephen Jay Gould, John Gribbin, Steve Harris, William Jarvis, Penn Jillette, Lawrence Krauss, Gerald Larue, Jeffrey Lehman, William McComas, John Mosley, Richard Olson, Donald Prothero, James Randi, Vincent Sarich, Eugenie Scott, Nancy Segal, Elie Shneour, Jay Stuart Snelson, Julia Sweeney, Carol Tavris, Teller, and Stuart Vyse. And thanks for the institutional support for the Skeptics Society at the California Institute of Technology goes to Dan Kevles, Susan Davis, Chris Harcourt, Jerry Pine, and Kip Thorn. Larry Mantle, Ilsa Setziol, Jackie Oclaray, Julia Posie, and Linda Othenin-Girard at KPCC 89.3 FM radio in Pasadena have been good friends and valuable supporters for promoting science and critical thinking on the air. Thanks to Linda Urban at Vroman's bookstore in Pasadena for her contributions to skepticism; to Robert Zeps and Gerry Ohrstrom, who has played an important role in professionalizing skepticism and critical thinking, and to Bruce Mazet, who has been a good friend to the skeptics and has influenced the movement in myriad unacknowledged ways. Finally, special thanks go to those who help at every level of our organization: Yolanda Anderson, Stephen Asma, Jaime Botero, Jason Bowes, Jean Paul Buquet, Adam Caldwell, Bonnie Callahan, Tim Callahan, Cliff Caplan, Randy Cassingham, Shoshana Cohen, John Coulter, Brad Davies, Janet Dreyer, Bob Friedhoffer, Jerry Friedman, Gene Friedman, Nick Gerlich, Sheila Gibson, Michael Gilmore, Tyson Gilmore, Andrew Harter, Laurie Johanson, Terry Kirker, Diane Knudtson, Joe Lee, Bernard Leikind, Betty McCollister, Liam McDaid, Tom McDonough, Sara Meric, Tom McIver, Frank Miele, Dave Patton, Brian Siano, Tanja Sterrmann, and Harry Ziel.

Charles Darwin once remarked that half his (geological) thoughts had come out of Charles Lyell's brain. With regard to scientific history and psychobiography I cannot find a better parallel acknowledgment than to thank Frank

Sulloway, whose revolutionary ideas about applying scientific methods to the study of history have set new standards for historians and biographers. Frank's influence is most noticeable in the Prologue, but the cloven hoofprint of his work can be found throughout this volume. Whether historians and biographers take us up on the challenge of treating historical personages as human subjects, and thus subject to analysis by the best tools of the social sciences, remains to be seen. Either way, I remain committed to the values of scientific methods applied to all aspects of the human condition.

No less important to this work are the contributions of Kim Ziel Shermer, whose ability to decipher both handwriting and intent made the reading of thousands of letters to and from Wallace so much the better. Her participation in this biography, as well as the interactive nature of scholarship, are (for me) best illustrated in a story about the discovery of a letter we made at the Darwin Correspondence Project at Cambridge University. Pasted on a folio page of one of the correspondence volumes in the Darwin collection there is a "P.S." of a letter (unsigned and undated) to Darwin, in support of Wallace's heretical views on the evolution of man that led to so many important ideas and controversies in Wallace's career. The collection catalogue noted that the letter fragment was from Spencer, but a question mark accompanied the notation. Further, the letter did not sound like Spencer, who, according to one of the editors of the project (Mario DiGregorio), generally just talks about himself and references his own ideas, interacting very little with the reader. The postscript begins: "I quite agree with you that Wallace's sketch of natural selection is admirable," then continues later:

> I was therefore not opposed to his idea, that the Supreme Intelligence might possibly direct variation in a way analogous to that in which even the limited powers of man might guide it in selection, as in the case of the breeder and horticulturist. In other words, as I feel that progressive development or evolution cannot be entirely explained by natural selection, I rather hail Wallace's suggestion that there may be a Supreme Will and Power which may not abdicate its functions of interference, but may guide the forces and laws of Nature.

No one else at the Darwin project recognized the handwriting, which was quite legible, and most agreed it did not sound like Spencer. But who could it be? Mario suggested we check the correspondence to or from Darwin for anyone interested in human evolution. After about two hours of wondering, searching, and digging through books and boxes of correspondence, Kim found a letter from Darwin to Lyell dated May 4, 1869, in which he writes (regarding Wallace's paper on the evolution of the human mind): "What a good sketch of natural selection! but I was dreadfully disappointed about Man, it seems to me incredibly strange . . . and had I not known to the contrary, would have sworn it had been inserted by some other hand." The wording was too similar to be a coincidence. We then went to Lyell's *Life and Letters* for the month of May 1869 and, sure enough, there it was, May 5: "I quite agree with you that Wallace's sketch of natural selection," etc. And the handwriting? Lyell's was not typically very legible, but because his eyesight was quite poor late in his life he frequently dictated letters to his wife or a secretary. It was a trivial correction to a vast body of literature, but a proud moment of original contribution for us. It is to Kim with good reason that I dedicate this biography.

Schemata

In Darwin's Shadow is a narrative biography that employs quantitative and analytical techniques to get our minds around this complex man. To accomplish this there are five schemata that roughly outline the work:

1. *Wallace the Man.* This is, first and foremost, a biography of Alfred Russel Wallace, utilizing a number of never-before-used archival sources that bring to bear new interpretations on a number of theoretical issues. In this regard, much of this volume resembles the standard womb-to-tomb narrative style of most biographies; nevertheless, I occasionally break the narrative flow to consider some of the theoretical issues involved, particularly in the development of Wallace's scientific, quasi-scientific, and nonscientific ideas. On this level *In Darwin's Shadow* is an intellectual biography—a history of ideas, particularly those of Wallace and his contemporaries.

2. *Wallace and Darwin.* Within the biography are the two major points of intersection between Alfred Wallace and Charles Darwin: (1) the question of priority in the discovery of natural selection, presented and analyzed in light of the archival evidence seen through a different theoretical model—though the absence of hypothesis confirming evidence (a "smoking gun") makes any final resolution impossible, the dispute can be settled to almost everyone's satisfaction through a careful analysis of the extant data; (2) Wallace's heresy that Darwin found so shocking in the man who had become more Darwinian than himself—Wallace's belief that natural selection cannot account for the full development of the human mind and that this was further evidence of the existence of a higher intelligence (although not necessarily God in any traditional sense).

3. *Wallace the Heretic.* Wallace's specific heresy had much broader implications than Darwin realized, and the term "heretic scientist" applies to a great many eccentric and fringe causes championed by Wallace throughout his career. One full chapter is devoted to Wallace as a scientist among the spiritualists, and how and why these experiences became so important as

theory-confirming data for his scientistic worldview. Unlike other scholars who have interpreted Wallace's spiritualism as directly influencing his science, I argue that the causal vector is in the other direction. Wallace's scientistic worldview forced him to shoehorn his encounters, experiences, and experiments in spiritualism into his larger scientism. Two additional chapters examine other Wallace heresies from a psychological and historical perspective. From the moment he first published his modified views on human evolution in 1869, to the present, it has been a historical curiosity as to why Wallace, as co-founder of the theory and stout defender of Darwinism, was apparently unable to apply completely his own and Darwin's theory of natural selection to man and mind. Historical interpretations of this problem have been primarily monocausal, focusing on Wallace's spiritualist investigations, his naturalistic outlook wedded to socialist ideals, and his hyper-selectionism. In a multicausal model it can be demonstrated that Wallace's thoughts on this subject were complex and deeply influenced by certain aspects of his Victorian culture. A "heretic personality," coupled to his thoughts on hyper-selectionism, monogenism and polygenism, egalitarianism, and environmental determinism, were strongly governed by his culture through both spiritualism and phrenology, as well as other intellectual traditions such as teleological purposefulness, scientific communal support for his belief in the limited power of natural selection, exotic anthropological experiences in the Amazon and Malay Archipelago, and working-class associations. Although he was a revolutionary thinker of the highest order—a heretic scientist who often thought outside of the box—Wallace's intellectual style and personality were nevertheless shaped by these cultural forces that drew him toward supernatural and spiritual explanations for some of the deepest mysteries of evolution.

4. *Wallace and the Psychology of Biography.* To determine how and why Wallace thought as he did, I have applied the statistical methods and historical analyses of social scientist Frank Sulloway, whose groundbreaking work *Born to Rebel* presented the results of his numerous tests of historical hypotheses. For example, was the Darwinian revolution primarily a social class revolution as claimed by many historians of science? Can we test the hypothesis of Adrian Desmond and James Moore that Darwin was a "tormented evolutionist" (as they call him in the subtitle of their biography) because his upper-class background and aristocratic lifestyle conflicted with the working-class revolution he helped to lead? We can. Sulloway coded hundreds of people who spoke out publicly on evolution for a number of different variables, including socioeconomic status (SES). There was no statistically significant difference in SES between those who led and supported the Darwinian revolution and those who opposed it. This finding is borne out most obviously in the lives of the revolution's two leaders, Charles Darwin and Alfred Russel

Wallace, who could not be more dissimilar in their social and economic backgrounds, yet they shared other important personality-shaping variables, and historical contingencies, that led them down similar paths toward intellectual rebelliousness.

5. *Wallace and the Nature of History.* This work concludes with an epilogue that considers the nature of historical change, as well as why history should be treated as a science, through the life and theories of Wallace. We see in the conflation of his personality, thoughts, and culture—as Wallace himself observed in a late-life reflection—the dynamic interaction of contingencies and necessities. In the history of life, as well as the history of a life, conjunctures of past events compel certain courses of action by constraining prior conditions.

Finally, I have included a number of appendices primarily for use by scholars. APPENDIX I: Wallace Archival Sources summarizes the holdings of the various archives throughout England. Not only are several archives not listed in any other source, but the recording of the materials in the published ones is either incomplete or in error. I have provided as much information of what is housed at each archive as space would permit. APPENDIX II: Wallace's Published Works includes all 22 books and 747 articles, papers, essays, letters, reviews, and interviews. Finally, the Bibliography is the complete citation source for this biography, with the exception of Wallace's references.

Although this book is roughly divided between quantitative analysis in the Prologue and Epilogue, and narrative synthesis in the twelve chapters in between, the division is not always so neatly cleaved. Since this is a biography not just of the life but of the science of Alfred Wallace, it is necessary to periodically digress into sidebars about some of the major theoretical issues of the age in order to understand the origins of his scientific ideas. And since Wallace was, first and foremost, a world-class practicing scientist, it is necessary to devote a moderate amount of time to his major ideas, what they were and how he developed them, both in time and in logic. Also making appearances in a couple of the narrative chapters are debates among historians of science about Wallace and the scientific disputes of his age, particularly with regard to the Darwin–Wallace priority question and where Darwin and Wallace departed in their interpretations of the power of natural selection. Since this biography is driven as much by ideas as by chronology, the structure of the narrative is that of shingled roof tiles, slightly overlapping one another but with a general flow of linear time. Thus, the reader may occasionally find it necessary to jump slightly back or forward in time as the sequence picks up after a discussion of the development of an important idea or event, which I felt was better served by a full explication in one location rather than having it fragmented throughout the text with nuanced details lost between chapters.

The Skirts of Happy Chance

It was a long journey from collegiate seminar to doctoral dissertation, with many chance conjunctures along the way. So many contingencies, summed over time, add up to a single necessity. As it was for Darwin and Wallace, whose paths crossed so many times, so it is for all of us that so many bits and pieces of history make up a life. Wallace's friend and colleague, Edward B. Poulton, in the centenary of Wallace's birth in 1923, wrote a lengthy article in the *Proceedings of the Royal Society,* praising Wallace and putting him into a historical and scientific context with the other luminaries of the Victorian age. In this comparison Poulton identifies the role of contingency and necessity in the lives of these great men:

> It is a noteworthy fact that nearly all these men began their scientific careers by long voyages or travels—Darwin, Hooker and Huxley with the Navy, Wallace and Bates in the South American tropics. With most of them, and especially with Wallace, the way to science was long and difficult. But this was not all loss: strength grows in one who
>
> ". . . grasps the skirts of happy chance
> And breasts the blows of circumstance
> And grapples with his evil star."[2]

IN DARWIN'S SHADOW

PROLOGUE

The Psychology of Biography

In the December 1913 issue of *Zoologist,* Oxford University professor of zoology Edward B. Poulton wrote an obituary notice for the death of his friend and colleague Alfred Russel Wallace: "So much has been written about this illustrious man, both before and after his death on November 7th, that it is difficult to say anything fresh or arresting. Looking back over a warm friendship of more than twenty-five years, and reading again the numerous letters received from him, I do however recall memories and find striking statements which help to create a picture of the great personality now lost to the world."[1]

A great personality indeed. For ninety years now—as long as Wallace lived—he has largely been lost to us, partly as a result of the ravages of time, partly because of the dark shadow cast by his more famous contemporary Charles Darwin, and partly because of his own modesty. As Poulton recalled, "Ten years ago the Hon. John Collier generously offered to paint a portrait of Wallace. If the offer had been accepted we should have had a noble presentation of one of the greatest men of the last century—a splendid companion to the Darwin and Huxley we all know and love so well. But nothing would induce Wallace to sit."[2] The man was modest to a fault, a trait that would contribute to his eventual obscurity. But as Darwin's star has brightened, those in his orbit are beginning to glow in his reflected light. Wallace now presents himself for a well-illuminated portrait.

Darwin's Dictum and Wallace's Wisdom

In 1861, less than two years after the publication of Charles Darwin's *The Origin of Species,* in a session before the British Association for the

Advancement of Science devoted to discussing the theory of evolution by means of natural selection, a critic claimed that Darwin's book was too theoretical and that he should have just "put his facts before us and let them rest." In attendance was Darwin's friend Henry Fawcett, who subsequently wrote him to report on the theory's reception. (Darwin did not attend such meetings, usually due to ill health and family duties.) On September 18, Darwin wrote Fawcett back, explaining the proper relationship between facts and theory: "About thirty years ago there was much talk that geologists ought only to observe and not theorize, and I well remember someone saying that at this rate a man might as well go into a gravel-pit and count the pebbles and describe the colours. How odd it is that anyone should not see that all observation must be for or against some view if it is to be of any service!"[3]

Alfred Russel Wallace, Darwin's younger contemporary who lived in the penumbra of the eclipse created by the sage of Down, was no less savvy in his wisdom of the ways of science when he observed: "The human mind cannot go on for ever accumulating facts which remain unconnected and without any mutual bearing and bound together by no law."[4] Darwin's dictum and Wallace's wisdom encode a philosophy of science that dictates that if scientific observations are to be of any use, they must be tested against a theory, hypothesis, or model. The facts never just speak for themselves, but must be interpreted through the colored lenses of ideas—percepts need concepts. Science is an exquisite blend of data and theory, facts and hypotheses, observations and views. If we think of science as a fluid and dynamic way of thinking instead of a staid and dogmatic body of knowledge, it is clear that a data/theory stratum runs throughout the archaeology of human knowledge and is an inexorable part of the scientific process. We can no more expunge our biases and preferences than we can find a truly objective Archimedean point—a god's-eye view—of the human condition.

The challenge from these two historical scientists has been undertaken in this work, in a metahistorical sense, to understand how and why observation and theory came to be conjoined in Alfred Russel Wallace as it did. This book, then, is both biography and investigation, narrative and analysis, history and theory. It is a study in psychology and biography, as well as the psychology *of* biography. It asks of the past, what happened and why? It is a look at both *a* biography and biography itself. The data from the life of Alfred Russel Wallace do not speak for themselves, but instead are interpreted through theoretical models and are presented for or against specific views, so that the facts of this biography are bounded and of service.

A scientific analysis of a living human being requires extensive data not only on the generalities of human behavior, but on the details and intricacies, nuances and vagaries of the particular person under investigation. But as social scientists are discovering, along with their colleagues in the physical and biological sciences, unique pathways of history are more common and important in the development of individuals than previously suspected. From galaxies and planetary bodies to ecosystems and individual members of a species, the particular histories of the subject play at least as important a role as the governing laws of physical, biological, or social action. Alfred Kinsey knew this all too well, as prior to his research foray into the sexual behavior of men and women he was an entomologist, studying gall wasps for over two decades, publishing a number of important and pioneering works. What he discovered about wasps—a fairly homogeneous group compared to humans—prepared him for his subsequent realization of the nearly incomprehensible variation in human actions, as he explained in his first volume: "Modern taxonomy is the product of an increasing awareness among biologists of the uniqueness of individuals, and of the wide range of variation which may occur in any population of individuals." Extrapolating to humans, Kinsey noted: "Males do not represent two discrete populations, heterosexual and homosexual. The world is not to be divided into sheep and goats. Not all things are black nor all things white. It is a fundamental of taxonomy that nature rarely deals with discrete categories. Only the human mind invents categories and tries to force facts into separate pigeonholes. The living world is a continuum in each and every one of its aspects."[5] It is no different with historical personages. Dependently linked events operating through time shape each and every life in a unique pattern of contingencies that mold individual lives. We cannot understand *all* human action outside of examining *a* human action. The particular shapes the general. History counts.

In his classic 1943 study *The Hero in History,* Sidney Hook captured this primary struggle between the players and forces of history—between the individual and the collective, the freedom of choice and the determinism of law, the contingent and the necessary—when he drew a distinction between the *eventful man,* who was merely at the right place at the right time, and the *event-making man,* who helped create the events himself: "The event-making man finds a fork in the historical road, but he also helps . . . to create it. He increases the odds of success for the alternative he chooses by virtue of the extraordinary qualities he brings to bear to realize it."[6] The hero is a great individual who was at the right place and time. The hero is both a product and producer of culture.

If only it were a simple task to know when and where a historical figure affected change or was affected by change. Which historical variables are cause, and which are effect? Arguments about whether heroes in history are "great men" or products of their culture rest on a false dichotomy created, in part, by both the participants themselves and the historians who write about them. As historical scientists we must, like social scientists, tease out the dependent and independent variables—the effects and the causes—in the historical development of our subjects. We must determine which variables made a significant difference and which did not. And most important, like post-Skinnerians, we must recognize that the complexity of human thought and action cannot be reconstructed in a simple box with one or two intervening variables. If behavioral psychologists have failed to adequately explain and predict human actions with these simple models, then behavioral historians shall fair no better. We need a more complex explanatory model that approaches the complexity of human actions, yet remains simple enough to allow us to get our minds around the subject.

The Historical Matrix Model

The past may be constructed as a massively contingent multitude of linkages across space and time, where the individual is molded into, and helps to mold the shape of, those connections. Traditional interpretations of history (e.g., the iconographies of ladders and cycles; the forces of culture and class) do not adequately reflect the rich and radiantly branching nature of contingent historical change and the role of the individual within that system.[7] Social scientists use a statistical tool called *analysis of variance* to sort out the relative effects of several causal independent variables on a single dependent variable. A *multiple analysis of variance,* also known as a *factorial arrangement of the treatments,* or just a *factorial matrix,* is a more complex instrument that examines several independent variables and their relative influences on more than one dependent variable.[8] The more sophisticated the factorial matrix, the closer the model comes to representing the complexities of real life. If this is true of the present, then it is certainly true of the past. If the present does not change in straight lines along single variables, then certainly the past did not. A *factorial matrix of history,* complex but not too complicated, may more appropriately represent the past.

Alfred Russel Wallace was a complex man with a nontraditional background and an unconventional life, who lived nine full decades in a rich culture. When I first undertook this project I was overwhelmed by the amount of individual and cultural data surrounding Wallace's life and thoughts, so I constructed a matrix that I call the *Historical Matrix Model* (*HMM*) to differentiate influencing variables and sort them into relative importance in the shaping of his ideas, most particularly his controversial theory of human evolution. The HMM is a complex arrangement of historical factors in a three-dimensional structure of *internal forces* (thoughts) interacting with *external forces* (culture) over *time* (as they changed). The interaction of these variables over time requires a model that can handle their relative effects. "We say that two variables interact when the effect of one variable changes at different levels of the second variable," statistician G. Keppel explains. "Thus, an interaction is present when the simple main effects of one variable are not the same at different levels of the second variable." Variables may also be *additive* when "the effect of one variable simply adds to the effect of the second variable."[9] Such interactions, of course, make the model more complex, but a closer approximation to what really happens in human-culture interactions: "In an increasing number of experiments being reported in the literature, interactions not only are predicted but represent the major interest of the studies. The discovery or the prediction of interactions may lead to a greater understanding of the behavior under study. In short, then, if behavior is complexly determined, we will need factorial experiments to isolate and to tease out these complexities. The factorial allows us to manipulate two or more independent variables concurrently and to obtain some idea as to how the variables combine to produce the behavior."[10] The Historical Matrix Model (Figure P-1) helps us visualize the additive and interactive development of Wallace's thoughts within his culture, as they changed over time.[11]

The tension between what *can* be and what *must* be—between the power of the individual and the force of culture—encapsulates this biography and drives the analysis of Wallace's life and science. To what degree does culture mold and shape an individual's ideas and behaviors? To what extent does an individual's thoughts and actions affect change in a culture? What is the relative influence of thought and culture as they interact over time? These questions will be asked with specific regard to Wallace in the context of nineteenth-century Victorian culture, to attempt to understand the complex development of his ideas and those of his contemporaries, particularly with regard to science and evolutionary theory.

The Historical Matrix Model presents a $5 \times 5 \times 6$ factorial design, in which five *internal forces* (thoughts) interact with five *external forces* (culture) over six periods of time representing Wallace's life. The selection of five forces was somewhat subjective, though fewer would have eliminated significant events and more would have oversaturated the analysis. The six time periods are marked by significant changes in Wallace's life and career, such as schooling, explorations, major publications, intellectual battles, and personal travails. The five internal forces in decreasing order of influence on the left vertical axis represent the individual ideas and experiences of Wallace and include: (a) *Hyper-selectionism,* or the overemphasis of adaptationism in explaining the evolution of organisms, particularly the evolution of humans and the human mind; (b) *Mono-polygenism,* the great debate in Wallace's time over the origin of humans from either a single source (monogenism) or mul-

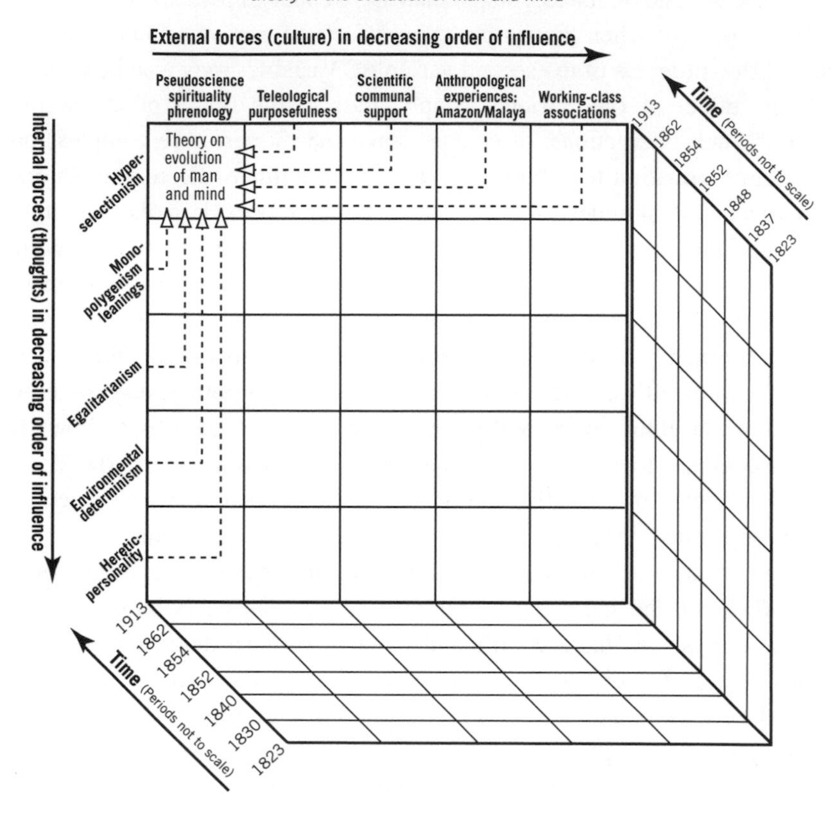

Figure P-1 The Historical Matrix Model. The Historical Matrix Model is a complex arrangement of historical factors in a three-dimensional array of *internal forces* (Wallace's thoughts) interacting with *external forces* (Wallace's culture) over *time* (as they changed throughout Wallace's long life of ninety years). The *interaction* of these variables over time allows us to study their relative effects. (Rendered by Michael Shermer and Pat Linse)

tiple sources (polygenism), derived partly from his anthropological studies and reflections after considerable travel in South America and the Malay Archipelago, and partly from the influence of other anthropologists in Victorian England; (c) *Egalitarianism,* Wallace's belief that people are inherently equal, derived from several sources including his unique blend of mono-polygenism, the cross-cultural experiences from extensive travel, as well as (d) *Environmental determinism,* Wallace's belief that since people are inherently equal, all apparent differences must be due to environmental differences; all of these variables were heavily influenced by Wallace's (e) *Personality,* which includes independence of thought, separatist tendencies, and especially an exceptionally high openness to experience that led him to make a number of revolutionary scientific discoveries, but with a concomitant level of gullibility that made Wallace more susceptible to many non-empirical and non-scientific assertions, most notably his uncritical endorsement of the spiritualism movement.

The five external forces in decreasing order of influence on the upper horizontal axis represent the cultural variables that affected Wallace's thinking throughout his intellectual life and include: (a) *Spiritualism and phrenology,* and other such quasi- and nonscientific phenomena popular in Victorian England throughout much of the nineteenth century; (b) *Teleological purposefulness,* the belief that all life is directed toward a goal by a higher force, pervasive throughout many fields of thought, but especially prevalent in both natural theology and natural history; (c) *Scientific communal support* in both the rejection of natural selection to the human mind, as well as for spiritualism and other nonscientific claims; (d) *Anthropological experiences* in Amazonia and the Malay Archipelago that directly influenced Wallace's hyper-selectionism in the controversy over the evolution of the human mind, as well as his social theories, attitudes, and activist causes; and (e) *Working-class associations* tethered to the populism of the Mechanics' Institutes through which Wallace gained his education, the socialism of Robert Owen in Wallace's younger years, and the worldview of Herbert Spencer in his later life, all of which directed his social activism in a distinct direction.

The Historical Matrix Model is used as a heuristic in the middle chapters of this biography, dealing with the historical variables in the matrix, their interaction, and how they interface with social and cultural events and the intellectual climate in which Wallace lived. The primary focus of the model is the problem of the evolution of the human mind, over which Wallace and Darwin broke their intellectual bonds and that so came to dominate Wallace's thoughts and actions. I have placed this problem in the upper left corner of the HMM in order to adjoin closely the two most significant internal and external causal variables—*Hyper-selectionism* and *Spiritualism* respectively. Broken-lined arrows represent *force-vectors,* or direct linkages from an influencing force, so that the farther each is from the problem, the weaker the link. The decision of relative position is based on textual analysis and cultural–historical interpretation.

It follows from this factorial design that the HMM could be used to examine the effects of the various forces on each other, since they do sometimes interact, and this is occasionally done in the biography. But the overriding flow of the *force-vectors* is from thought or culture to the thesis question, arguably a primary focal point in Wallace's life and career. Designing the

HMM after the factorial matrix was done to help explain the variance of the dependent variable—Wallace's theory of the evolution of man and mind. It does this, in part, by estimating the contributions to this variance, not by one or two independent variables, but by a 5 × 5 matrix of independent variables, interacting and influencing the dependent variable, over six units of time. The traditional monocausal explanations of Wallace's position on this question—for example, that his interest in spiritualism determined his belief—simply fail to consider not only other variables, but the interaction of these variables.

In this scientific approach to history, I take as a guide the work of social scientist and historian of science Frank Sulloway, who emphasizes the importance of examining both psychological and sociological considerations in historical settings in order to gain a greater understanding of one's subject, as in his statistical studies of historical figures: "Any multivariate model that can move Louis Agassiz from a 1.5 percent probability of supporting Darwinism to a 93 percent probability of supporting glaciation theory can only do so by encompassing a major role for historical context." In fact, Sulloway's justification for the utility of his own matrix model serves as a useful reinforcement for the application of the HMM in this biographical study:

> I believe that multivariate models are an underutilized resource in testing general claims about scientific change. Because most historical generalizations are inherently probabilistic in nature and because most historical outcomes are overdetermined, multivariate and epidemiological models are particularly suited to assessing the complicated interweave of influences associated with conceptual change. Given sufficient information, such models can also detect complex interaction effects between factors that would probably be missed using standard historical approaches. With regard to orthodoxy and innovation in science, multivariate models are able to simulate the considerable diversity of intellectual behavior that is actually observed in the history of science.[12]

Clearly scientific modeling has it limitations (see Epilogue), not the least of which is keeping the focus narrow enough to say something significant, without losing sight of the bigger picture. Thus, to get our minds around the general complexities of Alfred Russel Wallace and how he became a "heretic scientist," we must get to know the man himself. Without examining the unique particulars that molded his general personality, or the specific events that went into the construction of a composite lifetime, the compression of his story into the gridwork of a heuristic like the HMM would be analytic, but not synthetic. It is useful to break down a whole into its parts, but it is equally important to put the pieces back together into a complete historical narrative. The analysis is contained in the remainder of this chapter. The synthesis can be found in the narrative biography that follows.

Quantitative Biography

The measure of a life comes in many forms, from newspaper obituaries and encyclopedia entries, to potted biographies and womb-to-tomb narratives. The sources for reconstructing a life are numerous and include interviews, letters, notes, manuscripts, papers, articles, essays, reviews, books, photographs, diaries, and autobiographies. The tools of biography are varied, such as psychology, sociology, cultural history, oral history, the history of ideas, demographics, and statistics. But since humans are storytelling animals, the primary expression for biography is narrative.

One limitation of narrative biography, however, is what is known in cognitive psychology as the *confirmation bias,* where we tend to seek and find confirmatory evidence for what we already believe, and ignore disconfirmatory facts.[13] I discussed this problem in essay-length biographies of Carl Sagan[14] and Stephen Jay Gould,[15] and showed how quantitative methods can help the biographer navigate around the confirmation bias. In the case of the Cornell astronomer Carl Sagan, his biographers asked such questions as: Was he a tender-minded liberal or a tough-minded careerist? Was he a feminist or a misogynist? Was he a scientist of the first rank or merely a media-savvy popularizer? One way to answer such questions is to start off with a hunch and then comb through books, papers, notebooks, diaries, interview notes, and the like, pick out the quotes that best support the hypothesis, and draw the anticipated conclusion. This is the confirmation bias. In statistics it is called "mining the data."

One way to avoid the bias is to apply the tools of the social sciences. For example, I did a quantitative analysis of Sagan's curriculum vitae, which totals 265 single-spaced typed pages, classifying 500 scientific papers and 1,380 popular articles and essays by content and subject matter. Was Sagan politically and socially liberal? The data give us an unequivocal answer: one-third of everything he wrote or lectured on was on nuclear war, nuclear winter, environmental destruction, women's rights, reproductive rights, social freedoms, free speech, and the like. Was Sagan a mere science popularizer, but never really a serious scientist (now known as the "Sagan effect," where one's scientific output is thought to be inversely proportional to one's popular output)? To answer this question, I compared Sagan to several recognized eminent scientists, including Jared Diamond, Ernst Mayr, Edward O. Wilson, and Stephen Jay Gould. It turns out that Sagan falls squarely in the middle of this distinguished group in both total career publications (500) and average publications per year (12.5). Graphing Sagan's rate of publishing popular articles versus scientific papers over time revealed that the latter was unaffected by the former, even following the airing of *Cosmos* in 1980 and his sudden jump to superstardom. Throughout his career that began in 1957 and ended in December 1996, Sagan averaged a scientific peer-reviewed paper per month. The "Sagan effect" is a chimera, but only a quantitative analysis could have answered that question.

For a scientist whose literary output has been extraordinary, it is not possible to glean an overall subject emphasis and thematic focus without starting with such a large-scale quantitative analysis. From there the biographer can scale down from global trends to individual works, to see how the particular fits into the general. In the case of Harvard paleontologist Stephen Jay Gould, I not only did a quantitative analysis of his total literary corpus, I conducted a detailed content analysis of his 300 *Natural History* essays, classifying them by primary, secondary, and tertiary subject emphasis (e.g., evolutionary theory, history of science, zoology), as well as by thematic scheme (e.g., theory—data, contingency—necessity). In Gould's case, by initially dividing his 479 scientific papers into 15 specialties, it was possible to then collapse them into five related taxons of evolutionary theory, paleontology, history of science, natural history, and interdisciplinary. For his 300 essays that spanned 27 years and 1.2 million words, the history of science and science studies dominates at 148 essays, versus 78 on evolutionary theory. But when secondary and tertiary subjects are factored into the total, evolutionary theory totals 248 versus 212 for the history of science and science studies. In other words, a cursory look at Gould's essays leads to the conclusion that he is primarily doing the history of science, but a deeper examination reveals that Gould is, in fact, an evolutionary theorist who uses historical examples in the service of his personal theories of evolution—an overt example of Darwin's dictum and Wallace's wisdom.

In our time, Sagan and Gould have achieved a level of fame and influence in both science and culture matched by that of Darwin and Wallace in the nineteenth century. In the long history of science, however, only a handful of scientists stand out above the masses of rank-and-file researchers. They got there not just because of their important discoveries, but because they are synthetic thinkers on the grandest scale, integrating not only data and theory, but other themata that explore the deepest themes in all of Western thought. In both Sagan's and Gould's work, as well as that of Darwin and Wallace, one finds several large-scale themata, including: *Theory—Data* (how culture and science interact); *Time's Arrow—Time's Cycle* (unique historical change versus repetitive natural law); *Adaptationism—Nonadaptationism* (optimality versus suboptimality in the evolution of organisms); *Punctuationism—Gradualism* (catastrophic change versus uniformitarian change); *Contingency—Necessity* (directionless and purposeless change versus directed and purposeful change).[16]

Alfred Wallace, no less than Charles Darwin, integrated such themata into his work, which is one reason why, in his own time, he was as well known and nearly as influential as Darwin. In many ways, in fact, in conducting a quantitative analysis of Wallace's works we see that the scope of his intellectual interests far outstripped that of Darwin. Throughout this biography, we will see how Wallace integrated his many different scientific interests, as well as how these larger thematic pairs often formed an underlying substrate beneath the superficial issues at hand.

Quantifying Wallace

Alfred Russel Wallace led a remarkably long and productive life of nearly ninety-one years, having been born in 1823 just after Napoleon's death and dying in 1913 just before the Great War erupted. Wallace's influence was (and in many ways still is) pervasive. Numerous species of plants and animals carry his name. "Wallace's Line" (and "Wallacea") refers to the transitional zone between the Australian and Asian biogeographical regions. The planet's six basic biogeographic regions are sometimes referred to as "Wallace's realms." The "Wallace effect" involves the production of sterile hybrids in reproductively isolated populations. A plaque commemorating his life lies near Darwin's in Westminster Abbey. Although he never earned a doctorate nor was he a professor, he was referred to as "Dr. Wallace" and "Professor Wallace" in countless interviews and articles about him, and he did, in fact, receive honorary doctorates from the University of Dublin in 1882 and Oxford University in 1889, not to mention professional membership in all of Britain's major scientific societies, including the Royal Society. Even now his name carries weight around the world. In 2000 "Operation Wallacea Trust" was founded "to support activities that could directly contribute towards the conservation of biodiversity in the Wallacea region of eastern Indonesia," as well as the "Zoological Society Wallacea," a "new society for zoological research in South East Asia."[17]

In his final years, and on his death, Wallace was hailed as one of the greatest scientists to ever live. Press accounts refer to him as "England's greatest living naturalist" (1886); "[one of the two] most important and significant figures of the nineteenth century" (1904); "a mid-Victorian giant" (1909); "this greatest living representative of the Victorians" (1910); "the Grand Old Man of Science" (1911, 1913, 1913); "the last of the great Victorians" (1912); "the last of that great breed of men with whose names the glory of the Victorian era is inseparably bound up" (1913); "the acknowledged dean of the world's scientists" (1913); "one of the greatest naturalists of the nineteenth century" (1913); "We should not know where to look among the world's greatest men for a figure more worthy to be called unique" (1913); "Of all the great men of his time, or times, he was, with the single exception of Huxley, the most human" (1913); "Only a great ruler could have been accorded by the press of the world any such elaborate obituary recognition as was evoked by the death of Alfred Russel Wallace" (1914); "the last of the giants of English nineteenth-century science" (1914), and so on.[18]

Wallace archivist and historian Charles Smith has logged additional Wallacean references, including that the earth's moon features the "Wallace crater" and, for levity, the fact that cartoonist Scott Henson has chosen the pen name "Russ Wallace." More seriously, the Bristol Zoo Gardens in Clifton, Bristol, U.K., contains the "Wallace Aviary," an open-air facility for birds; the Department of Earth Sciences at Cardiff University in Wales houses the "Wallace Lecture Theatre"; the National Botanic Garden of Wales includes the "Wallace Garden" dedicated to genetics and evolution education; Kansas Wesleyan University offers an annual "Alfred Russel Wallace Award"; the "Wallace House" is a medical center in Broadstone, Dorset, near where Wallace once lived; the "Wallace Road" takes one to the site of his home (no longer extant) in Broadstone; on the site of another one of Wallace's homes at Old Orchard now stands an apartment complex called "Wallace Court"; the "Wallace Lecture Theatre" can be found at Bournemouth University and the "Wallace Room" is at the Bournemouth Natural Sciences Society; in 1985 the Royal Entomological Society of London and the Indonesian Department of Science instituted "Project Wallace," a year-long study of the Dumoga-Bone area of north Sulawesi in Indonesia; in Neath, Wales, is a library once used for Mechanics' Institutes lectures that was designed and built in 1846 by Wallace and his brother John and still stands in tribute to the primary source of Wallace's science education (as it was for so many others from the working class); and the "Wallace House" can be seen at 11 St. Andrew's Street in Hertford, on the grounds of the Richard Hale School (previously called the Hertford Grammar School that Wallace attended), and includes a circular concrete plaque over its door that reads: "In this house lived Alfred Russel Wallace OM. LLD. DCL. FRS. FLS. Born 1823—Died 1913. Naturalist, Author, Scientist. Educated at Hertford Grammar School."[19]

Wallace certainly was a naturalist, author, and scientist of the first rank. In his two most famous and productive expeditions—four years in the Amazon and eight in the Malay Archipelago—he undertook over a hundred specific collecting trips that logged in excess of 20,000 miles. Tragically, much of his collection from the first expedition was lost in a disaster at sea on the voyage home, but the Malay Archipelago produced an almost unimaginable 125,660 specimens, including 310 mammals, 100 reptiles, 8,050 birds, 7,500 shells, 13,100 butterflies, 83,200 beetles, and 13,400 other insects, over a thousand of which were new species. Yet, throughout a professional career that began with his first published paper in 1845 at the age of twenty-two, and continuing right up to the final month of his life with a book preface published in November 1913, Wallace was much more than a naturalist. Reviewing his 747 published articles, essays, reviews, commentaries, and letters we can globally divide them into 68 percent scientific and 32 percent social commentary, indicating that, although Wallace was a social activist of the highest profile, he was, first and foremost, a serious scientist whose output was more than two-to-one scientific to social. And, as we shall see, almost all his social causes were grounded in what he considered to be hard science. (His stance on anti-vaccination, for example, was almost entirely based on what he believed to be solid data proving that vaccination was—by that time—doing more harm than good.)

How serious a scientist was Wallace? Of his 508 scientific papers, a remarkable 191 (38 percent) were published in *Nature,* one of the most prestigious of all scientific journals. Wallace's papers, in fact, can be found in all the top journals of his time: 22 in the *Proceedings of the Entomological Society of London,* 21 in *Ibis,* 19 in the *Fortnightly Review,* 16 in the *Proceedings of the Zoological Society of London,* 14 in the *Annals and Magazine of Natural History,* 14 in *Zoologist,* 13 in the *Journal and Proceedings of the Royal Geographical Society,* 12 in the *Journal of the Anthropological Society of London,* and 6 in the *Quarterly Journal of Science.* Wallace's social commentaries and popular writings also found their way into influential publications: 24 in *Land and Labour,* 23 in *Vaccination Inquirer,* 22 in the *Daily Mail, Daily News, Daily Telegraph, Daily Chronicle,* and *Daily News and Leader,* 15 in *The Times* of London, 9 in *Light,* 9 in *The Spiritualist,* 9 in the *Journal of the Society for Psychical Research,* 8 in the *Reader,* 6 in *Spectator,* 6 in *Christian Commonwealth,* 4 in *Macmillan's Magazine,* 4 in *Echo,* and 2 in *Outlook.*

We can further classify Wallace's publications into numerous specialized subjects. In science Wallace published on ancient history and archaeology, animal behavior, astronomy and cosmology, botany, entomology, ethnography and ethnology, evolutionary ethics, evolutionary theory, exobiology, history of science and evolutionary thought, geography and geology, linguistics, the origins of life and the plurality of worlds, paleontology, phrenology, primatology, spiritualism, systematics and taxonomy, and zoology. In social commentary and activist causes Wallace published on agricultural economics, anti-vaccination, commerce, conservation of the environment, crime and punishment, economic theory and capitalism, education, equal opportunity, eugenics, labor, land nationalization, literature and poetry, museum design, poor laws, railroad nationalization, religion and the role of institutionalized churches, social justice, socialism, systematics, trade regulation, and women's rights and suffrage. Even these fields can be further subdivided. For example, in physical geography and geology Wallace published papers on glaciation, mountain formation, lake formation, and climatology. In evolutionary theory Wallace wrote on natural selection, sexual selection, the evolution of mind versus body, hybrid infertility, mimicry and protective coloration, variation and species divergence, instinct and behavior, and so on.

Wallace wrote on hundreds of different very specific subjects, so many, in fact, that such lists really do not tell us much beyond the obvious conclusion that Wallace was a polymath. We need to follow Wallace's wisdom here and bind together these accumulated facts. We begin with his books. Figure P-2 presents a classification of Wallace's twenty-two books by subject, in which we see that he focused on seven different general areas of study.

Wallace's books, however, tend to include many different subjects and thus are difficult to classify into single categories. Plus, there are a great many subjects that Wallace researched and published on that never made it into

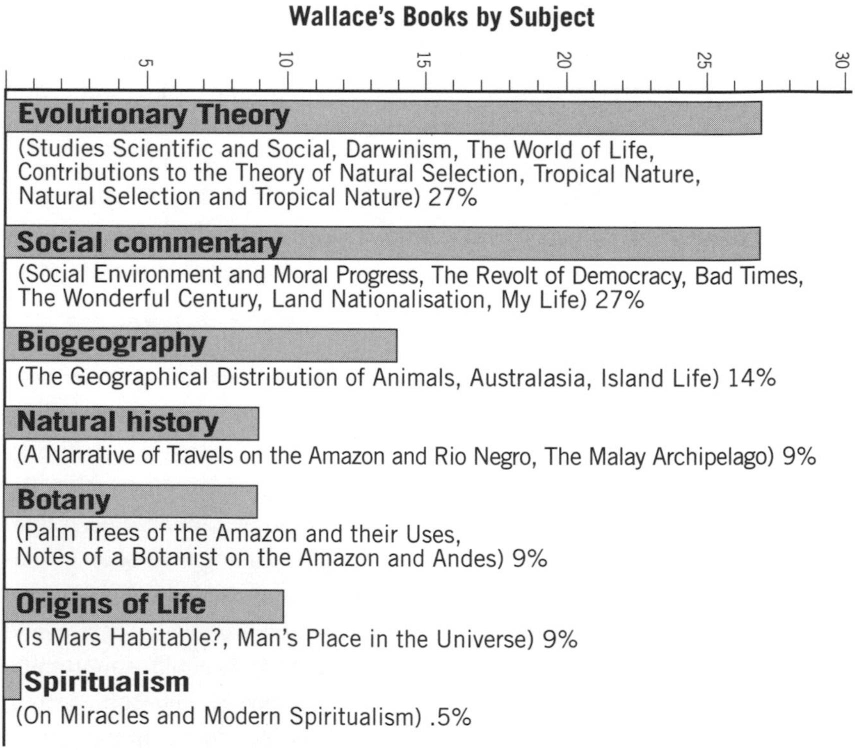

Wallace's Books by Subject

5 10 15 20 25 30

Evolutionary Theory
(Studies Scientific and Social, Darwinism, The World of Life, Contributions to the Theory of Natural Selection, Tropical Nature, Natural Selection and Tropical Nature) 27%

Social commentary
(Social Environment and Moral Progress, The Revolt of Democracy, Bad Times, The Wonderful Century, Land Nationalisation, My Life) 27%

Biogeography
(The Geographical Distribution of Animals, Australasia, Island Life) 14%

Natural history
(A Narrative of Travels on the Amazon and Rio Negro, The Malay Archipelago) 9%

Botany
(Palm Trees of the Amazon and their Uses, Notes of a Botanist on the Amazon and Andes) 9%

Origins of Life
(Is Mars Habitable?, Man's Place in the Universe) 9%

Spiritualism
(On Miracles and Modern Spiritualism) .5%

Figure P-2 Wallace's twenty-two books classified by subject.

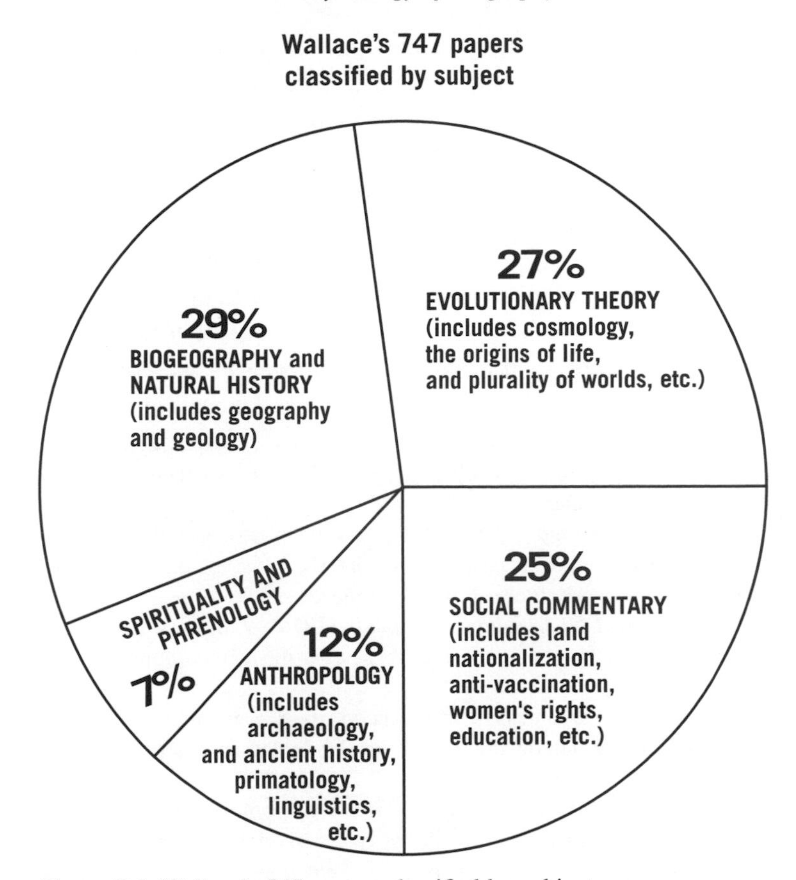

Figure P-3 Wallace's 747 papers classified by subject.

books, so we need to classify his 747 published articles, essays, reviews, commentaries, interviews, and letters into five major taxa, presented in Figure P-3, as a percentage of total publications.

As we shall see in the narrative biography, these five fields are distinctly defined by travels, events, and publications in Wallace's life. In his two major expeditions to the Amazon and Malay Archipelago, Wallace's focus in his day-to-day work was primarily on biogeography and natural history, secondarily on evolutionary theory (particularly in the Malay Archipelago), and tertiarily on anthropology, primatology, and linguistics. Although he dabbled in phrenology in his youth before his natural history excursions, it was not until his return to England in the early 1860s that he converted to spiritualism and made it an integral part of his scientistic worldview. And from roughly 1870 to his death in 1913, such social causes as land nationalization, anti-vaccination, women's rights, and education reform consumed roughly a quarter of his professional activities. So, although at first blush it appears that Wallace was all over the intellectual board, in fact he concentrated his energies on a handful of subjects and causes important to him and his unique worldview that, I shall argue, was scientistic (at least in Wallace's mind) to the core.

This quantitative taxonomy of Wallace's interests and work is borne out in an analysis of Wallace's most cited works in the *Science Citation Index, Social Sciences Citation Index,* and *Arts and Humanities Citation Index,* compiled by historian Charles Smith[20] and listed in rank order:

1. *Darwinism* (1889). This is Wallace's magnum opus and his definitive statement of his own views on evolutionary theory and how they are similar to and different from Darwin's.
2. *The Malay Archipelago* (1869). This is a travelogue and natural history of the archipelago from Wallace's travels from 1854 to 1862 and was Wallace's most successful work, literarily and commercially.
3. *The Geographical Distribution of Animals* (1876). Wallace was the founder of the science of biogeography and this is his most important work in this field.
4. *Island Life* (1880). Islands provide isolated experiments in evolution, and in this book Wallace demonstrates how insular biotas can cause rapid evolutionary change.
5. *Tropical Nature, and Other Essays* (1878). This book is a collection of Wallace's essays that did not appear in his earlier works in tropical biology.
6. *Contributions to the Theory of Natural Selection* (1870). In this volume Wallace clarifies what he did or did not contribute to evolutionary theory.
7. *My Life* (1905). Wallace's two-volume magisterial autobiography that includes many never-before-published letters and papers.

The following is a list of Wallace's most cited articles in rank order:

1. "The Origin of Human Races and the Antiquity of Man Deduced from the Theory of Natural Selection" (1864). In this paper Wallace applies the theory of natural selection to the problem of human racial diversity.
2. "On the Tendency of Varieties to Depart Indefinitely from the Original Type" (1858). This is the famous "Ternate essay" that Wallace sent to Darwin in March 1858 that contained his theory of natural selection that so resembled Darwin's own. This essay triggered Darwin to complete his big book on evolution that became *The Origin of Species,* published the following year.
3. "On the Zoological Geography of the Malay Archipelago" (1859). This paper presents Wallace's identification of the biogeographical break from Asia species that later became known as "Wallace's Line."
4. "On the Phenomena of Variation and Geographical Distribution as Illustrated by the Papilionidae of the Malayan Region" (1864). In this paper Wallace

presents his theories on species polymorphyism, mimicry, and protective coloration.

5. "On the Monkeys of the Amazon" (1852). In this early paper Wallace identifies the Amazon River and its many tributaries as reproductive isolating mechanisms that separate species and helps account for the biogeographical distribution patterns observed by naturalists.
6. "Sir Charles Lyell on Geological Climates and the Origin of Species" (1869). This is the paper that caused Darwin so much grief over Wallace's belief that natural selection cannot account for the human mind and that, therefore, a higher intelligence must have intervened. Within the scientific community this is Wallace's most controversial publication.
7. "On the Law Which Has Regulated the Introduction of New Species" (1855). This is the "Sarawak paper" in which Wallace states his belief in evolution as the theory best able to account for the geographical distribution of species.

Most striking in this ranking of Wallace's most cited works is the lack of a single nonscientific publication. There is not one of Wallace's many social commentaries that has survived into our time, and this is indicative of what in Wallace's work is currently important to us and what is not. For us, Wallace's science is what matters most, even though historically and in his own time, clearly these social issues were vital and compelling to both him and his contemporaries. To the extent that we remember Wallace (and that extent has been rather limited throughout most of the twentieth century), we do so primarily for his contributions to natural history and evolutionary theory. And that remembrance is primarily due to the fact that evolutionary theory has triumphed like no other overarching theory has since Newton united terrestrial and celestial mechanics into a cosmic worldview. Wallace's contribution to the monumental, pedestal-shattering evolution revolution was second only to Darwin's.

Themata

The deepest themes running throughout Wallace's many and diverse works are, as we saw earlier, those that concern most synthetic thinkers: *Theory—Data; Contingency—Necessity; Adaptationism—Nonadaptationism; Time's Arrow—Time's Cycle; Punctuationism—Gradualism.* In the narrative biography we will be exploring these in depth as they form the core of Wallace's thoughts, both scientific and social, and go a long way toward binding the varied details of his work with a handful of generalities. Some examples here, however, will suffice to show how important such themata were to Wallace.

Theory—Data concerns the interaction of culture and science, concepts and percepts. When Wallace was only twenty, for example, he wrote a paper entitled "The Advantages of Varied Knowledge," inspired by his experiences at the Mechanics' Institute through which he gained most of his early science education. In this essay, possibly given as a lecture at one of the Mechanics' Institutes, he argued for the importance of a varied education in giving one lots of specific examples of more generalized principles:

> There is an intrinsic value to ourselves in these varied branches of knowledge, so much indescribable pleasure in their possession, so much do they add to the enjoyment of every moment of our existence, that it is impossible to estimate their value, and we would hardly accept boundless wealth, at the cost, if it were possible, of their irrecoverable loss. . . . He who has extended his inquiries into the varied phenomena of nature learns to despise no fact, however small, and to consider the most apparently insignificant and common occurrences as much in need of explanation as those of a grander and more imposing character. He sees in every dewdrop trembling on the grass causes at work analogous to those which have produced the spherical figure of the earth and planets; and in the beautiful forms of crystallization on his window-panes on a frosty morning he recognizes the action of laws which may also have a part in the production of the similar forms of planets and of many of the lower animal types. Thus the simplest facts of everyday life have to him an inner meaning, and he sees that they depend upon the same general laws as those that are at work in the grandest phenomena of nature.[21]

The *Theory—Data* thematic pair can also be seen clearly expressed in a letter Wallace penned to his brother-in-law Thomas Sims on March 15, 1861, just three years after his discovery of natural selection. In this letter Wallace discusses the relationship between belief and evidence, particularly with regard to religion:

> You intimate that the happiness to be enjoyed in a future state will depend upon, and be a reward for, our belief in certain doctrines which you believe to constitute the essence of true religion. You must think, therefore, that belief is voluntary and also that it is meritorious. But I think that a little consideration will show you that belief is quite independent of our will, and our common expressions show it. We say, "I wish I could believe him innocent, but the evidence is too clear"; or, "Whatever people may say, I can never believe he can do such a mean action." Now, suppose in any similar case the evidence on both sides leads you to a certain belief or disbelief, and then a reward is offered you for changing your opinion. Can you really change your opinion and belief, for the hope of reward or the fear of punishment? Will you not say, "As the matter stands I can't change my belief. You must give me proofs that I am wrong or show that the evidence I have heard is false, and then I may change my belief"? It may be that you do get more and do change your belief. But this change is not voluntary on your part. It depends upon the force of evidence upon your individual mind, and the evidence remaining the same and your mental faculties remaining unimpaired—you cannot believe otherwise any more than you can fly.[22]

Contingency—Necessity concerns the thematic pairs of directionlessness and direction, or purposelessness and purpose in nature. At the age of eighty-seven, Wallace voiced his belief in the latter interpretation in an interview, referencing his forthcoming book whose title alone tells us where Wallace stood on this thematic pair: *The World of Life; A Manifestation of Creative Power, Directive Mind and Ultimate Purpose:*

> Ah, we come to a great question. I deal with it in a book which Chapman and Hall are to publish this winter. In some ways this book will be my final contribution to the philosophic side of evolution. It concerns itself with the great question of Purpose. Is there guidance and control, or is everything the result of chance? Are we solitary in the cosmos, and without meaning to the rest of the universe; or are we one in "a stair of creatures," a hierarchy of beings? Now, you may approach this matter along the metaphysical path, or, as a man of exact science, by observation of the physical globe and reflection upon visible and tangible objects. My contribution is made as a man of science, as a naturalist, as a man who studies his surroundings to see where he is. And the conclusion I reach in my book is this: That everywhere, not here and there, but everywhere, and in the very smallest operations of nature to which human observation has penetrated, there is Purpose and a continual Guidance and Control.[23]

Closely related is the thematic pair *Adaptationism—Nonadaptationism,* or the optimality and suboptimality of organisms as designed by nature. This theme was succinctly expressed by Wallace in an 1856 paper he wrote during his Malay Archipelago expedition entitled "On the Habits of the Orang-utan of Borneo," in which he rejected the idea of a first cause "for any and every special effect in the universe," and yet embraced the idea that there is a "general design" behind nature. (This paper was written shortly after his 1855 Sarawak paper in which he outlined the fundamentals of his theory of evolution, but well before his 1858 Ternate paper in which he introduced the evolutionary mechanism of natural selection.) Here we see that Wallace has clearly rejected creationism and accepted evolution, but was still grappling with the problem of determining the relative influence of contingency and necessity and adaptation and nonadaptation in nature's design:

> Naturalists are too apt to imagine, when they cannot discover, a use for everything in nature: they are not even content to let "beauty" be a sufficient use, but hunt after some purpose to which even that can be applied by the animal itself, as if one of the noblest and most refining parts of man's nature, the love of beauty for its own sake, would not be perceptible also in the works of a Supreme Creator. The separate species of which the organic world consists being parts of a whole, we must suppose some dependence of each upon all; some general design which has determined the details, quite independently of

> individual necessities. We look upon the anomalies, the eccentricities, the exaggerated or diminished development of certain parts, as indications of a general system of nature, by a careful study of which we may learn much that is at present hidden from us.[24]

In time, however, Wallace became, as he once said, more Darwinian than Darwin in his zeal to apply natural selection and find the adaptive significance of every structure and function of an organism. Such hyper-adaptationism (or hyper-selectionism) led to Wallace's biggest blunder in the eyes of Darwin, when he rejected natural selection as the primary influence in the development of the human mind, in favor of a direct involvement of a higher intelligence. His first public statement of this was in 1869, yet as early as 1853 in only his second book, *A Narrative of Travels on the Amazon and Rio Negro,* Wallace speculated on this themata: "In all works on Natural History, we constantly find details of the marvelous adaptation of animals to their food, their habits, and the localities in which they are found. But naturalists are now beginning to look beyond this, and to see that there must be some other principle regulating the infinitely varied forms of animal life."[25] Of course, natural selection was that principle he went on to discover in 1858, and then reject in specific cases where it failed to give a full account.

In the thematic pair *Time's Arrow—Time's Cycle* we see the interplay of unique historical events and repeating law-governed forces—history as one thing after another versus history as the same thing over and over. On the one hand, Wallace recognized the contingently unique nature of history and the unrepeatability of such complex entities as intelligence, which led him to conclude "that all the available evidence supports the idea of the extreme unlikelihood of there being on any star or planet revealed by the telescope—I don't say life, but any intelligent being, either identical with or analogous to man."[26] This statement was made in an interview given following the 1903 publication of Wallace's *Man's Place in the Universe; A Study of the Results of Scientific Research in Relation to the Unity or Plurality of Worlds.* Here Wallace clearly expressed his preference for a contingent, time's arrow model of history in arguing for humanity's uniqueness based on the extreme improbability that every contingent step of evolutionary change from basic bacteria to big brains could have been repeated elsewhere: "The ultimate development of man has, therefore roughly speaking, depended on something like a million distinct modifications, each of a special type and dependent on some precedent changes in the organic and inorganic environments, or in both. The chances against such an enormously long series of definite modifications having occurred twice over . . . are almost infinite."[27]

On the other hand, from the time of his earliest writings Wallace believed in the inherent perfectibility of man, linking several of these themata into one theory that emphasized cyclical but necessary progress that would lead to a utopian socialist society brought about through equality of opportunity for all. Here we also find the final thematic pair, *Punctuationism—Gradualism,* or catastrophic versus uniformitarian change. In natural history Wallace was usually a gradualist, emphasizing the principle of uniformitarianism through the actions of natural selection—slow and steady wins the race. Likewise for human history, where Wallace argued that although social change may come about through both revolution and evolution, he expressed his emphasis on the latter, fighting, for example, for women's suffrage because the vote was how he envisioned his utopian society coming to fruition gradually and legally.

These thematic tensions can be seen in many of Wallace's writings on spiritualism and socialism that, especially in his later years, were deeply integrated. In his 1875 monograph *On Miracles and Modern Spiritualism,* Wallace endeavored to present a "Theory of Human Nature," which he described as follows:

> 1. Man is a duality, consisting of an organised spiritual form, evolved coincidently and permeating the physical body, and having corresponding organs and development.
> 2. Death is the separation of this duality, and effects no change in the spirit, morally or intellectually.
> 3. Progressive evolution of the intellectual and moral nature is the destiny of individuals; the knowledge, attainments, and experience of earth-life forming the basis of spirit-life.[28]

How best to bring about this moral destiny? In his 1898 address to the International Congress of Spiritualists, appropriately entitled "Spiritualism and Social Duty," Wallace connected his spiritualism and his socialism, starting with a confession of his political preference: "As many of my friends here know, I myself, against all my early prepossessions, have come to believe that some form of Socialism is the only complete remedy for this state of things; and I define Socialism as simply the organisation of labour for the highest common good." Why should Spiritualists adopt socialism? "As Spiritualists we must uphold justice; and equality of opportunity for all is but bare justice. Knowing that the life here is the school for the development of the spirit, we must feel it our duty to see that the nascent spirit in each infant has the fullest and freest opportunity of developing all its faculties and powers under the best conditions we can provide for it." In fact, Wallace argued, Spiritualists more than any other organized body should become social activists:

> In this movement for justice and right, Spiritualists should take the lead, because they, more than any other body, know its vital importance both for this world and the next. The various religious sects are all working, according to their lights, in the social field; but their forces are almost exclusively directed to the alleviation of individual cases of want and misery by means of charity in various forms. But this method has utterly failed even to diminish the mass of human misery everywhere around us, because it deals with symptoms only and leaves the causes untouched. . . . But let us Spiritualists take higher ground. Let us demand Social Justice. This will be a work worthy of our cause, to which it will give dignity and importance. It will show our fellow-countrymen that we are not mere seekers after signs and wonders, mere interviewers of the lower denizens of the spirit-world; but that our faith, founded on knowledge, has a direct influence on our lives; that it teaches us to work strenuously for the elevation and permanent well-being of all our fellow men. In order to do this our watchword must be—NOT CHARITY ONLY BUT JUSTICE.[29]

All in a day's work for a heretic personality like Alfred Russel Wallace.

Heretic Personality

In Wilma George's 1964 study of Wallace she offered this disclaimer: "No attempt has been made to study Alfred Russel Wallace the man, nor to investigate the psychological reasons for his being both spiritualist and founder of zoogeography."[30] This biography does, with an examination of Wallace's personality to see how his seemingly disparate intellectual interests and his scientific and spiritualistic beliefs relate to one another. Wallace, in short, was a heretic personality. But what does that mean? A heretic is "one who maintains opinions upon any subject at variance with those generally received or considered authoritative," and personality is a "unique pattern of traits," in which "a trait is any distinguishable, relatively enduring way in which one individual differs from others."[31]

Personality, however, can be a fuzzy concept. Just what do we mean by personality, or a personality trait? The personality psychologist J. Guilford explained the confusion this way: "One does not need to read very far in the voluminous literature on personality to be struck by the fact that there is a somewhat bewildering variation in treatments of the subject. One might even conclude that there is confusion bordering on chaos." Nevertheless, Guilford prefaces his definition of personality with "an axiom to which everyone seems agreed: each and every personality is unique." Uniqueness means differences from others (though "similar in some respects"), known in the trade as "individual differences," which allows Guilford to conclude that "an individual's personality is his unique pattern of traits" and that a trait is "any distinguishable, relatively enduring way in which one individual differs from others."[32]

We may, based on this analysis, construct a composite (and modified) definition of personality: *The unique pattern of relatively permanent traits that makes an individual similar to but different from others.* Therefore a heretic personality is: *The unique pattern of relatively permanent traits that makes an individual maintain opinions upon any subject at variance with those considered authoritative.* In other words, a heretic-personality is an individual who is different from others in his or her tendency to accept and support ideas considered heretical, although similar to those who also maintain such anti-authoritarian, pro-radical tendencies. The assumption is that these traits, in being "relatively permanent," are not provisional states, or conditions of the environment, the altering of which changes the personality. The heretic personality tends to be heretical in most environmental settings, throughout much of a lifetime. This definition fits Wallace, who routinely maintained opinions on a variety of subjects typically at odds with the received authorities.

Today's most popular trait theory is what is known as the Five Factor model, or the "Big Five": (1) *Conscientiousness* (competence, order, dutifulness), (2) *Agreeableness* (trust, altruism, modesty), (3) *Openness to Experience* (fantasy, feelings, values), (4) *Extroversion* (gregariousness, assertiveness, excitement seeking), and (5) *Neuroticism* (anxiety, anger, depression).[33] To measure Wallace's personality Frank Sulloway and I had ten historians of science and Wallace experts rate him on a standardized Big Five personality inventory of forty descriptive adjectives using a nine-step scale. For example:

I see Alfred Russel Wallace as someone who was . . .
Ambitious/hardworking 1 2 3 4 5 6 7 8 9 Lackadaisical
Tough-minded 1 2 3 4 5 6 7 8 9 Tender-minded
Assertive/dominant 1 2 3 4 5 6 7 8 9 Unassertive/submissive
Organized 1 2 3 4 5 6 7 8 9 Disorganized
Rebellious 1 2 3 4 5 6 7 8 9 Conforming

Figure P-4 presents the results for Wallace in percentile rankings relative to Sulloway's database of over a hundred thousand subjects. Even though most of our expert raters expressed skepticism about the validity and reliability of this measurement on a historical figure, we computed an interrater reliability score for the ten raters of .59, a very respectable measure of reliability.[34] That is, whatever it is we were measuring, these ten experts were very consistent in their measurements. The validity of the scale will be considered next and in Chapter 10, in which Wallace's personality is explored further.

This cluster of traits befits a heretic personality. Although a heretic personality could be low on *extroversion,* or high on *neuroticism,* the key is high

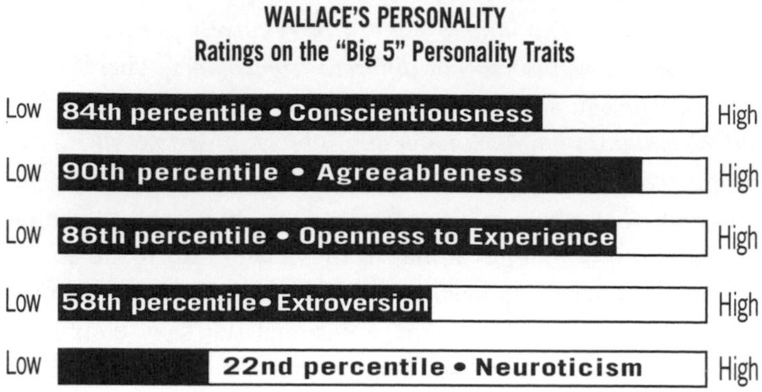

WALLACE'S PERSONALITY
Ratings on the "Big 5" Personality Traits

Low **84th percentile • Conscientiousness** High

Low **90th percentile • Agreeableness** High

Low **86th percentile • Openness to Experience** High

Low **58th percentile • Extroversion** High

Low **22nd percentile • Neuroticism** High

Figure P-4 Wallace's personality profile as rated by ten historians of science and Wallace experts and based on forty descriptive adjective pairs relative to a database collected by Frank Sulloway of over 100,000 subjects. (Rendered by Michael Shermer and Pat Linse)

openness to experience, which makes one more receptive to radical ideas and change. An exceptionally high *conscientiousness* makes one more conforming to the status quo and thus more intellectually conservative. Darwin, for example, scored in the 99th percentile on *conscientiousness* as well as in the 99th percentile on *openness to experience.* These personality profiles go a long way toward unraveling the mystery and apparent paradox of Wallace the scientist and Wallace the spiritualist, and the break with his more conservative colleagues like Darwin in accepting so many and different radical ideas. Darwin's high *conscientiousness* kept his high *openness* in check, helping him find that exquisite balance between orthodoxy and heresy. For a number of heretical claims Wallace did not have that personality brake. And he was agreeable to a fault. Wallace was simply far too conciliatory toward almost everyone whose ideas were on the fringe. He had a difficult time discriminating between fact and fiction, reality and fantasy, and he was far too eager to please, whereas his more tough-minded colleagues (Huxley especially) had no qualms about not suffering fools gladly. Although the theory of evolution was a moderately radical idea, it had the support of many in the scientific community before 1859 and many quickly converted shortly after, so Darwin's personality was well suited to it. By contrast, Wallace's other heresies such as phrenology and spiritualism never found mainstream support and remained on the intellectual fringes, precisely where a heretic personality like Wallace enjoys residing.

Birth Order and Heretical Science

The next obvious question to ask for the psychobiographer is what determines personality? Obviously sociocultural forces of the historical time period play a powerful role, and these we will explore in detail in the narrative biography. But it is constructive to consider the general origins and development of such a personality in all heretic-scientists throughout the history of science. This has been done by Frank Sulloway in his study of orthodoxy and innovation in science,[35] and especially in his book *Born to Rebel,* a study of "birth order, family dynamics, and creative lives."[36] Sulloway confirmed the importance of a number of factors in the development of a personality willing to explore and ultimately accept heretical ideas. Sulloway conducted a multivariate correlational study examining the tendency toward rejection or receptivity of a new scientific theory based on such variables as "date of conversion to the new theory, age, sex, nationality, socioeconomic class, sibship size, degree of previous contact with the leaders of the new theory, religious and political attitudes, fields of scientific specialization, previous awards and honors, three independent measures of eminence, religious denomination, conflict with parents, travel, education attainment, physical handicaps, and parents' ages at birth."[37] In 2,784 participants in 28 diverse scientific controversies that spanned over 400 years of history Sulloway discovered, "using multiple regression models with these and other variables, that birth order consistently emerges as the single best predictor of intellectual receptivity."[38]

Consulting over a hundred historians of science, Sulloway had them "judge the stances taken by the participants in these debates," which included the Copernican revolution, relativity, phrenology, quantum mechanics, the indeterminacy principle, Freudian psychoanalysis, Semmelweiss and puerperal fever, Lister and antisepsis, mesmerism, Hutton's theory of the earth, Harvey and the circulation of the blood, spontaneous generation, Lyell and uniformitarianism, and many more, including the Darwinian revolution. For the collected twenty-eight controversies, spanning in dates from 1543 to 1967, Sulloway found that only 34 percent firstborns supported the new ideas, compared to 64 percent laterborns. Specifically, Sulloway discovered that the likelihood of a laterborn accepting a revolutionary idea was 3.1 times higher than a firstborn, and for radical revolutions the likelihood was 4.7 times higher. Using a Fisher's one-tailed exact test of significance, Sulloway found this laterborn tendency for acceptance to be significantly greater than the firstborns at the .0001 level, which means the probability of this happening by chance is virtually zero.[39] Historically speaking, this indicates that "laterborns have indeed generally introduced and supported other major conceptual transformations over the protests of their firstborn colleagues. Even when the principal leaders of the new theory occasionally turn out to be firstborns—as was the case with Newton, Einstein, and Lavoisier—the opponents as a whole are still predominantly firstborns, and the converts continue to be mostly laterborns."[40] Children without siblings, a "control group" of sorts, were sandwiched in between firstborns and laterborns in their percentage of support for radical theories.

Relevant to our purposes here, Sulloway also assessed the attitudes of over 300 scientists toward the Darwinian revolution between 1859 and 1870. The criteria for acceptance or rejection of Darwinism were based on three premises: "(1) that evolution takes place, (2) that natural selection is an important (but not an exclusive) cause of evolution, and (3) that human beings are descended from lower animals without supernatural intervention." Acceptance of all three makes one a Darwinian. The results for this particular controversy were consistent with those of the general model—83 percent Darwinians were laterborns and 55 percent non-Darwinians were firstborns, a statistically significant difference at $p<.0001$. Included in those who rejected Darwinian evolution by natural selection were Louis Agassiz, Charles Lyell, John Herschel, and William Whewell, all firstborns. Counted as full supporters were Joseph Hooker, Thomas Henry Huxley, Ernst Haeckel, and, of course, Charles Darwin and Alfred Russel Wallace, all laterborns.[41]

Plotting over 1,000,000 data points (which also confirm the effect in nonscientific revolutions such as the Protestant Reformation and the French and American Revolutions), Sulloway found that the degree of radicalness of the new theory was also correlated with birth order. Laterborns prefer probabilistic views of the world, such as Darwin's and Wallace's theory of natural selection, to a more mechanical and predictable worldview preferred by firstborns. Finally, Sulloway found that when firstborns did accept new theories, they were typically the most conservative of the bunch, "theories that typically reaffirm the social, religious, and political status quo and that also emphasize hierarchy, order, and the possibility of complete scientific certainty."[42] Louis Agassiz, for example, opposed Darwin (predicted at a 98.5 percent probability by the model), but he supported glaciation theory, explained by the fact that relative to the ideological implications of Darwinism, glaciation was conservative, as well as being linked with the already accepted theories of catastrophism and creationism. The theory of evolution by means of natural selection, by contrast, did not reaffirm the status quo of any social institution and was thus especially appealing to laterborn radicals like Wallace.

A deeper subject to probe in this context is *why* firstborns are more conservative and influenced by authority, while laterborns are more liberal and receptive to ideological change. What is the causal connection between birth order and personality? One hypothesis is that firstborns, being first, receive substantially more attention from their parents than laterborns, who tend to enjoy greater freedom and less indoctrination into the ideologies of and obedience to authorities. Firstborns generally have greater responsibilities, including the care and liability of their younger sibs. Laterborns are frequently a step removed from the parental authority, and thus less inclined to obey and adopt the beliefs of the higher authority. Sulloway summarizes the birth order–personality connection:

> Sandwiched between parents and younger siblings, the firstborn child occupies a special place within the family constellation and, for this reason, generally receives special treatment from parents. Moreover, as the eldest, firstborns tend to identify more closely with parents, and through them, with other representatives of authority. This tendency is probably reinforced by the firstborn's frequent role as a surrogate parent to younger siblings. Consistent with these developmental circumstances, firstborns are found to be more respectful of parents and other authority figures, more conforming, and more conscientious, conventional, and religious. Laterborns, who tend to identify less closely with parents and authority, also tend to rebel against the authority of their elder siblings.[43]

There is independent corroboration of this hypothesis in the field of developmental psychology. J. S. Turner and D. B. Helms, for example, report that "firstborns become their parents' center of attention and monopolize their time. The parents of firstborns are usually not only young and eager to romp with their children but also spend considerable time talking to them and sharing their activities. This tends to strengthen bonds of attachment between the two."[44] Quite obviously this attention would include more rewards and punishment, further reinforcing obedience to authority and controlled acceptance of the "right way" to think. Adams and Phillips[45] and Kidwell[46] report that this excessive attention causes firstborns to strive harder for approval than laterborns. Markus has discovered that firstborns tend to be anxious, dependent, and more conforming than laterborns.[47] Hilton, in a mother–child interactive experimental setting with twenty firstborn, twenty laterborn, and twenty only children (four years of age), found that firstborns were significantly more dependent on, and asked for help or reassurance from their mothers, than the laterborn or only children.[48] In addition, mothers of firstborns were significantly more likely to interfere with their child's task (constructing a puzzle) than were the mothers of laterborn or only children. Finally, it has been shown by Nisbett[49] that laterborns are far more likely to participate in relatively dangerous sports than are firstborns, which is linked to risk taking, and thus "heretical" thinking.[50]

Birth order alone, of course, does not determine the ideological receptivity to radically new ideas. Instead, Sulloway explains, birth order is a proxy for other influencing variables, such as age, sex, and socioeconomic class, which in turn influence openness to new ideas. For example, although "social class itself exerts absolutely no direct influence on the acceptance of new scientific ideas," Sulloway discovered "it is only through a triple-interaction effect with birth order and parental loss that social class plays a subtle but significant role in attitudes toward scientific innovation."[51]

In Sulloway's book *Born to Rebel* he presented a summary of 196 controlled birth-order findings classified according to the "Big Five" personality dimensions. The results are as follows:

> *Conscientiousness:* Firstborns are more responsible, achievement oriented, organized, and planful.
> *Agreeableness:* Laterborns are more easygoing, cooperative, and popular.
> *Openness to experience:* Firstborns are more conforming, traditional, and closely identified with parents.
> *Extroversion:* Firstborns are more extroverted, assertive, and likely to exhibit leadership.
> *Neuroticism/emotional instability:* Firstborns are more jealous, anxious, neurotic, fearful, and likely to affiliate under stress.[52]

Look again at Figure P-4. This birth-order-driven personality profile perfectly matches the Big Five findings on Wallace by expert raters. Consider the influencing variables in this multivariate matrix. Alfred Wallace was laterborn (the eighth of nine children), was in the middle/lower class (in Sulloway's classification system), and was separated from his parents at age fourteen—the triple-interaction effect that generates the greatest amount of support for radical scientific theories. According to Sulloway's multivariate model, which includes twelve predictors and their interaction effects, Wallace possessed a 99.5 percent probability of championing the theory of evolution.[53]

Wallace's parents began in the middle class, but soon deteriorated into the working class. Wallace's formal education (and thus indoctrination into conserving traditional beliefs) was a minimal seven years, and his father went bankrupt when Wallace was only thirteen, at which time he went to live with his brother John, and then later with his brother William. Wallace rarely returned home, and by the time he completed his four-year trip to the Amazon, at age twenty-nine, his father had died. By Sulloway's analysis, Wallace was almost destined to be a radical scientist because of his heretic personality. Other factors make this even more likely, as Sulloway explains:

> In my multivariate model, Wallace is, of course, a laterborn, in the most liberal, political, and religious cohorts of the model, already somewhat acquainted with Darwin before 1859, and relatively young at that time (36). What differentiates

> him from Darwin is primarily his greater degree of political radicalness. Darwin falls in the third rather than the fourth category for political and religious beliefs, since he was a Whig and a deist (averaging about 3.6 on my 5-point scales), whereas Wallace was a clear radical and a deist (about a 4.4).[54]

(To demonstrate the interactive nature of both historical contingencies as well as his multivariate model, Sulloway notes that Bishop "Soapy Sam" Wilberforce, who vehemently opposed Darwinism in the now famous debate against Thomas Huxley, was himself a laterborn, but other mitigating factors such as differences in age, religiosity, politics, and personal contact with Darwin, made "the likelihood of these two individuals agreeing with one another . . . almost nil."[55])

Of the Big Five personality traits, *openness to experience* is most sensitive to birth order effects, and here both Wallace and Darwin score exceptionally high. Given this concatenation of psychological, social, and cultural variables, and their multivariate interactions, we cannot fail to glean a deeper understanding of Wallace. And, following Darwin's dictum, we should not fail to apply such psychobiographical methods to glean a deeper understanding of all humans.

Adding to the Fund of Instruction

What end, then, will this biography serve? The same as that of any work of history—that we may learn from those who came before. History is primarily for the present, secondarily for the future, and tertiarily for the past. In 1843 a very youthful Alfred Russel Wallace composed a lecture on "The Advantages of Varied Knowledge" in which he reflected on this very question and answered it eloquently:

> Is it not fitting that, as intellectual beings with such high powers, we should each of us acquire a knowledge of what past generations have taught us, so that, should the opportunity occur, we may be able to add somewhat, however small, to the fund of instruction for posterity? Shall we not then feel the satisfaction of having done all in our power to improve by culture those higher faculties that distinguish us from the brutes, that none of the talents with which we may have been gifted have been suffered to lie altogether idle? And, lastly, can any reflecting mind have a doubt that, by improving to the utmost the nobler faculties of our nature in this world, we shall be the better fitted to enter upon and enjoy whatever new state of being the future may have in store for us?[56]

1

UNCERTAIN BEGINNINGS

The year 1889 was a historically interesting one, although no more nor less than most others in the latter half of this rapidly changing century. Adolf Hitler was born, Brazil was proclaimed a republic, Benjamin Harrison became the twenty-third president of the United States, the British South Africa Company was given a Royal Charter, Barnum and Bailey's circus opened in London, Vincent van Gogh painted his *Cypress Tree* landscape, Alexander Gustave designed the Eiffel Tower, Richard Strauss penned his poem "Don Juan," and Gilbert and Sullivan produced "The Gondoliers." In science Ivan Pavlov began his research into the digestive system that would lead to his discovery of classical conditioning, Francis Galton introduced the concept of the correlation coefficient as a tool for the scientific study of the heritability of human abilities, and George Fitzgerald anticipated Einstein when he formulated the principle that objects shrink slightly in the direction they are traveling.

In May of that year a British naturalist published a panoramic summary of evolutionary theory in which he outlined his heretical views on the evolution of the human mind and the spiritual purposefulness of all evolutionary progress: "To us, the whole purpose, the only *raison d'être* of the world—with all its complexities of physical structure, with its grand geological progress, the slow evolution of the vegetable and animal kingdoms, and the ultimate appearance of man—was the development of the human spirit in association with the human body."[1] The book was entitled *Darwinism, An Exposition of the Theory of Natural Selection with Some of Its Applications,* but its author was not Charles Darwin, who was already seven years interred at Westminster Abbey. It was the definitive statement of Alfred Russel Wallace who, at seventy-six years of age, was bringing together in a consilience of inductions (as his colleague William Whewell called this process of convergence from many different sources) a lifetime of data and theory, observations and generalizations, from a surfeit of fields both scientific and social. It was a "theory of everything"—as a later generation of physicists would call their search for a grand unifying principle behind the cosmos—a synthesis of knowledge that tied together the physical, biological, and social sciences.

Whatever became of that grand theory? It died with the death of its chief defender. What was the *raison d'être* of this man, with all his complexities, progress, and slow evolution to the ultimate appearance of his spiritual and bodily purpose? To find out, we return to the birth of the author and the theory, a consilience of creator and culture (since theories are constructed out of brains in conjunction with environments) that begins, appropriately enough for this subject, with a date of uncertain origin.

Thrown on Our Own Resources

For the majority of his exceptionally long life, Alfred Russel Wallace thought he was born in 1822. But in 1903, while researching his autobiography, he confided to his friend and colleague at Oxford, Edward B. Poulton, that he had recently come "into possession of an old Prayer Book in which the date of birth of my father is given by his father, & of all my brothers & sisters in my father's handwriting, & there I am put down as born in 1823 Jan. 8th. & the date is repeated for my baptism Feb. 16th. 1823." It must have been a pleasant surprise for a man in the twilight of his life to discover he was "younger than I had supposed."[2]

Wallace's birthplace was never in doubt (Figure 1-1). He was born in Usk, Monmouthshire, Wales, the eighth child of Thomas Vere Wallace and Mary Anne Greenell (Figure 1-2), both devout members of the Church of England. Although young Alfred never met his grandparents on either side, through a 1723 prayer book and old birth registries, he was able to trace his heritage back to "that famous stock" of Sir William Wallace of Hanworth, Middlesex. On his mother's side his grandfather was John Greenell of Hertford, where he would later spend several years of his childhood.[3] On both sides of the family a number of Wallace's relations practiced law, including his father Thomas, who began by apprenticing to a solicitor preparing legal cases and then "was duly sworn in as an Attorney-at-Law of the Court of King's Bench" and for a brief time earned a respectable middle-class income of 500 pounds a year.[4]

In time, however, Thomas left the law to try his hand at many a business venture—tutoring, operating a small subscription library, publishing an illustrated magazine—but each in turn failed, thrusting the family into the working classes where the young Alfred spent most of his childhood in difficult eco-

Figure 1-1 The birthplace of Alfred Russel Wallace, Kensington Cottage, Usk. (From *My Life,* 1905, v. I, 20)

Figure 1-2 Mary Anne Wallace (née Greenell), Alfred's mother. Thomas Vere Wallace, Alfred's father. (From *My Life*, 1905, v. I, 22)

nomic circumstances in which "we were all of us very much thrown on our own resources to make our way in life." But as was typical through a life strewn with setbacks—financial, physical, and psychological—Wallace's sanguine temperment led him to conclude that even though he "inherited from my father a certain amount of constitutional inactivity or laziness, the necessity for work that our circumstances entailed was certainly beneficial in developing whatever powers were latent in us; and this is what I implied when I remarked that our father's loss of his property was perhaps a blessing in disguise."[5] Unfortunately, not all of his siblings would get the opportunity to test their mettle. One unnamed sister died at five months; two others, Mary Anne and Emma, passed at ages six and eight years, respectively. His brother William died from a lung infection that he caught in a freezing train ride from London to South Wales, while his brother Herbert died in South America shortly after traveling there to join Alfred on his expedition. Those Wallaces who survived childhood, by contrast, lived long lives. Alfred's brother John moved to California where he lived to age seventy, and his sister Frances, who was involved in his investigations into the spirit world and married a photographer name Thomas Sims who became a lifelong friend of Alfred, survived eighty-one years. Ever the number-crunching naturalist, Wallace computed a mean survival age of seventy years for the Wallaces and seventy-six years for the Greenells, not counting infancy or childhood deaths.[6]

Much has been made of the sharp contrast between Darwin's upper-crust Cambridge education and Wallace's lower-grade grammar school training. In fact, the story is more complex. Since his father "was fond of reading, and through reading clubs or lending libraries we usually had some of the best books of travel or biography in the house," the young Alfred listened to the poetic homilies of Lear and Cordelia, Hamlet and Lady Macbeth. He even recalled that his father wrote a history of Hertford (never published) and tried his hand at antiquities and heraldry, sketching and poetry. A quatrain from one poem, entitled "On the Custom observed in Wales of dressing the Graves with Flowers on Palm Sunday," proved most prophetic for the tragedies that befell so many families of that age:

> That place of rest where parents, children, sleep,
> Where heaves the turf in many a mould'ring heap
> Affection's hand hath gaily decked the ground
> And spring's sweet gifts profusely scatter'd round.[7]

Even before his limited education began, Alfred was a tinkerer. One day his parents read to him an Aesop's fable about a thirsty fox who could not reach the water at the bottom of a pitcher and solved the problem by dropping pebbles into it until the water rose to a drinkable level. The youngster was puzzled. It seemed like magic to him. How could stones cause water to rise? To find out, the boy who would grow up to become one of the greatest empirical observers of his age performed his first experiment—a simple one involving a bucket of water and a pile of small stones. He soon got his answer. "I could not see that the water rose up as I thought it ought to have done. Then I got my little spade and scraped up stones off the gravel path, and with it, of course, some of the soft gravel, but instead of the water rising, it merely turned to mud." The young scientist finally grew tired and gave up, concluding "that the story could not be true." It was his first refutation of a conjecture that "rather made me disbelieve in experiments out of story-books."[8] His skepticism of what authorities published in books would grow far more serious over the decades.

This anecdote was emblematic of Wallace's educational experiences. The only formal schooling he received was seven years at Hertford Grammar School, a quaint structure built in 1617 that featured a graveyard next to the long entry drive (see Figure 1-3). He remembered it as a largely valueless experience, with the sole exception of gaining a working knowledge of Latin and French, useful in later scientific writings. Ever the resourceful laterborn seeking to find his own path in life separate from his older and more tradition-bound siblings, Alfred did not let his mediocre schooling interfere with his education, particularly as he grew more and more interested in the natural

Figure 1-3 The grammar school at Hertford, the place of Alfred's only formal education. (From *My Life*, 1905, v. I, 34)

world. It was during this period, he later recalled, that he "began to feel the influence of nature and to wish to know more of the various flowers, shrubs, and trees I daily met with but of which for the most part, I did not even know the English names."[9] Although his Hertfordian experiences were largely forgettable, his impact on the town was not—the year after his death, Stephen Austin and Sons, Ltd., of Hertford, published a short biography on Wallace entitled *A Great Hertfordian,* written by G. W. Kinman, M.A., in which the author placed Wallace on the English Olympic team, "if Olympic contests were of an intellectual character."[10]

Perhaps some of that character was built on the floggings with a cane delivered to disobedient children who sometimes deserved it, but more likely such punishment, often doled out for nothing more than intellectual shortcomings, instilled in Alfred a sense of inequity that needed redressing. More than anything in these limited years at the Hertford Grammar School, however, was the sense he received that most education involved the rote memorization of unconnected facts—data without theory. He was forced to memorize geographical names, but provided no cultural or historical context. History was a mind-numbing sequence of names and dates, kings and queens, wars and rebellions, without a hint as to why history unfolded as it did. Most significantly, nature was treated as a static entity to be named, but not understood, pinned with labels, but without the connective string to tie it together.

These formative years, although seemingly uneventful and for the most part forgotten by all but a handful of historians of science, were critical in one respect: the dynamics of his family structure. It is not unimportant that Wallace was eighth born, that his parents had limited resources, and that like all children of large families he had to compete with his siblings to find a niche. Like the principle of diversification he would later come to discover during his explorations of the multifarious flora and fauna of the Amazon and Malay Archipelago, as a youngster Alfred learned to diversify his interests and activities, and this diversification led him down a path radically different from those of most of his siblings and friends.

God's Works and Words

By the time Wallace turned thirteen the family's resources were all but depleted and his parents could no longer afford his school's tuition. Alfred managed to squeeze in another year, covering the tuition by tutoring his classmates in writing and reading. Through this experience he not only developed the skill of communicating concepts in a clear fashion to others, which would serve him well decades later in his popularization of science, he honed an entrepreneurial independence that would carry him through a working life that never saw a guaranteed steady income. By 1837 (the year of Victoria's accession to the throne), the Wallace family financial structure utterly collapsed, and with it their ability to support Alfred. The now six-foot-tall, brown-haired, blue-eyed "little Saxon" (as he was called by intimates) was sent to London to live and work with his older brother John, who at age nineteen was apprenticed to a master builder.

In reflecting back on his schooling and early education, Wallace called the educational system an "utter failure" in the teaching of science. There were several reasons for this, including "the notion that any good can result from the teaching of such a large and complex subject to youths who come to it without any preliminary training whatever, and who are crammed with it by means of a lesson a week for perhaps one year." Worse still were the teachers who lacked "the whole range of subjects" necessary to teach science, as well as "how to communicate to others the knowledge they themselves possess."[11] On this latter front Wallace would mount a lifelong assault through two dozen books and hundreds of articles, many of which were written not just for his fellow scholars and scientists, but for anyone interested in the subject and willing to exert a modicum of effort to master the material.

For Alfred, as for so many budding youngsters interested in dabbling in nature, science was struggling to find an identity. The term "scientist" had not yet been invented, and most of those doing what we might think of as science would have thought of it as natural philosophy or natural theology. In the previous century philosophers and theologians adopted the new mechanical philosophy that sprung out of Newton's synthesis of Copernicus, Brahe, Kepler, and Galileo, and united it with the natural theology of correlating the words and works of God. Far from this new mechanical philosophy removing God's providence in the world, most understood it to mean that by its very nature of lifelessness and passivity, such mindless matter needed God periodically to energize it. Not only did science not remove God from the picture, it gave natural philosophers and theologians a way to distinguish between God's *ordinary providence,* reflected in the normal cycles of nature, and God's *special providence,* witnessed in the unique events of nature associated with miracles. It was the themata of time's cycle versus time's arrow, necessity versus contingency, the universal and the particular.[12]

Before Darwin and Wallace gave to science a mechanism by which nature can create apparently intelligent design without an intelligent designer (thereby cleaving science and religion into two entirely separate spheres of knowledge), the study of nature and theology were two sides of the same coin. The study of nature was a supplement to scripture, and the study of scripture was enhanced by a deeper understanding of nature. Theology liberated those interested in nature to study her secrets in this words-to-works system. The chemist Robert Boyle put it well in his 1662 book *On the Usefulness of Natural Philosophy:* "When . . . I see in a curious clock, how orderly every wheel and other part performs its own motions, and with what seeming unanimity they conspire to show the hour, and accomplish the other designs of the artificer: I do not imagine that any of the wheels, etc., or the engine itself is endowed with reason, but commend that of the workman, who framed it so artifically."[13] This sentiment was echoed by his countryman Walter Charleton, who noted, "Just as a watch cannot run without a mainspring, so the world cannot run without God as an 'energetical principle' or as the *Spring* in the Engine of the world."[14]

British natural theology was given a boost by a group of liberal thinkers called the *Latitudinarians*—so named for their greater "latitude" in the acceptance of heretical ideas—when they devised a natural religion that not only included logic and reason as part of its doctrinal platform, but eschewed inspiration and personal revelation as a justification of belief. The *Latitudinarians* argued that empiricism is essential to religious belief, and in so doing they opened the floodgates for the study of nature. They believed that events in nature, particularly those out of the ordinary (earthquakes, volcanic eruptions, comets, not to mention political upheavals), were signs of God's involvement in the earthly domain.[15] The *Latitudinarians* were reinforced by the *Physico-Theologians,* such as John Ray and William Derham, who insisted that even in the ordinary structure and workings of nature there was ample evidence of design that proved the existence of a designer.[16] Even Newton, that paragon of mechanical philosophy and textbook hero of empirical science, saw in his "system of the world" the workings of the Almighty: "When I wrote my treatise upon our Systeme I had an eye upon such Principles as might work with considering men for the beliefe of a Deity and nothing can rejoyce me more than to find it usefull for that purpose."[17] Newton corresponded regularly with the Reverend Thomas Burnet, who was himself developing a theory of the earth that included deep implications for natural theology. In one letter, for example, Newton explained to Burnet what can be inferred from something as mundane as the earth's rotation: "Where natural causes are at hand God uses them as instruments in his works, but I do not think them alone sufficient for the creation and therefore may be allowed to suppose that amongst other things God gave the earth its motion by such degrees and at such times as was most suitable to the creatures."[18]

Deists, though less enthusiastic about discovering God's continuing providence in nature, did feel that nature reveals the evidence of God's existence as the creator. Furthermore, a popular group of Millenarians, who believed that the second coming of Christ would occur with the millennium (of various calculations), held that the destructive events in nature, particularly those geologic, were signs of the end times, as predicted by the prophets of the Bible. Anglican millenarianism in particular (of which Newton and Burnet were a part) held that the eventual destruction and reconstruction of God's kingdom on earth would place the Church at its triumphant head.[19] Because the end of the millennium and the second coming would be punctuated by such "signs of the times" as earthquakes, volcanos, comets, and colliding planets, a theory to explain these events was critical to the millenarian model of the world.

The Reverend Burnet's book *Sacred Theory of the Earth* provided one such model.[20] "We are not to suppose that any truth concerning the natural world can be an enemy to religion; for truth cannot be an enemy to truth, God is not divided against Himself."[21] But Burnet added something unique to the works-to-words system: a system running on its own without constant maintenance was superior to one that required regular attention because "it is no detraction from Divine Providence that the course of Nature is exact and regular, and that even in its greatest changes and revolutions it should still conspire and be prepar'd to answer the ends and purposes of the Divine Will."[22] This regularity was made possible by a clockwork world that does not require "extraordinary consourse and interposition off the First cause." Burnet believed that this analogy "is clear to everyman's judgment. We think him a better Artist that makes a Clock that strikes regularly at every hour from the springs and Wheels which he puts in the work, than he that hath so made his Clock that he must put his finger to it every hour to make it strike."[23]

Thomas Burnet was no out-of-time evolutionist, and his "sacred theory" emphasized a cycle of time more in tune with biblical creationism than an arrow of contingent evolutionary time, but he did introduce a model of a changing, evolving earth: "Since I was first inclin'd to the Contemplation of Nature, and took pleasure to trace out the Causes of Effects, and the dependance of one thing upon another in the visible Creation, I had always, methought, a particular curiosity to look back into the first Sources and ORIGINAL of Things; and to view in my mind, so far as I was able, the Beginning and Progress of a RISING WORLD. . . . there is a particular pleasure to see things in their Origin, and by what degrees and successive changes they rise into that order and state we see them in afterwards, when compleated."[24]

Burnet's "sacred theory," along with Newton's "system of the world," established a research paradigm that included nature as an essential component of study for all learned men that carried the day throughout the eighteenth century and set the stage for the final removal of the Deity from nature in the nineteenth. The scriptural geologists preceding and concurrent with Newton and Burnet were primarily concerned with the *when* and the *why* of the world. Newton and Burnet wanted to know the *how,* or the mechanics of the origin, development, and eventual destruction of the earth. To Burnet, the study of matters terrestrial was just as important as matters celestial: "The greatest objects of Nature are, methinks, the most pleasing to behold; and next to the great Concave of the Heavens, and those boundless Regions where the Stars inhabit, there is nothing that I look upon with more pleasure than the wide Sea and Mountains of the Earth. There is something august and stately in the Air of these things that inspires the mind with great thoughts and passions."[25]

A century and a half later it was a philosophy of science that had fully come of age and was shared by a young naturalist who would find endless pleasure, thoughts, and passions in his voyages on the seas and across the mountains of the earth, and would discover there a principle that would make Newton's and Burnet's Deity unnecessary.

Working-Class Science

For the young Alfred even at its best classroom-based and book-bound science education harbored serious shortcomings. What he later called hands-on "nature-knowledge" he felt was "the most important, the most interesting, and therefore the most useful of all knowledge. But to be thus useful it must be taught properly throughout the whole period of instruction from the kindergarten onwards, always by means of facts, experiments, and outdoor observation, supplemented, where necessary, by fuller exposition of difficult points in the classroom." Wallace believed, not surprisingly (as it describes him), that the best teachers are those who "have largely taught themselves by personal observation and study" because "they alone know the difficulties felt by beginners; they alone are able to go to the fundamental principles that underlie the most familiar phenomena, and are thus able to make everything clear to their pupils." These teachers "should be carefully sought for and given the highest rank in the teacher's profession."[26]

Decades later, through thousands of public lectures and popular writings, Wallace would pioneer his own teaching style outside of the classroom, educating the masses on knowledge in and the virtues of the various sciences. Degreeless, and without an all-important institute affiliation, Wallace nevertheless carved a path through life, and the life sciences, in which he would leave his indelible stamp. It was his good fortune to have been born at a time when both practically and culturally this could be accomplished. If he had been born a century earlier, such a path would likely have been closed off by the restrictions of social class and the lack of a market for such self-taught men. Born a century later, he would have come to scientific age just as science was becoming "Big Science," yet before the university educational system was open to all. As always, timing is everything.

The Victorian England into which Alfred Russel Wallace was coming of age was young, with almost half the population under twenty by mid-century. Her workmen were described by one continental traveler as having "superior persevering energy" and "whose untiring, savage industry surpasses that of every other country I have visited, Belgium, Germany, and Switzerland not excepted." Another commentator noted that "the people of England, which calls herself Old, are younger than the people of many other countries, and certainly younger than the people of the countries of stagnation."[27] It was an age of visible and dramatic transition, not from the immediate previous age, but from long before. As one historian has noted: "To Mill and the Victorians the past which they had outgrown was not the Romantic period and not even the eighteenth century. It was the Middle Ages."[28] All cultures change. But the mode and tempo of this change was significant and the participants knew it.

Victorian England was also a stratified society, with the landed gentry on top, the middling class (as the name suggests) in the upper middle and middle, artisans and some working class in the lower middle, and the rest of the working class, farmers, and "deserving poor" on the bottom. Although the two would eventually come to share near equal cultural status in the sciences, Charles Darwin and Alfred Wallace were separated by a socioeconomic abyss. Although Alfred's father occasionally propelled himself and his family into a middle-class station, they just as quickly sank when another business venture failed. For the Darwins and his close relations the Wedgewoods, investments in property, agriculture, and industry afforded their children the finest in classical education. These upper-class children, who spent seventeen of twenty-two hours per week on the classics of Greece and Rome, were not trained to become scholars but well-educated men of the world (and it was almost exclusively men) in order to develop "character."[29] And nothing built character more than outdoor sports, most notably hunting and shooting, two of Darwin's youthful favorites.

While the gentry owned land, farmers rented it and laborers worked it. But this was changing, as "individual effort, backed by austerity of life, would propel any man, no matter what his origins, to success in this world and if reinforced by the right brand of piety, to salvation in the next."[30] Rapid industrialization and technological innovation propelled the English economy to become the most powerful in the world.[31] Wallace was the beneficiary of this social trend, and he would epitomize Samuel Smiles's *Self Help* biographies of self-made men filled with heroic individualism and enterprising industry. He would make it in the world of science, and he would do so with self-sufficiency and pride, joined by many others who shared his station.

Fundamental Principles

These fellow travelers on the road to universal and hitherto unaccessible knowledge could be found in the night schools of the nineteenth century—the Mechanics' Institutes. Wallace's earliest significant intellectual pursuits came from evenings spent with his brother John "at what was then termed a 'Hall of Science' . . . really a kind of club or mechanics' institute for advanced thinkers among workmen, and especially for the followers of Robert Owen, the founder of the Socialist movement in England."[32] The Mechanics' Institutes, founded in London and Glasgow the year Wallace was born, were custom-designed for intellectually restless working-class young adults. It was here that the intellectual foundation was built for Wallace's career in science. His philosophical predilections, religious skepticism, and political–economic speculations also found their roots at the Institutes. "Here we sometimes heard lectures on Owen's doctrines, or on the principles of secularism or agnosticism, as it is now called." (The latter term was not coined until 1869 by Thomas Huxley.) The intellectual explorations knew no bounds, as this is when "I also received my first knowledge of the arguments of sceptics, and read among other books Paine's *Age of Reason*."[33] Wallace directly acknowledged this influence on his mature sociopolitical thoughts when he noted that "my introduction to advanced political views, founded on the philosophy of human nature, was due to the writings and teachings of Robert Owen and some of his disciples."[34]

Further, Wallace's lifelong belief in and commitment to the position that the environment shapes our behavior more than biology had its start here when he was introduced to the always controversial debate on the relative roles of nature and nurture through Owen. "His great fundamental principle, on which all his teaching and all his practice were founded," Wallace explained, "was that the character of every individual is formed *for* and not *by* himself, first by heredity, which gives him his natural disposition with all its powers and tendencies, its good and bad qualities; and secondly, by environment, including education and surroundings from earliest infancy, which always modifies the original character for better or for worse." Wallace believed that theories of biological determinism had failed to implement effective social change: "The utter failure of this doctrine, which has been followed in practice during the whole period of human history, seems to have produced hardly any effect on our systems of criminal law or of general education; and though other writers have exposed the error, and are still exposing it, yet no one saw so clearly as Owen how to put his views into practice."[35]

Owen's "fundamental principle" would be inculcated into Wallace's scientistic worldview that spanned and integrated both biological and social systems, and the Owenian influence was pervasive throughout Wallace's life. Born in Newtown, Montgomeryshire, Wales, in 1771, Robert Owen was a shop-hand, manufacturer, factory reformer, educator, trade union leader, utopian socialist, and founder of numerous social movements and experimental social programs. In his fellow Welshman, Wallace could see his past as well as his future. Owen's was a working-class background, his father a saddler and ironmonger. At age nine he left school and became a shop-boy at a local factory. A series of jobs in London and Manchester prepared him to enter business on his own, which eventually positioned him to become a partner in the New Lanark Mills in 1800. Self-educated, Owen worked his way into the literati of Manchester through its Literary and Philosophical Society.

In the quarter of a century that Owen operated the New Lanark Mills, he built it into a well-controlled human community founded on his ideals, which included the belief that owners should provide their workers not only a steady and respectable income, but decent homes and quality food and clothing at affordable prices. One of the reasons New Lanark was so successful, in fact, was that Owen's workers were paid better wages, worked shorter hours, and were provided with better equipment than most of his competitors. He *chose* not to employ young children below the age of ten (where others used children six years old and younger) and was not coerced into so doing. He even constructed a school system for his worker's children. For all of this he was rewarded with greater profits as well as national and international recognition (the philosopher and founder of utilitarianism, Jeremy Bentham, became a partner in the firm). The New Lanark experiment was duplicated by others, and Owen even tried one in America in New Harmony, Indiana.

Like Wallace, Owen not only lived a long life (eighty-seven years, dying in 1858 when Wallace was in the Malay Archipelago), he wrote voluminously until the end. His most influential writings centered on his theories and actions at New Lanark, including his classic *New View of Society,* which carried the appropriately descriptive subtitle *Essays on the Principle of the Formation of the Human Character, and the Application of the Principle to Practice.*[36] The idea was to take the model of New Lanark and expand it to an entire society. A country could be structured in individual communities consisting of 500 to 3,000 people each, based primarily on agriculture, and then expand the principle outward until the entire world was filled with these self-contained social units. It was all possible, Owen believed, because of the pliability of human nature. The opening statement on the title page of the first edition of the *New View of Society* sums up his socialistic belief in human malleability (and could have been written by Wallace half a century later): "Any character, from the best to the worst, from the most ignorant to the most enlightened, may be given to any community, even to the world at large, by applying certain means; which are to a great extent at the command and under the controul [*sic*], or easily made so, of those who possess the government of nations."[37]

Though his utopian socialism ultimately failed, Owen remained steadfast until the end when he, like Wallace, embraced spiritualism. Wallace never forgot the lesson and on many levels his life, particularly in his later years, was a recapitulation of Owen's. In 1905 Wallace recalled that "I have always looked upon Owen as my first teacher in the philosophy of human nature and my first guide through the labyrinth of social science. He influenced my character more than I then knew, and now that I have read his life and most of his works, I am fully convinced that he was the greatest of social reformers and the real founder of modern Socialism."[38]

The Republic of Science

Wallace's participation in the Mechanics' Institutes was emblematic of a growing class of individuals without money or formal education who were determined to enter the technological and scientific trades. "Why are the avenues of science barred against the poor just because they are poor?" It was a rhetorical question asked by Dr. George Birkbeck, who likely conceptualized the idea of a proprietary school as early as 1799 while teaching at Anderson's Institution in Glasgow, a school offering courses for artisans.[39] For these "mechanics," it was Birbeck's goal "to improve extensively habits and functions." But there was another agenda afoot. Birbeck also hoped the Institutes would "advance the arts and sciences, and to add largely to the power, resources and prosperity of the country."[40] Thus, the Mechanics' Institutes originally served a dual function as avenue for individual achievement and a means of social control. The working classes were assumed to be populated by simple minds with simple pleasures, expressed through pedestrian recreations such as drinking, gambling, and sexual promiscuity. Scientific and technological pursuits, designed into the curriculum, would raise them above "the grossness of sensuality" and provide them "with safe and rational recreation, which might otherwise be sought in scenes of low debauchery." Further, such an education "had the effect of promoting the strength and prosperity of the country in general." Inculcated into their scientific and technological education was a "ready acceptance of the industrial system and their place in it."[41]

The Mechanics' Institutes, then, were clearly not intended to be a mechanism of social leveling. Nevertheless, regardless of their underlying social significance, the pure sciences were not only well represented, they were among the most popular courses. At the Hudersfield Mechanics' Institute, for instance, Physiography (physical geography), Animal Physiology, Elementary Botany, and Geology outdrew Applied Mechanics and Metallurgy in attendance.[42] Science periodicals of the time reflected this new "Republic of Science," where every man became his own scientist, adding a brick here and some mortar there in the overall edifice of knowledge and truth, as science was then seen in this progressivist perspective. Amateurs could participate and share in the unsullied joy of discovery previously monopolized by the minority elite. An issue of *Popular Science Review* of 1862 explained this relationship between amateur and professional: "The principle of combination, whilst it has lightened the labours of the student, has aided materially to enrich our stores of knowledge, and the greater the harvest becomes, the more numerous will be the husbandmen."[43]

This nascent form of night school would be a primary avenue for those whose curiosity knew no boundaries and who also had the courage to be upwardly mobile. The church, the rich, and the few supported and participated in privately funded schools, leaving a gaping hole in the educational needs of a rapidly expanding industrial society. Before compulsory mass education, entrepreneurs rushed in to meet the demand in the form of these Mechanics' Institutes. The figures are not insignificant. The Yorkshire Union, founded in 1837, within thirteen years controlled 109 Institutes catering to the needs of nearly 18,500 enrollees. By 1851 there were over 700 "Literary and Mechanics' Institutes" in Great Britain and Ireland, catering to the needs of over 120,000 students.[44] Most of them were from the working classes (though there were some from the petite bourgeoisie), and they believed that knowledge was for everyone and not, as the *Edinburgh Review* of 1825 described it, for a "few superior minds, any more than being able to read or write now constitutes, as it once did, the title to scholarship." In time, such useful knowledge even became a mark of social grace such that, as described in Stewart's *Philosophical Essays of 1810–1811,* a "man can scarcely pass current in the informed circles of society, without knowing something of political economy, chemistry, mineralogy, geology, and etymology,—having a small notion of painting, sculpture, and architecture,—with some sort of taste for the picturesque, and a smattering of German and Spanish literature, and even some idea of Indian, Sanskrit and Chinese learning and history,—over and above some little knowledge of trade and agriculture."[45]

Wallace drank from this Pierian spring, but his thirst knew no quenching point until his life was interrupted by tragedy, as it so often would be, forcing him once again to return to practical matters.

To Know the Cause of Things

In 1845 William Wallace died, forcing Alfred to forgo his education and resign a temporary teaching post he had procured, in order to straighten out his brother's business and personal affairs. Ever anxious to turn tragedy into opportunity, Alfred further educated himself in the science of surveying (which he had learned informally in 1840; for the next two years he worked as a surveyor during the railroad-building bubble from 1846 to 1848). He called this interruption "the most important in my early life," and it gave the young man a chance to become financially solvent and accumulate some savings that could be applied to his science.[46]

Despite his obvious precocity in the face of deprivation, with characteristic modesty Wallace later recalled that "I do not think that at this time I could be said to have shown special superiority in any of the higher mental faculties."[47] It was blatant and false modesty, but unimportant in the sense that all scientists are smart; what sets one apart from another is something else, something Wallace was willing to publicly admit: "I possessed a strong desire to know the causes of things . . . [and] if I had one distinct mental faculty more prominent than another, it was the power of correct reasoning from a review of the known facts in any case to the causes or laws which produced them."[48]

Wallace's religious views in these early years, almost conspicuous by their absence in his later ventures into the spiritual and supernatural worlds, were already nontraditional. "What little religious belief I had," he recalled, "very quickly vanished under the influence of philosophical or scientific scepticism." His parents, like most in the neighborhood, belonged to the Church of England, and when Alfred was young they escorted him to church every Sunday, taking in both the morning and evening services. A voracious reader, Thomas allowed only limited reading material on Sundays. Fortunately for the precocious Alfred, that included such literary classics as *Pilgrim's Progress* and *Paradise Lost.* Not insignificantly, Alfred's father had a good many Quaker and Dissenter friends who periodically led to Alfred's attending alternative services with his parents, gaining him valuable exposure to the beliefs of those outside the religious mainstream. Nevertheless, Alfred could find "no sufficient basis of intelligible fact or connected reasoning to satisfy my intellect," and he shortly therafter lost what little religion he had. His skepticism was subsequently reinforced in the time spent with his brothers, particularly William, where, "though the subject of religion was not often mentioned, there was a pervading spirit of scepticism, or free-thought as it was then called, which strengthened and confirmed my doubts as to the truth or value of all ordinary religious teaching."[49]

This skepticism had several sources. One was his experiences at the Mechanics' Institutes and his reading of Owen, which led him to believe that "the only true religion is that which preaches service to humanity and brotherhood of man." Owen preached (and that is the right word) that people were a product of their social circumstances and should not be held responsible for their failings, as traditional doctrines of the Church of England proclaimed. This appealed to Wallace's humanity. Another influence was Thomas Paine and the other religious skeptics of the Enlightenment and post-Enlightenment period who identified and went public with such theological conundrums as the problem of evil—if God is omnipotent and omnibenevolent, then whence and whyfore evil? "Is God able to prevent evil but not willing? Then he is not benevolent. Is he willing but not able? Then he is not omnipotent. Is he both able and willing? Whence then is evil?" Theodicy hit the young Wallace hard, so he turned to his father for advice, "expecting he would be very much shocked at my acquaintance with any such infidel literature. But he merely remarked that such problems were mysteries which the wisest cannot understand, and seemed disinclined to any discussion of the subject." To a young heretic, however, this was nothing more than an evasion, and one that seemed to him "to prove that the orthodox ideas as to His [God] nature and powers cannot be accepted."[50]

Finally, Wallace's general commitment to science sounded the death knell of already weak religious convictions. "In addition to these influences my growing taste for various branches of physical science and my increasing love of nature disinclined me more and more for either the observances or the doctrines of orthodox religion, so that by the time I came of age I was absolutely non-religious, I cared and thought nothing about it, and could be best described by the modern term 'agnostic.' "[51]

Work with a Purpose

Whether the Mechanics' Institutes succeeded or failed in their stated goals remains problematic as far as the collective effects on the whole are concerned.[52] But individually there is no doubt that they provided opportunities where none previously existed, and there were those who took advantage of them, and none more than Wallace. There is no question that these working-class affiliations were radically different from the privileged upbringing and education experienced by many of Wallace's later colleagues, such as Darwin and Lyell, and would play an influential role in the development of his ideas, particularly as they diverged from these other thinkers.

Wallace and these gentlemen-scientists did share one thing in common in their respective youths—the love of nature. As a teenager his passion for the great outdoors indicated the type of life Wallace would lead and what kind of science he would practice over the next seven decades. A week after his seventeenth birthday he wrote to George Silk, his boyhood friend, of the joys of land surveying that involves "half in doors and half out doors work." It was the latter that made the former all the more rewarding:

> It is delightful on a fine summers day to be cutting, about over the country, following the chain & admiring the beauties of nature, breathing the fresh and pure air on the Hills, or in the noontime heat eating a piece of bread and cheese in a pleasant valley by the side of a rippling brook. Sometimes indeed it is not quite so pleasant on a cold winters day to find yourself like a monument on the top of an open hill with not a house within a mile and the wind and sleet ready to cut you through—but it is all made up for in the evening, when those who sit at home all day cannot have any idea of the pleasure there is in setting down to a good dinner and being hungry enough to eat plates, dishes and all.[53]

This physical work ethic would help carve a rugged individualism into Wallace that he would later need in the harsh years of exploring and collecting in the tropical rain forests of the Amazon and Malay Archipelago. "As to health & life, what are they compared with peace & happiness," he asked his future brother-in-law Thomas Sims from the Malayan island of Ternate years later, "& *happiness* is admirably described in the *Fam. Herald* as obtained by 'work with a purpose,' & the nobler the purpose the greater the happiness." To which he added a final passionate ingredient: "So far from being angry at being called an Enthusiast it is my pride & glory to be worthy to be so called. Who ever did anything good or great who was not an enthusiast?"[54] Such enthusiasm had its origins in these early life experiences.

Now completely on his own, Wallace learned the basics of such sciences as mechanics and optics from yet another venue of learning for the potential rank-and-file of Britain's nonprivileged working scientists—the *Society for the Diffusion of Useful Knowledge.* Through the Society Wallace learned to construct a crude telescope, by which he determined the meridian by equal altitudes of the sun, as well as by the pole star at its upper or lower culmination. This early interest in astronomy would, in the twilight of his career, be tethered to his lifelong interest in the evolution of life in a book *Man's Place in the Universe,* in which he argued that because of the complexity and delicately balanced state of the cosmos, a higher mind must be pervasive throughout, though the *embodiment* of this universal mind happened only once, and that was in the enlarged brain of *Homo sapiens.* Many of Wallace's earliest interests that hatched from these societies and institutes were carried through to the end of his life. In the summer of 1838, for example, when he went to Barton-by-the-sea to learn surveying with his brother William, Alfred discovered the science of geology. And from surveying and geology he discovered paleontology. "My brother, like most land-surveyors, was something of a geologist, and he showed me the fossil oysters."[55] It was his introduction to the geological history of life on earth.

During this period Wallace also bought a small paperback book, "the title of which I forgot," on the structure of plants. More important, he began to consider the structure of nature itself. "This little book was a revelation to me and for a year was my constant companion. I began to realize for the first time the order that underlay all the variety of nature." Interested in learning more, but unable to afford books, Wallace frequented a bookstore whose owner was generous enough to allow him to make copious notes while in the store. Wallace read of Ray, Jussieu, and other early naturalists concerned with the problem of the nature and mutability of species. On the question of species transmutation, for example, Wallace was sufficiently intrigued to quote in his notes the following passage from John Lindley's *Elements of Botany:* "As nature never passes from one extreme to another, except by something between the two, so she is accustomed to produce creations of an intermediate and doubtful condition, which partake of both extremes."[56]

Here the seeds of his own evolutionary thought were sown. "If true [this] would obviously go to the annihilation of all definitions of natural history; for if every object is admitted to pass into some other object by an insensible gradation, it would be necessary to admit also that no real limits are to be found between one thing and another and that absolute distributions can have no existence."[57] Wallace saw that the clear and distinct definition of a species becomes blurred at the edges as they blend from one into another, making such distinctions apparently artificial. Further, if species are mutable, the change must be slow and gradual, not sudden and catastrophic. The early transmutationists believed that *natura non facit saltus* (nature does not make leaps), a doctrine that would find both favor and evidentiary support in the subsequent decades by both Wallace and especially Darwin, who made the Latin motto a central organizing principle of his theory of evolution.

Simple Facts and Inner Meanings

Alfred now began to develop an educational philosophy that prized diversity as much as specialization. In 1843 he turned twenty and sported tiny wire-rimmed glasses framed by a thick mop of flaxen hair. He was still physically immature, as he was intellectually, but his boldness and intelligence shone through nonetheless in a short lecture he gave on "The Advantages of Varied Knowledge," written (as his first published paper, reprinted in his autobiography) "in opposition to the idea that it was better to learn one subject thoroughly than to know something of many subjects." This is because "we see the advantage possessed by him whose studies have been in various directions, and who at different times has had many different pursuits, for whatever may happen, he will always find something in his surroundings to interest and instruct him."[58]

Such diversification of interest led Wallace to develop a deep appreciation of science, and of the power of a few general principles to enrich one's perception of the many varied and seemingly unconnected phenomena in nature. "Many who marvel at the rolling thunder care not to inquire what causes the sound which is heard when a tightly-fitting cork is quickly drawn from a bottle, or when a whip is cracked, or a pistol fired," he told the audience, "and while they are struck with awe and admiration at the dazzling lightning, look upon the sparks drawn from a cat's back on a frosty evening and the slight crackle that accompanies them as being only fit to amuse a child; yet in each case the cause of the trifling and of the grand phenomena are the same." The well-honed and mature skills of a thinker and writer of only twenty years of age must have surprised the listeners, as Wallace drew to his grand conclusion:

> He who has extended his inquiries into the varied phenomena of nature learns to despise no fact, however small, and to consider the most apparently insignificant and common occurrences as much in need of explanation as those of a grander and more imposing character. He sees in every dewdrop trembling on the grass causes at work analogous to those which have produced the spherical figure of the earth and planets; and in the beautiful forms of crystallization on his window-panes on a frosty morning he recognizes the action of laws which may also have a part in the production of the similar forms of plants and of many of the lower animal types. Thus the simplest facts of everyday life have to him an inner meaning, and he sees that they depend upon the same general laws as those that are at work in the grandest phenomena of nature.[59]

This study of the principles behind observations undoubtedly came from his reading of the astronomer John Herschel's *Preliminary Discourse on the Study of Natural Philosophy*. Published in 1830 as a volume in Lardner's *Cabinet Cyclopedia,* the book was written for the amateur scientist and promoted the virtues of theory: "To a natural philosopher there is no natural object unimportant or trifling. From the least of nature's works he may learn the greatest lessons."[60] It was a philosophy of science not lost on Wallace, who, over the next two decades of collecting, learned the greatest of lessons from nature's tropics. He would become a natural philosopher who backed the grandest lessons with mountains of facts, and subscribe to Herschel's belief that "we may still continue to speak of causes . . . as those proximate links which connect phenomena with others of a simpler, higher, and more general or elementary kind."[61] It became Wallace's philosophy of science throughout his career.

Alfred called this time of scientific exploration "the turning point of my life." While he grew intellectually rich, however, he remained economically impoverished. In 1843 his father died and the family scattered. His mother took a job as a housekeeper, his sister emigrated and became a teacher in Georgia, and his brothers continued practicing their tradesman crafts. Early the following year, Alfred applied for and received a teaching post at Reverend Abraham Hill's Collegiate School at Leicester, where he met the soon-to-be-famous entomologist Henry Walter Bates. Both of modest means, Bates and Wallace took a liking to each other and developed a close friendship that would culminate in a joint venture to South America.

While still in England, however, Bates introduced Wallace to the importance of variety in nature, particularly the abundant diversity of insect species just within the local area—an estimated 10,000 varieties in a circle of just ten miles! Wallace added literary discoveries to his entomological finds, including Humboldt's *Personal Narrative of Travels in South America,* Prescott's *History of the Conquests of Mexico and Peru,* Darwin's *Voyage of the Beagle,* and "perhaps the most important book I read," Malthus's *Essay on Population,* the "main principles" of which "remained with me as a permanent possession."[62] Like most people interested in natural history, Wallace also read the 1844 anonymously published *Vestiges of the Natural History of Creation* (by Robert Chambers) and was intrigued by the author's hypothesis that "the simplest and most primitive type, under a law to which that of like-production is subordinate, gave birth to the type next above it, that this again produced the next higher, and so on to the very highest, the stages of advance being in all cases very small." The highest, of course, was man, placed there ultimately by "Providence."[63] No pure materialist was Chambers, who stated unequivocally the fact that "God created animated beings, as well as the terraqueous theatre of their being, is a fact so powerfully evidenced, and so universally received, that I at once take it for granted."[64] The idea grabbed Wallace's attention. It was a theory of evolution without a mechanism. Already a synthetic thinker with a bold personality, perhaps Wallace was already thinking that he might be the one to find that mechanism.

Bates, along with most everyone else in the scientific community, including and especially Darwin, lambasted *Vestiges* for being too speculative and lacking observational support. Chambers accepted the French naturalist Jean Baptiste Lamarck's theory of the inheritance of acquired characteristics, noting, for example, that the children of parents that habitually lie to them will grow up to be habitual liars themselves, in turn passing on the trait to their children. Reflecting the cultural attitudes of the day, Chambers explained that this problem was especially prevalent among the poor. He also cited experiments allegedly supporting the doctrine of spontaneous generation. The larvae of *Oinopota cellaris,* for example, was apparently found nowhere but in beer and wine during fermentation, leading Chambers to conclude that they had been generated spontaneously only after the invention of the process of fermentation. The book was so speculative—and panned viciously for being so—that it sent a message to Darwin that anyone attempting a grand synthesis had better amount so many facts in support that even the most skeptical of critics would collapse under their weight. (When another French naturalist named Frédéric Gérard published a book proffering an evolutionary theory, entitled *On Species,* which was also shredded by the critics, Darwin's botanist friend Joseph Hooker warned him that no one should "examine the question of species who has not minutely described many."[65] Darwin promptly turned to an extensive multiyear study of barnacles. This diversion allowed the younger Wallace, as it were, to catch up to him.)

It sent a different message to Wallace. Always more open to such fringe ideas than his colleagues, he wrote to Bates three days after Christmas 1845 and told him "I have rather a more favourable opinion of the 'Vestiges' than you appear to have. I do not consider it a hasty generalization, but rather as an ingenious hypothesis strongly supported by some striking facts and analogies, but which remains to be proved by more facts and the additional light which more research may throw upon the problem." Recognizing already the importance of having both data and theory—observations are useful only when held up to some unifying principle—Wallace opined to Bates: "It furnishes a subject for every observer of nature to attend to; every fact he observes will make either for or against it, and it thus serves both as an incitement to the collection of facts, and an object to which they can be applied when collected."[66]

If Wallace found a weakness in the book, it was Chambers's lack of distinction between varieties and species. As he told Bates, "An animal which differs from another by some decided and permanent character, however slight, which difference is undiminished by propagation and unchanged by climate and external circumstances, is universally held to be a distinct *species;* while one which is not regularly transmitted so as to form a distinct race, but is occasionally reproduced from the parent stock (like Albinoes), is generally, if the difference is not very considerable, classed as a *variety.*" But "if the theory of the 'Vestiges' is accepted, the Negro, the Red Indian, and the European are distinct species of the genus Homo," a distinction that Wallace later rejected. For now, however, it stimulated him to think hard on the question.

This "species question"—what is the difference between a variety and a species, and if a variety varies enough from its original type, can it become a new species?—had a long pedigree that Wallace would inherit and take with him to the tropics. Clearly *Vestiges* had an impact on Wallace, since he immediately began speculating on the relationship between geography and change within and between both varieties and species. In fact, he became an evolutionist shortly after reading *Vestiges,* and shortly before heading for South America on his first voyage. Although he was too early in his career for his evaluation of the *Vestiges* to have any impact on the professional scientists who mostly rejected it, Wallace was influenced by it more than his more mature and future colleagues. Contrary to how science is usually portrayed as careful deductions from copious facts, Wallace began with theory, then turned to data. He converted to evolution, then went out into the world to become a naturalist. It would take two extensive expeditions before he would wed the two.

2

The Evolution of a Naturalist

Whereas the fifteenth through the eighteenth centuries were the age of geographical exploration in the expansion of the world, the nineteenth century was the age of geological, zoological, and botanical exploration in the expansion of world scientific knowledge. Between 1839 and 1843 Joseph Hooker classified the flora of Antarctica as surgeon-botanist on the H.M.S. *Erebus* and H.M.S. *Terror;* from 1846 to 1850 Thomas Huxley explored marine specimens in the South Pacific and the Great Barrier Reef on board the H.M.S. *Rattlesnake;* and, of course, from 1831 to 1836 Charles Darwin circumnavigated the globe on the H.M.S. *Beagle,* the most famous of such voyages and a story known to all. Alfred Wallace had two extended expeditions, the first from 1848 to 1852 in the Amazon, and the second from 1854 to 1862 in the Malay Archipelago.

The social and economic contrast between Darwin and Wallace could not be sharper than in a comparison of their voyages of scientific discovery. Where Darwin's invitation to join the *Beagle* was through Professor John Henslow at Cambridge, Wallace had no invitation and hitched a ride for a fee. Where Darwin's father paid his expenses for the entire five years, Wallace was self-financed through the sale of specimen collections mailed back to an agent in England. Where Darwin's trip was sanctioned by the Royal Navy, and his correspondence and collections read and examined by the leading naturalists and geologists of the day, Wallace struggled for years in virtual anonymity. And where Darwin had the prospect of a published book on the voyage to justify his scientific musings, Wallace could only hope that his notes and his specimens, not to mention he himself, would make it safely back to England. As it would happen, on his first excursion even this outcome was doubtful.

Where Endless Summer Reigns

Throughout the early 1840s Wallace moved often and gained considerable experience in a number of odd jobs he undertook to make ends meet. Although the homes and jobs were uneventful, the variety of people and places he experienced instilled in him an openess to experience and flexibility to changing situations, a temperament well suited for travels to exotic locals.

In 1847, while in London to tie up some loose ends in the business affairs of his brother, Wallace visited the Insect Room of the British Museum of Natural History and commented in a letter to Bates of his dissatisfaction with "mere local" collections. "I should like to take some one family and study it thoroughly, principally with a view to the theory of the origin of species. I firmly believe that a full and careful study of the facts of nature will ultimately lead to a solution of the mystery."[1] He could not have known that a mere three years prior Darwin had penned (but did not publish) an essay on natural selection that, when refined over the course of the next decade and a half, would provide just such a solution. But where Darwin had settled in at Down House, coalescing his data on barnacles and reworking editions on his voyage on the *Beagle*, Wallace was just beginning his career as a naturalist. He had already read with great care Darwin's *Journal of Researches into the Natural History and Geology of the Countries visited during the Voyage of H.M.S. Beagle round the World,*[2] and with the imagination of a young man brimming with vitality envisioned how he might make his own mark in this blossoming science of natural history.

Where Darwin served as inspiration for Wallace, two other books gave him a target destination and a purpose. Alexander von Humboldt's *Personal Narrative of Travels to the Equinoctial Regions of the New Continent*[3] was "the first book that gave me the desire to visit the tropics."[4] Even more significant was William H. Edwards's *A Voyage Up the River Amazon,* published the same year as Darwin's book. Edwards wrote of the Amazon that the "vast numbers of trees add their tribute of beauty, and the flower-domed forest from its many coloured altars ever sends heavenward worshipful incense. Nor is this wild luxuriance unseen or unenlivened. Monkeys are frolicking through festooned bowers, or chasing in revelry over the wood arches. Squirrels scamper in ecstasy from limb to limb, unable to contain themselves for joyousness . . . Birds of the gaudiest plumage flit through the trees."[5] Such descriptions stirred Wallace's sense of adventure. "This little book was so clearly and brightly written, described so well the beauty and the grandeur of tropical vegetation, and gave such a pleasing account of the people, their kindness and hospitality to strangers, and especially of the English and American merchants in Para," he later pleasantly recalled, while also noting the all-important

Figure 2-1 A daguerreotype from 1848 of Alfred Russel Wallace at age twenty-five, shortly before his departure for the Amazon. (From *My Life,* 1905, v. I, 264)

financial factor in the equation, "while expenses of living and of travelling were both very moderate, that Bates and myself at once agreed that this was the very place for us to go to if there was any chance of paying our expenses by the sale of our duplicate collections."[6]

Wallace had visited Paris with his sister Fanny and, of course, had thoroughly explored the environs in and around London, but his was a call to undomesticated nature. "An earnest desire to visit a tropical country, to behold the luxuriance of animal and vegetable life said to exist there, and to see with my own eyes all those wonders which I had so much delighted to read of in the narratives of travellers, were the motives that induced me to break through the trammels of business and the ties of home, and start for 'Some far land where endless summer reigns.' "[7] He approached Bates, who recalled that Wallace "proposed to me a joint expedition to the river Amazons, for the purpose of exploring the natural history of its banks." But they were not to be mere collectors. Rather, they would "gather facts, as Mr. Wallace expressed it in one of his letters, 'towards solving the problem of the origin of species,' a subject on which we had conversed and corresponded much together."[8]

Figure 2-2 Henry Walter Bates, Wallace's traveling companion for the first part of the Amazon expedition. Bates went on to become one of the world's foremost entomologists and an expert on protective coloration and mimicry. (Courtesy of Richard Milner)

Edward Doubleday, curator of butterflies at the British Museum, assured them that "the whole of northern Brazil was very little known" and that "if we collected all orders of insects, as well as landshells, birds, and mammals, there was no doubt we could easily pay our expenses."[9] After carefully going through the collections and noting what was needed, Bates and Wallace chanced a meeting with William Edwards, the author of *A Voyage Up the River Amazon,* which had inspired them initially, and procured a letter of introduction from him. They also met with Dr. Thomas Horsfield, a curator at the India Museum, who showed them how best to pack specimens into boxes for safe shipment home, and secured a sales agent to whom they would ship their collections in one Mr. Samuel Stevens, whose reputation was reinforced through his brother who was an established natural history auctioneer. It was a business relationship that would last throughout Wallace's two lengthy voyages totaling a dozen years, for which he could not "remember that we ever had the least disagreement about any matter whatever."[10]

Wallace could have had no inkling that this would be the beginning of a prodigious literary and exploratory career spanning nearly seven decades and including over 20,000 miles of travel and 10,000 pages of published documents.

Mischief in the Amazon

On April 20, 1848, Henry Walter Bates and Alfred Russel Wallace left England on a relatively small, 192-ton square-rigger barque, appropriately named H.M.S. *Mischief.* Although it was rated A1 with Lloyds of London, the ship was small enough to be battered about in high winds, which rose soon after launch "with waves that flooded our decks, washed away part of our bulwarks, and was very near swamping us altogether." Like so many landlubbers, Darwin included, Wallace spent much of the first part of the voyage "in my berth prostrate with sea-sickness." After about a week, however, the weather and his body turned tranquil and Wallace went up on deck in time to witness the passing through the celebrated Sargasso Sea where he began his observations in the floating seaweed of "great numbers of small fish, crabs, mollusca, and innumerable low forms of marine life."[11]

Twenty-nine days later he and Bates landed in Pará, on the Brazilian coast below the mouth of the Amazon. "The city of Para is a curious, outlandish looking place, the best part of it very like Baulogne, the streets narrow and horribly rough—no pavement. The public buildings handsome, but out of repair or even ruinous. The squares and public places covered with grass and weeds like an English common." Tellingly, for he would come to write a book about them, one of his first observations on land was of "palm trees of many different kinds, bananas and plantains abundant in all the gardens, and orange trees innumerable, most of the roads out of the city being bordered on each side with them."[12] Wallace's exposure to a variety of races of peoples was immediate. Blacks, whites, browns, and "between these a hundred shades and mixtures, which it requires an experienced eye to detect," at first overwhelmed his percepts, he recalled in a short volume published after his return. Nevertheless, he was disappointed that the people, the weather, and the vegetation were not "the glowing picture I had conjured up in my imagination, and had been brooding over during the tedium of a sea-voyage."[13]

At the age of twenty-five and with his entire life savings of £100 in his pocket, Wallace began his exploration of the legendary rain forest on the 23rd of June. With Bates by his side during the first leg of the journey, a typical day found the naturalist-explorers up at 6:00 A.M. and collecting for a solid six hours from 8:00 A.M. to 2:00 P.M., in the heat of the midday sun, causing the locals to question their sanity. A bath at 3:00 P.M. followed by tea revived them for several more hours of cleaning, sorting, cataloguing, and classifying their catch. "The constant hard exercise, pure air, and good living, notwithstanding the intense heat, kept us in the most perfect health, and I have never altogether enjoyed myself so much" he wrote.[14] A letter written by Wallace and Bates to their collections agent Stevens (published the next year in the *Annals and Magazine of Natural History* as "Journey to Explore the Province of Pará") gave a description of "Messrs. Wallace and Bates, two enterprising and deserving young men . . . on an expedition to South America to explore some of the vast and unexamined regions of the province of Para, said to be so rich and varied in its productions of natural history." It is one of the few extant descriptions of their adventures that survived the trip because it was sent back with specimens. Wallace's description of a typical leg of the journey is as informative as it is colorful:

> We had the usual difficulties of travellers in this country in the desertion of our crew, which delayed us six or seven days in going up; the voyage took us three weeks to Guaribas and two weeks returning. We reached a point about twenty miles below Arroya, beyond which a large canoe cannot pass in the dry season, from the rapids, falls and whirlpools which here commence and obstruct the navigation of this magnificent river more or less to its source; here we are obliged to leave our vessel and continue in an open boat, in which we were exposed for two days, amply repaid however by the beauty of the scenery, the river (here a mile wide) being studded with rocky and sandy islets of all sizes, and richly clad with vegetation; the shores high and undulating, covered with a dense but picturesque forest; the waters dark and clear as crystal; and the excitement in shooting fearful rapids, &c. acted as a necessary stimulant under the heat of an equatorial sun, and thermometer 95° in the shade.[15]

Once they were well up the Amazon, conditions grew worse and danger lurked around every bend in the river or behind every tree. They ate ants, turtle and alligator meat, were constantly harrassed by insects, and "jaguars I knew abounded here, deadly serpents were plentiful, and at every step I almost expected to feel a cold gliding body under my feet, or deadly fangs in my leg." The evenings "were dull and dreary," except for the occasional attack by a vampire bat. Hacking their way through this tropical morass made Wallace dream prophetically that "the whole glory of these forests could only be seen by sailing gently in a balloon over the undulating flowery surface above: such a treat is perhaps reserved for the traveller of a future age."[16] For relief from the heat and humidity they replenished their fluids with fresh oranges and pineapples, and, as he told Stevens on September 12, 1849, "The Tapajoz here is clear water with a sandy beach, and the bathing is luxurious; we bathe here in the middle of the day, when dripping with perspiration, and you can have no idea of the excessive luxury of it." The pleasures of being in a glorious foreign land clearly outweighed the difficulties of travel: "The more I see of the country, *the more I want to,* and I can see *no end* of, the species of butterflies when the whole country is well explored."[17]

One of Wallace's earliest experiences that resulted in a scientific publication in the prestigious *Proceedings of the Zoological Society of London*—"Monkeys of the Amazon"—began one morning while walking through the forest. Wallace "heard a rustling of the leaves and branches . . . and expected every minute to see some Indian hunter make his appearance, when all at once the sounds appeared to be in the branches above, and turning up my eyes there, I saw a large monkey looking down at me, and seeming as much astonished as I was myself." The monkey retreated and the next day Wallace and a hunting companion glimpsed a troop in the same area. As "one approached too near for its safety," it was shot by the other hunter and the "poor little animal was not quite dead, and its cries, its innocent-looking countenance, and delicate little hands were quite childlike." Not unusual for naturalists for the day, Wallace ate his catch for breakfast, exclaiming that it "resembled rabbit, without any very peculiar or unpleasant flavour."[18] Gastric interest aside, comparing primate and human anatomy and behavior got Wallace cogitating about the relationship between the two.

By 1850 the men were a thousand miles up the Amazon (Figure 2-3) and, on March 26, they split up and went their separate ways, Bates toward the Andes to explore the Solimoens, or Upper Amazon, Wallace toward Venezuela up the Rio Negro and the unknown Uaupés. They figured it would be more profitable if they were collecting different specimens from different areas, and it seemed natural to explore both of these important branches of the great river. An Englishman accustomed to perceiving the Thames as a large river, Wallace noted that even "an insignificant tributary of the Amazon was wider than the Thames," which is dwarfed into insignificance by the Amazon itself. Wallace mapped the river as best he could with crude instruments. Sixty years later Dr. Hamilton Rice of the Royal Geographical Society, after mounting a full-scale expedition with sophisticated equipment and astronomical observations, declared that Wallace's work "still holds good."[19] It was in these formative first two years that Wallace found his niche, both in nature and in life. He was a brave and hardy explorer and a talented observer and collector. He knew it and so did Bates, who told Stevens on the final day of the year 1850: "Mr. Wallace, I suppose, will follow up the profession, and probably will adopt the track I have planned out to Peru; he is now in glorious country,

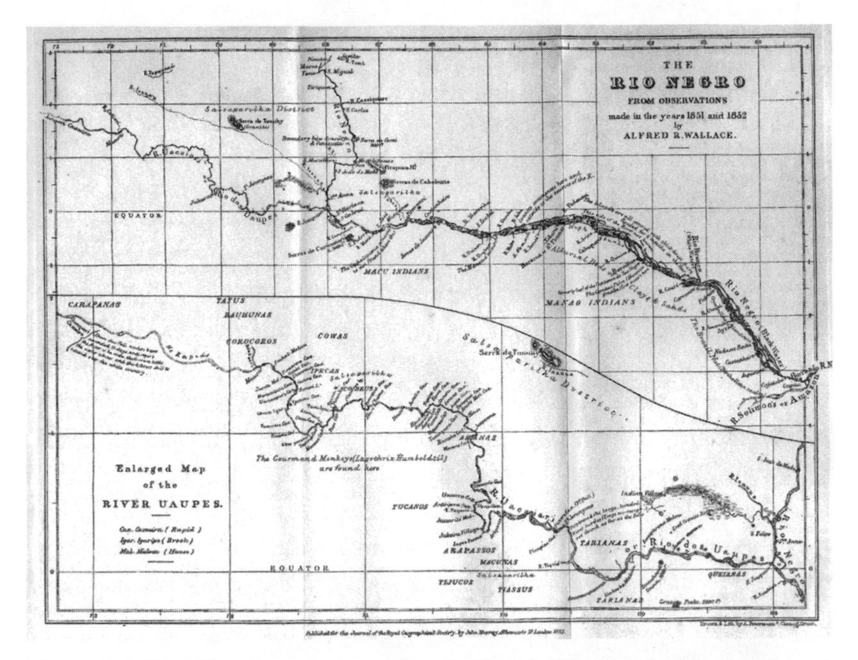

Figure 2-3 The Rio Negro, mapped by the Royal Geographical Society based on Wallace's observations and descriptions. After exploring the Amazon for several years together, Bates and Wallace split up, Bates taking the Upper Amazon while Wallace ascended the Rio Negro and the virtually unexplored Uaupés. This was Wallace's first expedition and it lasted a total of four years. (From *My Life,* 1905, v. I, 320)

and you must expect great things from him. In perseverance and real knowledge of the subject, he goes ahead of me, and is worthy of all success."[20]

Wallace did follow up the profession, in more ways than one. He became a profitable collector, and "because I am so much interested in the country and the people that I am determined to see and know more of it and them than any other European traveller," he told Stevens from Guia, Rio Negro, in January 1851. "If I do not get profit, I hope at least to get some credit as an industrious and persevering traveller." But success in collecting for both naturalists was beyond the wildest of their imaginations, enabling them to continue exploring indefinitely. The first shipment to England consisted of 400 butterflies, 450 beetles, and 1,300 other assorted insects. Stevens was able to sell most of them through advertisements, such as the following that appeared on the inside cover in the *Annals and Magazine of Natural History:*

TO NATURALISTS, &c.

SAMUEL STEVENS, Natural History Agent, No. 24 Bloomsbury Street, Bedford Square, begs to announce that he has recently received from South America

> *Two* beautiful Consignments of INSECTS of all orders in very fine Condition, collected in the province of Pará, containing numbers of very rare and some new species, also a few LAND and FRESHWATER SHELLS, and some BIRD SKINS and several small parcels of Insects, &c., from New Zealand, New Holland, India, and the Cape, all of which are for Sale by Private Contract.[21]

The pecuniary reward soon followed and the expedition was given its first of many financial boosts.

Following his departure from Bates, Alfred was joined by his brother Herbert, who, unable to find financial security in England, had heard of his brother's collecting success and thought he would try his hand at it. The brothers enjoyed each other's company and on several occasions they "mesmerized" willing native subjects, a skill Wallace had learned in Leicester in 1844 from Spencer Hall (and one he would find useful in later odysseys into the spiritual world). As Herbert "was the only one of our family who had some natural capacity as a verse-writer," while Alfred collected Herbert wrote such verses as these, which captured the poetry of the rain forest and her flora and fauna:

> And now upon the Amazon,
> the waters rush and roar—
> The noble river that flows between
> A league from shore to shore;
> Our little bark speeds gallantly,
> The porpose [*sic*], rising, blows,
> The gull darts downward rapidly
> At a fish beneath our bows,
> The far-off roar of the onça,
> The cry of the whip-poor-will—
> All breathe to us in whispers
> That we are in Brazil.[22]

Unfortunately for Herbert he did not take well to the stress and harsh physical conditions of tropical living, and in this weakened state he succumbed to yellow fever on June 8, 1851, shortly after departing from his brother and just before leaving for home from Para. (Bates and Wallace were also stricken with the disease, but had much hardier constitutions to ward off the ill effects of the condition.) Wallace got word that his brother was seriously ill, but for a time was unable to determine the final outcome. It was an anxiety-ridden and "suspenseful" period before he received the "bad news."[23] His brother was dead, shocking Wallace into the realization that this was a risky business.

Latent Sparks of Genius

Wallace headed back up the Rio Negro on a second trek to the headwaters of the Rio Uaupés. He was in his element—a naturalist's dream country—as he recounted in a somewhat rambling stream of consciousness penned in a lengthy letter to the members of the Mechanics' Institutes, to whom he had promised a report from abroad: "There is, however, one natural feature of this country, the interest and grandeur of which may be fully appreciated in a single walk: it is the 'virgin forest.' Here no one who has any feeling of the magnificent and the sublime can be disappointed; the sombre shade, scarce illumined by a single direct ray even of the tropical sun, the enormous size and height of the trees, most of which rise like huge columns a hundred feet or more without throwing out a single branch, the strange buttresses around the base of some, and spiny or furrowed stems of others, the curious and even extraordinary creepers and climbers which wind around them, hanging in long festoons from branch to branch, sometimes curling and twisting on the ground like a great serpent, then mounting to the very tops of the trees, thence throwing down roots and fibres which hang waving in the air, or twisting round each other form ropes and cables of every variety of size and often of the most perfect regularity."[24]

Pressing on up the Rio Negro with a native assistant in tow helping out with the movement of supplies and the capturing and packing of specimens, Wallace veered north to the small village of Javita, Venezuela, where he "saw a large jet-black animal come out of the forest about twenty yards before me, which took me so much by surprise that I did not at first imagine what it was." It was a black jaguar. Wallace raised the gun and prepared to drop the predator when he suddenly remembered he had loaded buckshot in both barrels "and that to fire would exasperate without killing him," so the naturalist stood tall but silent, so in admiration of the magnificent beast that he could not even feel fear.[25]

In one of his earliest anthropological studies Wallace spent forty days as the only white man in the village, writing poetically of the contrasts with life in England that by now struck him as stilted and confining:

> The children of small growth are naked, and
> The boys and men wear but a narrow cloth,
> How I delight to see those naked boys!
> Their well form'd limbs, their bright, smooth, red-brown skin,
> And every motion full of grace and health;
> And as they run, and race, and shout, and leap,
> Or swim and dive beneath the rapid stream,
> I pity English boys; their active limbs

> Cramp'd and confined in tightly-fitting clothes;
> I'd be an Indian here, and live content
> To fish, and hunt, and paddle my canoe,
> And see my children grow, like young wild fawns,
> In health of body and in peace of mind,
> Rich without wealth, and happy without gold![26]

Wallace was no twenty-first-century politically correct liberal, but he was far ahead of most of his contemporaries in showing respect for indigenous peoples whose Otherness, for the most part, filled him with admiration. "They were all going about their own work and pleasure which had nothing to do with white man or their ways; they walked with the free step of the independent forest-dweller," he wrote in his narrative. The contrast with European colonialists was especially striking to him: "I could not have believed that there would be so much difference in the aspect of the same people in their native state and when living under European supervision. The true denizen of the Amazonian forests . . . is unique and not to be forgotten."[27]

In some cases such supervision included slavery, which Wallace railed against even at this early stage of his career, appealing not to a moral argument, but a pragmatic one grounded in what appears to be a nascent evolutionary argument of what is natural and unnatural. In observing the slave system on a large plantation on the lower Amazon, in which the slaves were well kept by their owner, Wallace began by noting that his evaluation was of slavery in "its most favourable light." Despite such conditions in which families were kept intact, the ill were treated medically, and working conditions were tolerable, Wallace asks, "Can it be right to keep a number of our fellow-creatures in a state of adult infancy, of unthinking childhood?" To achieve maturity an individual, like a species, must exercise its full powers against nature. "It is the responsibility and self-defence of manhood that calls forth the highest powers and energies of our race. It is the struggle for existence, the 'battle for life,' which exercises the moral faculties and calls forth the latent sparks of genius. The hope of gain, the love of power, the desire of fame and approbation, excite to noble deeds, and call into action all those faculties which are the distinctive attributes of man." Slavery, then, withholds this natural evolution by keeping the individual in a state of perpetual immaturity. "Childhood is the animal part of man's existence, manhood the intellectual; and when the weakness and imbecility of childhood remain, without its simplicity and pureness, its grace and beauty, how degrading is the spectacle! And this is the state of the slave when slavery is the best it can be."[28]

These were some of Wallace's earliest anthropological observations, out of which he also branched off into and began yet another field of study—philology. When he began his Amazonian journey Wallace had to learn Spanish and Portuguese. As he worked his way up the river he found that most of the "tame Indians" spoke the lingoa geral, a creole-type language developed by Jesuit missionaries to communicate with the Indians. But by the time Wallace branched off the Amazon and went up the Rio Negro and Rio Uaupés, he encountered Indians who spoke nothing but their native language, so he began to construct syllabaries and synonymies. He composed ten in all, published in the first edition of his narrative volume, but dropped in subsequent editions.

There Is No Part of Natural History More Interesting

Even more important to Wallace than the people of the Amazon was the physical geography, which "surpasses in dimensions that of any other river in the world." The amount of water pouring into the Atlantic Ocean is "far greater than that of any other river; not only absolutely, but probably also relatively to its area, for as it is almost entirely covered by dense virgin forests, the heavy rains which penetrate them do not suffer so much evaporation."[29] Wallace continued with careful measurements of the width, velocity, depth, color, currents, rise and fall, and eddies of the Amazon. He noted, for example, its "enormous width" of "twenty and thirty miles wide, and, for a very great distance, fifteen to twenty."[30]

Most disappointing to Wallace on the geology of the river valley was the lack of fossils. He was unable "to find any fossil remains whatever,—not even a shell, or a fragment of fossil wood, or anything that could lead to a conjecture as to the state in which the valley existed at any former period." But he was able to give detailed descriptions of rock formations, mountain ranges, stratigraphic arrangements, concluding "that here we see the last stage of a process that has been going on, during the whole period of the elevation of the Andes and the mountains of Brazil and Guiana, from the ocean." For how long this had been ongoing Wallace could not ascertain, "till the country has been more thoroughly explored, and the organic remains, which must doubtless exist, be brought forward, to give us more accurate information respecting the birth and growth of the Amazon."[31]

The botanical variety and splendor of the valley surpassed that of its geology, for, compared to here, Wallace observed, "perhaps no country in the world contains such an amount of vegetable matter on its surface." The entire valley "is covered with one dense and lofty primeval forest, the most extensive and unbroken which exists upon the earth." The "thinly wooded plains" of Asia and the "trifling" forests of Central Europe were no match for "the great variety of species of trees" found in the Amazon. In particular, Wallace noted the "medicinal properties" of trees producing fruit that "supply the whole population" of an area. Biogeographically speaking, Wallace contrasted his own observations with those of Humboldt, Spence, and Darwin, and concluded that, compared to the temperate zones, "there is a much greater number of species" in the tropical regions. In reverse, however, he wrote that the flowers of the temperate regions had more "brilliant colouring and picturesque beauty . . . than in the tropical regions." This was due to the fact that "in the tropics, a greater proportion of the surface is covered either with dense forests or with barren deserts, neither of which can exhibit many flowers." Nevertheless, the "endless carpet of verdure, with masses of gay blossoms, the varying hues of the foliage, and the constant variety of plain and forest, meadow and woodland, more than individual objects, are what fill the beholder with delight."[32] But Wallace was more than merely delighted. He was curious. He wanted to know why the biosphere should be divided as it was.

Where the Amazonian trees surpassed those of all other forests within his purview, the mammals Wallace found to be smaller in number "both of species and individuals," and the valley to be altogether "deficient in large animals" compared to "any other part of the world of equal extent, except Australia." These, however, are compensated for by the birds which, being "so numerous and striking, that it is impossible here to do more than mention a few of the most interesting and beautiful." Wallace alone collected more than 500 species, "a greater number than can be found all over Europe." Reptiles were equally abundant to the birds (and considerably more dangerous), with anacondas upwards of thirty to forty feet long (and "sometimes from sixty to eighty feet long"), quite capable of killing cattle and horses. Alligators (actually caimans) also reportedly nabbed several children a year. Although the natives killed the smaller alligators, "the larger often devour them in return."[33]

More important than animal anatomy and behavior for Wallace, however, was their geographical distribution. "There is no part of natural history more interesting or instructive." For species to become distinct they cannot "intermix," Wallace reasoned, and therefore geographic isolation could be the key to preventing such interbreeding. Geographical zones—tropical versus temperate, for example—contain plants and animals peculiar to their area. Even similar zones in separate continents "have scarcely an animal in common." Europe and North America, for example, are separated by "a wide extent of sea . . . which few animals can pass over; so that, supposing the animal productions to have been originally distinct, they could not well have become intermixed." Further, within a single country populations are divided into smaller groups "and that almost every district has peculiar animals found nowhere else." Oceans and rivers act as isolating mechanisms, as do "great mountain-chains" such as the Rockies, which "separate two distinct zoological districts; California and Oregon on the one side, possessing plants, birds, and

Figure 2-4 This series of hand sketches was in the Collection of Wallace Manuscripts at the Linnean Society of London, labeled by Wallace as: "Some of My Original Sketches on the Amazon." These are interesting for several reasons. First, they show yet another talent of Wallace's—the ability to draw in fine detail, an important skill for any naturalist throughout most of the first half of the nineteenth century before photographic equipment became practical to carry.

(Continued Next Page)

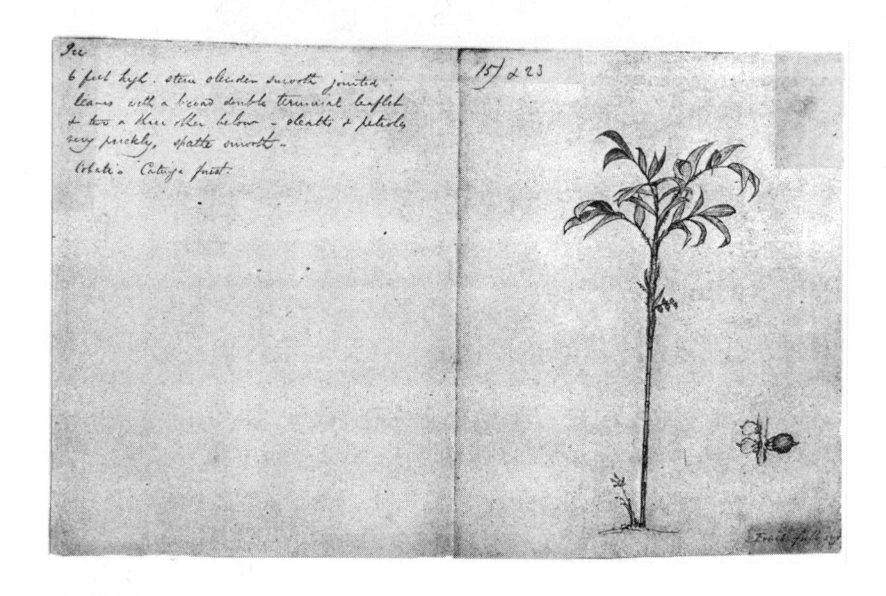

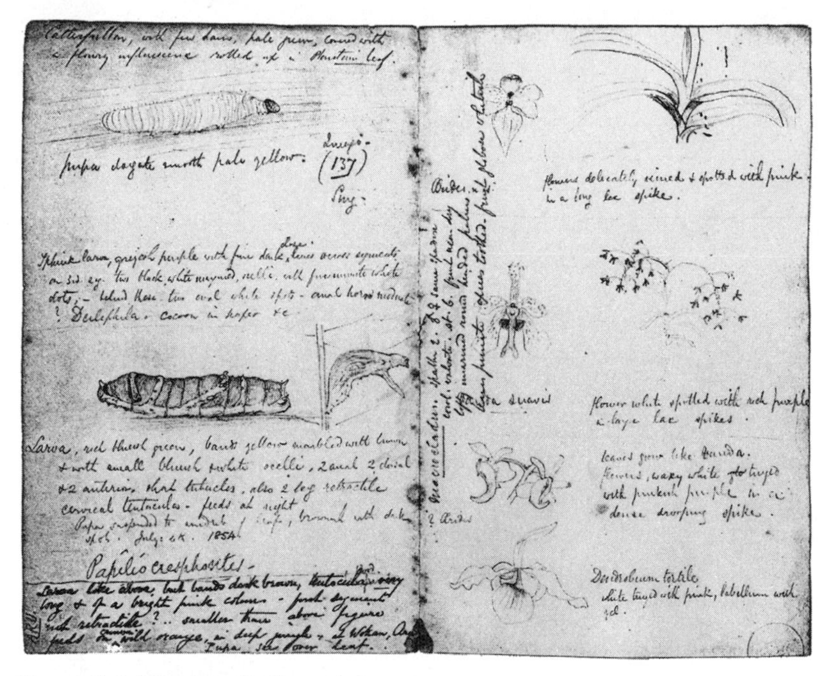

Figure 2-4 (Continued) Second, they demonstrate the range of Wallace's interest in the exploration of nature (page 69): fruit trees and general landscape; rain forest and geological processes and fossilization; and (this page) developmental stages of insects and botanical specimens. (Courtesy of the Linnean Society of London)

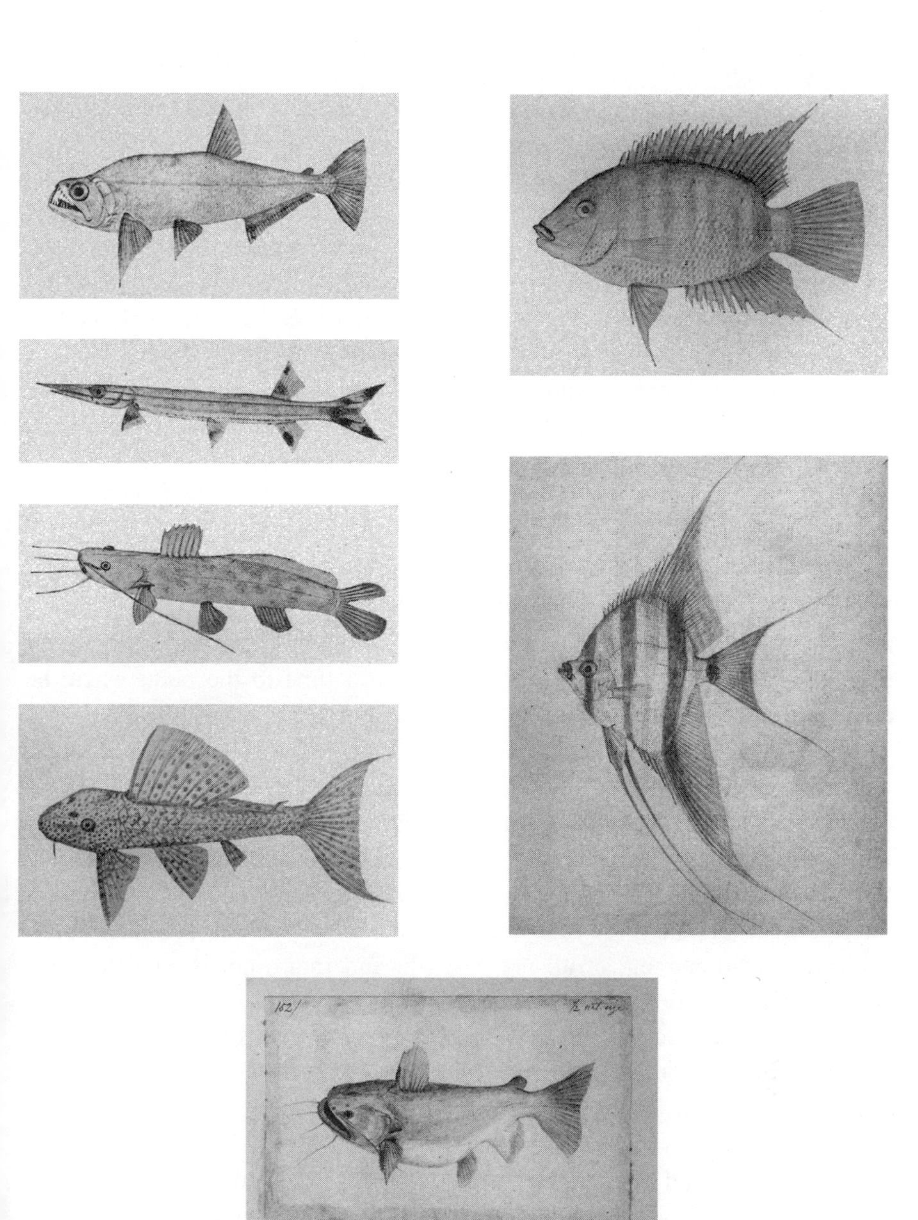

Figure 2-4 (Continued) The fish of the Amazon in Wallace's hand. These drawings are among the few items that survived the fire that sank Wallace's ship on the return voyage and include the well-known angelfish (center right), labeled by Wallace as the butterfly fish, *Pterophyllum scalara,* now popular among aquarium aficionados, as well as (from top to bottom) *Cichlosoma severum, Cynodon scombroides, Xiphostoma lateristriga, Pimelodus holomelas, Plecostomus guacari,* and the ubiquitous catfish, *Asterophysus batrachus.* (From *My Life,* 1905, v. I, 285–87; and the Linnean Society of London)

insects, not found in any part of North America east of that range." Even besides these obvious restrictions to the movement and cross-fertilization of species, Wallace noted that there are other, nonphysical, and much more subtle barriers. He recorded that "places not more than fifty or a hundred miles apart often have species of insects and birds at the one, which are not found at the others." Therefore, he concluded, "there must be some boundary which determines the range of each species; some external peculiarity to mark the line which each one does not pass."[34] The explanation for and consequences of these delimiting factors would come to Wallace in bits and pieces over the next six years. But first he had to survive the remainder of his voyage.

Irrecoverably Lost at Sea

Despite his youth and hearty disposition, after nearly four years of travel Wallace was wearing down from the daily assaults of the jungle. Foremost among these were the tropical diseases of malaria, dysentery, and yellow fever that not infrequently laid him out for days at a time to the point where he could not eat and struggled just to keep down juice. He recalled that one day "I found myself quite knocked up, with headache, pains in the back and limbs, and violent fever." Forcing down quinine and cream of tartar water in hopes of recovering, he "was so weak and apathetic that at times I could hardly muster resolution to move myself to prepare" his specimens. In such a state the mind plays tricks, distorting judgment and perception, and dulling one's motivation to continue. "While in that apathetic state I was constantly half-thinking, half-dreaming, of all my past life and future hopes, and that they were perhaps all doomed to end here on the Rio Negro."[35]

The last straw came at the end of 1851 when a bout of yellow fever left him prostrate for three months. By February 1852 he began thinking of returning to England. While recuperating he collected his thoughts and notes that would form the final four chapters of the narrative of his travels. Physically exhausted, and with enough material for a book, Wallace decided to cut his trip short by a year and head for home. It took several months to wrap up his affairs, and in July he boarded the 235-ton brig *Helen* of Liverpool, destined for England and loaded with India-rubber, cocoa, and a variety of other plant commodities. What happened next Wallace recounted on his arrival in England in the pages of the *Zoologist*.[36] At around 9:00 A.M. on August 6, after about three uneventful weeks at sea, at latitude 30°30'N, longitude 52°W, "smoke was discovered issuing from the hatchways."

The *Helen* had caught fire. The captain ordered the ship abandoned as "the smoke became more dense and suffocating, and soon filled the cabin, so as to render it very difficult to get any necessaries out of it." These necessaries included, tragically, his notes, journals, and collections. In one final desperate plunge into the lower decks and his cabin, Wallace managed to salvage his watch, some shirts, and a tin box containing drawings of trees, plants, landscapes, Indian tools and artifacts, and fish. By noon the flames had spread on deck and the crew began to man the lifeboats, which, "being much shrunk by exposure to the sun, required all our exertions to keep them from filling with water."

When Wallace lowered himself from the ship into the lifeboat by rope, he slipped and suffered excruciating rope burn on his hands, which, upon hitting the salt water, caused "a most intense smarting and burning on my scarified fingers." Watching from afar and bailing continuously, the men witnessed the inferno light up the night sky, by which time "the masts had fallen, and the deck and cargo was one fierce mass of flame." Wallace observed to his horror that a number of his live specimens, including parrots and monkeys, were trapped and retreated to the bowsprit to await their fate. One parrot escaped by flying to the lifeboat, but the monkeys perished in the fire.

By the following morning the *Helen* had slipped beneath the waves. The men hoisted a small sail on the lifeboat and steered for the nearest body of land, the island of Bermuda, a full 700 miles away. For two days an easterly kept them on course, but a change in wind on the third day left them soaked, overheated, and feeling hopeless. "We suffered much from the heat by day; and being constantly wet with the spray, and having no place to lie down comfortably, it may be supposed that we did not sleep very soundly at night." For nourishment they had to force down stale biscuits and raw salt pork, but after a week supplies were dwindling and water rations had to be severely restricted. "And as we now were in a part celebrated for squalls and hurricanes, every shift in the wind and change of the sky was most anxiously watched by us." Finally, after ten days of constant bailing, starvation, sunburnt skin, cracked and blistered lips, and sleep deprivation, one of the men yelled out, "Sail ho!" The tired lot pulled hard at the oars and were rescued by the London-bound *Jordeson* from Cuba. After a week and a half of sailing and rowing they were still over 200 miles from Bermuda.

Amazingly, the adventure was not over. The *Jordeson,* with her crew complement now doubled, began to run short of food and water. She "encountered three very heavy gales, which split and carried away some of the strongest sails in the ship, and made her leak so much that the pumps could with difficulty keep her free." Finally, after eighty days on the open ocean they made port, Wallace having managed to save "my watch, my drawings of fishes, and a portion of my notes and journals." The tragedy was that "most of my journals, notes on the habits of animals, and drawings of the transformation of insects were lost." Not to mention a collection of ten species of river tortoises, a hundred species of Rio Negro fishes, skeletons and skins of an anteater and cowfish (*Manatus*), and living monkeys, parrots, macaws, and other birds, all "irrecoverably lost":

> It was now, when the danger appeared past, that I began to feel fully the greatness of my loss. With what pleasure I had looked upon every rare and curious insect I had added to my collection! How many times, when almost overcome with the ague, had I crawled into the forest and been rewarded by some unknown and beautiful species! How many places, which no European foot but my own had trodden, would have been recalled to my memory by the rare birds and insects they had furnished to my collection! How many weary days and weeks had I passed, upheld only by the fond hope of bringing home many new and beautiful forms from those wild regions. . . . And now everything was gone, and I had not one specimen to illustrate the unknown lands I had trod, or to call back the recollection of the wild scenes I had beheld![37]

There is no way to know how much sooner Wallace might have discovered the mechanism of species transformation had he had his detailed notes and specimens on his return. On the other hand, perhaps he would not have made the discovery at all had he not suffered this great loss and been thereby motivated to travel to the East, where he did, in fact, make his great find.

Speculations on Species

Lacking a means of supporting himself on his return, Wallace was fortunate that Stevens had insured his collections, giving him £200 to begin anew and secure lodging for him. Wallace literally had only the shirt on his back, so Stevens took him to a ready-made clothes shop and bought him a suit, then to his haberdasher to outfit him for the season. Stevens's mother opened her home to Wallace until he was able to find housing for himself. The time he spent in London also afforded him the opportunity to develop his new scientific acquaintances who had been reading about his exploits and adventures in the pages of the *Annals & Magazine of Natural History* and the *Proceedings of the Zoological Society of London.*

Within a month Wallace returned to writing, reconstructing what he could of the trip from journal fragments and memory shards. From these he wrote his first two books, *Palm Trees of the Amazon* and *A Narrative of Travels on the Amazon and Rio Negro,* both in 1853. *Palm Trees* was published at his own expense and had an initial print run of only 250 copies, the sales of which barely covered the cost. As a consequence of haste in preparation and a dearth of materials, *Palm Trees* was not received uncritically. Lyell had some qualms, as did the director of the Royal Botanic Gardens at Kew, Sir William Hooker, who felt that "this work is certainly more suited to the drawing room table than to the library of a botanist."[38] Richard Spruce, a botanist with whom Wallace had done some collecting in the Amazon, echoed Hooker's assessment of the book, although he began with faint praise: "You asked me about Wallace's *Palms.* He has sent me a copy—the figures are very pretty, and with some of them he has been very successful. I may instance the figs of *Raphia taedigera* and *Acrocomia sclerocarpa.*" But from there Spruce savaged the book: "The worst figure in the book is that of *Iriartea ventricosa.* The most striking fault of nearly all the figˢ of the larger species is that the stem is much too thick compared with the length of the fronds, and that the latter has only half as many pinnae as they ought to have. The descriptions are worse than nothing, in many cases not mentioning a single circumstance that a botanist would most desire to know; but the accounts of the uses are good. His *Leopoldinia piassaba* and *Mauritia carana* are two magnificent new palms, both correctly referred to their genus; but the former has been figured from a stunted specimen."[39]

A Narrative of Travels on the Amazon and Rio Negro was published by arrangement with a small publisher named Lovell Reeve, who agreed to split the profits fifty/fifty with the author. Although a modest 750 copies were printed, a decade later only 500 had sold with no profits left to divide. It was an outcome emblematic of Wallace's struggle for financial security that eluded him until the final decades of his long life. *Narrative of Travels* was criticized by Darwin and others for also being thin in the data department, but most readers were understanding of the fact that Wallace lost much of his data in the sea disaster, and the book gained momentum and found a readership in the burgeoning travel literature of the period. Macmillan & Co. publishers picked up the copyright and it was kept in print for nearly three-quarters of a century, eventually earning Wallace a small profit.

What was Wallace able to accomplish from his four years in the Amazonian rain forest? Collectively for the trip, he and Bates boasted a remarkable assortment of 14,712 *species* (not just individuals) of insects, birds, reptiles, and other variegated biological items, approximately 8,000 of which had never before been seen in Europe. With such staggering empirical evidence of nature's abundance and variety, qualified by geographic limitation and its effect on varieties of species, Wallace's earliest scientific papers laid the foundations for what was to become the science of biogeography, of which he was a founder: "On the Umbrella Bird (*Cephalopterus ornatus*), 'Ueramimbé,' " "On the Monkeys of the Amazon," and "On some Fishes Allied to *Gymnotus*" in the *Proceedings of the Zoological Society of London;* "Some Remarks on the Habits of the Hesperidae" in the *Zoologist*; "On the Rio Negro" in the

Journal for the Royal Geographical Society; "On the Insects Used for Food by the Indians of the Amazon" and "On the Habits of the Butterflies of the Amazon Valley" in the *Transactions of the Entomological Society of London*.[40] Obviously, not all was lost with the ship.

It would appear by all reconstructions of the Amazon trip, including his own, that the discovery and description of natural selection as the primary mechanism of species transmutation was made on his second great collecting expedition, not this one. Though the origin of species still evaded him, the *limits* of species did not. The biogeographical boundaries delimiting the range of species was the deeper scientific achievement of his journey up the Amazon. In his paper on the range and distribution of monkeys he concluded that "on the accurate determination of an animal's range many interesting questions depend."[41] Probing the walls of the species citadel, Wallace wondered: "Are the very closely allied species ever separated by a wide interval of country? What physical features determine the boundaries of species and of genera? Do the isothermal lines ever accurately delimit the range of species, or are they altogether independent of them? What are the circumstances which render certain rivers and certain mountain ranges the limits of numerous species, while others are not?" The Amazon, for example, because of its extensive width, is a reproductive isolating mechanism, and thus a cause of speciation. "The native hunters are perfectly acquainted with this fact, and always cross over the river when they want to procure particular animals, which are found even on the river's bank on one side, but never by any chance on the other." Moving up smaller tributaries, however, "they cease to be a boundary, and most of the species are found on both sides of them."[42]

In his paper on the behavior of Amazonian butterflies, Wallace presented this discovery of the relationship between species geographical limitation, modification, and time: "All these groups are exceedingly productive in closely related species and varieties of the most interesting description, and often having a very limited range; and as there is every reason to believe that the banks of the lower Amazon are among the most recently formed parts of South America, we may fairly regard those insects, which are peculiar to that district, as among the youngest species, the latest in the long series of modification which the forms of animal life have undergone."[43] Innocent biogeographical questions and observations had hidden in them the seeds of the isolation and transformation of species. The relationship between geography and species would become the foundation of his theory of transformation that he would begin in 1855 and complete in 1858. But in the interim he had business to attend to back home.

3

Breaching the Walls of the Species Citadel

When Charles Darwin returned home from his five-year circumnavigation of the globe he settled into domestic life, starting his family, securing his finances, and opening his notebooks on the transmutation of species, on which he worked for over twenty years before publishing. He never again left England. By contrast, when Wallace returned home from his four years in the Amazon, he was restless, anxious to finish what he felt he had only begun in his studies of tropical nature. He was young and strong (Figure 3-1), had just turned thirty, and still felt the pang of wanderlust. From numerous scientific society meetings, extensive reading of travel literature, and in-depth review of the research of other naturalists (as well as a brief excursion to Switzerland to make glacial observations), it became apparent to Wallace that, unlike Darwin, he was not ready to settle down into domesticity and desk-bound reflection. Wallace had the data, but not the theory. He had evolved into a world-class observer, but had yet to develop the synthetic thinking of a theoretician. He was an evolutionist and a naturalist, but he had yet to connect the two. It would take another extensive voyage to breach the walls of the species citadel and solve that mystery of mystery—the origin of species.

Everything Except Pay

During his eighteen months at home Wallace became an intellectual insider—a member of the scientific club—with invitations to attend regular meetings of many scientific organizations, such as the Entomological and Zoological Societies. He was beginning to circulate among the present and future scientific luminaries of his age. He saw Darwin "for a few minutes in the British

Figure 3-1 Alfred Russel Wallace in 1853 at age thirty, just after returning from the Amazon and just before leaving for the Malay Archipelago. (From *My Life,* 1905, v. I, 324)

Figure 3-2 Charles Darwin in 1854 at age forty-five, on the eve of completing his manuscript on natural selection and at the height of his intellectual powers. (From F. Darwin, 1887, 205)

Museum"; he heard Thomas Huxley speak and was "particularly struck with his wonderful power of making a difficult and rather complex subject perfectly intelligible and extremely interesting." He was also taken aback by a comment Huxley made that would ring true throughout much of his life: "Science in England does everything—except pay." Wallace and Huxley maintained a casual correspondence for decades to follow. Though Huxley took a similar rough road to the top of the scientific establishment, Wallace apparently never saw himself as his scientific equal. "From that time I always looked up to Huxley as being immeasurably superior to myself in scientific knowledge, and supposed him to be much older than I was." Years later Wallace was chagrined "to find that he was really younger."[1]

In a literature search one day, Wallace's eye paused on a page in Goodrich's 1851 *Universal History,* in which the author made this observation about the Malay Archipelago: "It seems like a new world for its vegetable as well as animal kingdom and is unlike that of all other countries." With the exception of the island of Java, the natural history of the entire archipelago was largely unexplored. As always for Wallace, however, practical obstacles loomed as large as scientific ones. He would have to first manage the finances required for equipment and passage. Wallace's paper presentations at the Royal Geo-

Figure 3-3 Wallace in his late twenties or early thirties, with his mother, Mary Anne, and sister Frances. (Courtesy of Alfred John Russel Wallace and Richard Russel Wallace)

graphical Society caught the attention of its president, Sir Roderick Murchison, who arranged for Wallace a first-class ticket on a steamship bound for Singapore and the tropics of the South Pacific. His "chief object," he said, was "the investigation of the Natural history of the Eastern Archipelago in a more complete manner than has hitherto been attempted." In addition, he solicited "Astronomical & Meteorological instruments as are required to determine the position in Latitude, Longitude & Height above the Sea Level" for his chief research locations.[2]

In preparation for ornithological riches the archipelago promised, Wallace acquired a copy of Prince Lucien Bonaparte's 800-page bird catalogue *Conspectus Generum Avium,* featuring all known species of birds up to 1850, with demographic data and identifying descriptions. It was one of Wallace's most important purchases for the trip, as the book's wide margins allowed him to copy "out in abbreviated form such of the characters as I thought would enable me to determine each, the result being that during my whole eight years' collecting in the East, I could almost always identify every bird already described, and if I could not do so, was pretty sure that it was a new or unde-

Figure 3-4 Wallace playing chess with his sister shortly after his return from the Amazon. (Courtesy of Alfred John Russel Wallace and Richard Russel Wallace)

scribed species."[3] Wallace knew that in order to develop a theory to explain the transmutation of species he would need as complete a database as possible from which to work. Well equipped and endorsed by the Royal Geographical Society and "her Majesty's Government" (who, at the height of her imperialistic reaches requested reports, sketches, and maps in return), Wallace boarded the brig *Euxine* at Portsmouth (after being delayed in port for two months on another vessel that proved unseaworthy—an omen of things to come) and sailed into scientific history in March 1854.

The Least Known Part of the Globe

What the Galápagos Islands were to Charles Darwin the Malay Archipelago was to Alfred Wallace—an evolutionary mecca where the cathedrals of life housed transepts and spandrels revealing the genesis of species. Wallace was now thirty-one years old and a seasoned veteran, about to invest eight years of intense observation and reflection in a variety of locales. (Darwin was twenty-two when he departed England on his five-year voyage.) It was in Asia that Wallace began his serious study of what Darwin called "the mystery of mysteries"—the origin of species. For this period of time, "which constituted the central and controlling incident of my life," Wallace would later write, "the question of *how* changes of species could have been brought about was rarely out of my mind."[4]

Like Darwin, Wallace's reading of Lyell's *Principles of Geology* and Malthus's *Essay on the Principle of Population* shaped his thoughts on the transmutation of species. "In order to refresh my memory I have again looked through Malthus' work," he recalled half a century later, "and I feel sure that what influenced me was not any special passage or passages, but the cumulative effect of chapters III to XII of the first volume (and more especially chapters III to VIII) occupying about 150 pages." Given enough time, Malthus reasoned, populations will outstrip food supplies, causing their natural rate of growth to be severely curtailed. While Malthus was concerned with humans, Wallace saw how it also applied to animals: "In these chapters are comprised very detailed accounts from all available sources, of the various causes which keep down the population of savage and barbarous nations, in America, Africa, and Asia, notwithstanding that they all possess a power of increase sufficient to produce a dense population for any of the continents in a few centuries."[5]

In addition to better equipment, the sponsorship from the Royal Geographical Society also provided Wallace a collecting assistant—a sixteen-year-old carpenter's son named Charles Allen. The voyage to the Far East, like that to South America, was not without mishaps and tribulations. The ship was to leave England in January, but was delayed until March with the eruption of the Crimean War. Sailing through the Mediterranean to Alexandria, the group took the overland route to Suez (there was as yet no canal), which was littered with hundreds of camel skeletons that no doubt raised a brow of concern. Horse-drawn omnitrains carried the passengers while camels hauled the baggage. In Suez they picked up the steamer *Bengal* and finally arrived in Singapore on April 20. By the time they landed, Wallace "had obtained sufficient information to satisfy me that the very finest field for an exploring and collecting naturalist was to be found in the great Malay Archipelago." Even more than a naturalist's dream, Malaya had the added virtue of being relatively unexplored. Wallace would expand the cartographical maps as well as the zoological and botanical ones. "To the ordinary Englishman, this is perhaps the least known part of the globe."[6]

Before long Wallace introduced the neophyte Allen to the rigors of natural history. Up at 5:30 A.M. every day, they awakened their senses with a cold bath and hot coffee. The early morning was spent sorting through the previous day's catch, mending nets, refilling pincushions, cataloguing specimens, and logging it all in carefully kept journals (that survived the trip and are in the holdings of the Linnean Society of London). Five hours of collecting from 9:00 A.M. to 2:00 P.M. were followed by another bath, dinner at 4:00, and bed by 8:00 or 9:00. Eight years of such a disciplined lifestyle resulted in over 14,000 miles covered in ninety-six transitions between every major and minor group of islands from Malaya to New Guinea. When Wallace was done, the natural history of the region was fairly well known.

On the Law

One year into his eight-year voyage, Wallace found himself in Sarawak, the northwest region of the large island of Borneo, where he collected 320 different species of beetles in fourteen days, with a single-day best of 76 varieties, 34 of which were "new to me!" Many of these species still carry his name, such as *Ectatorphinus wallacei* and *Cryiophalpus wallacei.* More important, it was here that Wallace formulated the *Sarawak Law,* presented in his first theoretical paper, entitled "On the Law which has Regulated the Introduction of New Species."[7] The Sarawak article was written in February 1855, grafting a theoretical model onto a reasonably vast body of observational facts from Wallace's field research. It was the wet season, curtailing the amount of collecting that could be done in the field, thus affording him the time for thoughtful reflection on matters theoretic. Mulling over the fundamental differences between western and eastern tropics and considering the ideas of other naturalists such as Swainson, Humboldt, Chambers (the *Vestiges* in any case), and Lyell, "it occurred to me that these facts had never been properly utilized as indications of the way in which species had come into existence."[8] The paper was important. It set the stage for Wallace's 1858 article that challenged Darwin's priority in discovering and describing natural selection as the mechanism of evolutionary change, and established him as more than just a cataloger of nature's diversity.

The paper is divided into nine parts, with a central focus on the law as deduced from biogeographical data: "Many of these facts are quite different from what would have been anticipated, and have hitherto been considered as highly curious, but quite inexplicable. None of the explanations attempted from the time of Linneaus are now considered at all satisfactory; none of them have given a cause sufficient to account for the facts known at the time, or comprehensive enough to include all the new facts which have since been, and are daily being added." Among the new facts were geological discoveries "which have shown that the present state of the earth and of the organisms now inhabiting it, is but the last stage of a long and uninterrupted series of changes which it has undergone."[9]

Reflecting his study of Lyell's *Principles of Geology,* Wallace gives a brief account of Earth's history, including the formation of islands and continents, the elevation of mountain ranges, and how this must have happened gradually over an immense period of time. But not just geological changes, "organic life of the earth has undergone a corresponding alteration. This alteration also has been gradual, but complete; after a certain interval not a single species existing which had lived at the commencement of the period. This complete renewal of the forms of life also appears to have occurred several times:—That from the last of the geological epochs to the present or historical epoch, the change of organic life has been gradual: the first appearance of animals now existing can in many cases be traced, their numbers gradually increasing in the more recent formations, while other species continually die out and disappear, so that the present condition of the organic world is clearly derived by a natural process of gradual extinction and creation of species from that of the latest geological periods."

So far so conventional. The idea of evolution was not new to naturalists, who had been tinkering with the idea for over a century. The "mystery of mystery" was the mechanism (and, of course, adequate evidence). Although Wallace wrote the paper as if he were working on strict Baconian induction, reasoning from data to theory, in reality, by his own admission, the generalization came first, the facts second. It is a confession that reveals the true nature of the scientific enterprise, which is anything but inductive: "It is about ten years since the idea of such a law suggested itself to the writer of this essay, and he has since taken every opportunity of testing it by all the newly-ascertained facts with which he has become acquainted, or has been able to observe himself." That law, "deduced from well-known geographical and geological facts" (but, leaving himself a convenient out, constructed "in a place far removed from all means of reference and exact information"), is presented in nine facts and one deduction, divided into two subsections:

Geography

1. Large groups, such as classes and orders, are generally spread over the whole earth, while smaller ones, such as families and genera, are frequently confined to one portion, often to a very limited district.
2. In widely distributed families the genera are often limited in range; in widely distributed genera, well marked groups of species are peculiar to each geographical district.
3. When a group is confined to one district, and is rich in species, it is almost invariably the case that the most closely allied species are found in the same locality or in closely adjoining localities, and that therefore the natural sequence of the species by affinity is also geographical.

4. In countries of a similar climate, but separated by a wide sea or lofty mountains, the families, genera and species of the one are often represented by closely allied families, genera and species peculiar to the other.

Geology

5. The distribution of the organic world in time is very similar to its present distribution in space.
6. Most of the larger and some small groups extend through several geological periods.
7. In each period, however, there are peculiar groups, found nowhere else, and extending through one or several formations.
8. Species of one genus, or genera of one family occurring in the same geological time, are more closely allied than those separated in time.
9. As generally in geography no species or genus occurs in two very distant localities without being also found in intermediate places, so in geology the life of a species or genus has not been interrupted. In other words, no group or species has come into existence twice.
10. The following law may be deduced from these facts:—*Every species has come into existence coincident both in space and time with a pre-existing closely allied species.*[10]

So no one could miss his conclusion, Wallace punched it up with italics. Closely allied species come into existence not only near one another in space, but *from* one another in *time*. Adopting the metaphor of the tree of life, Wallace suggested how we might view these closely allied species by noting the difficulty of discerning whether species are similar because they are allied in space or time. He wrote of "the difficulty of arriving at a true classification, even in a small and perfect group;—in the actual state of nature it is almost impossible, the species being so numerous and the modifications of form and structure so varied, arising probably from the immense number of species which have served as antitype for the existing species, and thus produced a complicated branching of the lines of affinity, as intricate as the twigs of a gnarled oak or the vascular system of the human body. Again, if we consider that we have only fragments of this vast system, the stem and main branches being represented by extinct species of which we have no knowledge, while a vast mass of limbs and boughs and minute twigs and scattered leaves is what we have to place in order, and determine the true position each originally occupied with regard to the others, the whole difficulty of the true natural system of classification becomes apparent to us."[11] The power of the metaphor is unmistakable, and Wallace used it to full effect.

Wallace next attempts to test (more like confirm) his hypothesis by an extensive consideration of the geographical distribution of organisms around the world. "If in any case the antitype had an extensive range, two or more groups of species might have been formed, each varying from it in a different manner, and thus producing several representative or analogous groups. The Sylviadae of Europe and the Sylvicolidae of North America, the Heliconidae of South America and the Euploeas of the East, the group of Trogons inhabiting Asia, and that peculiar to South America, are examples that may be accounted for in this manner." But islands, not continents, were the key to Wallace's theory. Islands provide a laboratory of evolution, a time-lapse experiment affording observers the opportunity to study the relationships of organisms with a minimal of interventing variables. The classic case is that of the Galapagos Islands, of course, "which contain little groups of plants and animals peculiar to themselves, but most nearly allied to those of South America. . . . They must have been first peopled, like other newly-formed islands, by the action of winds and currents, and at a period sufficiently remote to have had the original species die out, and the modified prototypes only remain. In the same way we can account for the separate islands having each their peculiar species, either on the supposition that the same original emigration peopled the whole of the islands with the same species from which differently modified prototypes were created, or that the islands were successively peopled from each other, but that new species have been created in each on the plan of the pre-existing ones."[12]

Wallace still did not have the mechanism to explain how these closely allied species came to be, but he came ever so close in his identification of such reproductive isolating events as the rise of a mountain chain: "When a range of mountains has attained a great elevation, and has so remained during a long geological period, the species of the two sides at and near their bases will be often very different, representative species of some genera occurring, and even whole genera being peculiar to one side, as is remarkably seen in the case of the Andes and Rocky Mountains. A similar phenomenon occurs when an island has been separated from a continent at a very early period. The shallow sea between the Peninsula of Malacca, Java, Sumatra and Borneo was probably a continent or large island at an early epoch, and may have become submerged as the volcanic ranges of Java and Sumatra were elevated." And here is the test of the hypothesis: "The organic results we see in the very considerable number of species of animals common to some or all of these countries, while at the same time a number of closely allied representative species exist peculiar to each, showing that a considerable period has elapsed since their separation." In characteristic understatement Wallace draws the deduction. "The facts of geographical distribution and of geology may thus mutually explain each other in doubtful cases, should the principles here advocated be clearly established."

Wallace believed they were, and he provided additional examples just in case the reader missed it, once again from islands: "In all those cases in which an island has been separated from a continent, or raised by volcanic or coralline action from the sea, or in which a mountain-chain has been elevated in a recent geological epoch, the phaenomena of peculiar groups or even of single representative species will not exist. Our own island is an example of this, its separation from the continent being geologically very recent, and we have consequently scarcely a species which is peculiar to it; while the Alpine range, one of the most recent mountain elevations, separates faunas and floras which scarcely differ more than may be due to climate and latitude alone."

After another half-dozen examples Wallace then draws the deduction like a seasoned theoretician: "The question forces itself upon every thinking mind,—why are these things so? They could not be as they are had no law regulated their creation and dispersion. The law here enunciated not merely explains, but necessitates the facts we see to exist, while the vast and long-continued geological changes of the earth readily account for the exceptions and apparent discrepancies that here and there occur." In the Popperian mode of conjecture and refutation, Wallace instructs his readers on how to analyze his data. "The writer's object in putting forward his views in the present imperfect manner is to submit them to the test of other minds, and to be made aware of all the facts supposed to be inconsistent with them. As his hypothesis is one which claims acceptance solely as explaining and connecting facts which exist in nature, he expects facts alone to be brought to disprove it, not *a priori* arguments against its probability."[13] Wallace's plea was more prescriptive than descriptive.

In closing out this theoretical paper Wallace returns "to the analogy of a branching tree, as the best mode of representing the natural arrangement of species and their successive creation." He follows this with a discussion of "rudimentary organs" (today called vestigial organs), such as "the minute limbs hidden beneath the skin in many of the snake-like lizards, the anal hooks of the boa constrictor, the complete series of jointed finger-bones in the paddle of the Manatus and whale," and asks the heretical question: "If each species has been created independently, and without any necessary relations with pre-existing species, what do these rudiments, these apparent imperfections mean? There must be a cause for them; they must be the necessary results of some great natural law." That great natural law, of course, was Wallace's own, which he equates in his final sentence with those describing the operations of the cosmos, an overt attempt to elevate natural history to a science on par with the highly regarded physical sciences, particularly astronomy: "Granted the law, and many of the most important facts in Nature could not have been otherwise, but are almost as necessary deductions from it, as are the elliptic orbits of the planets from the law of gravitation."[14]

Nothing Very New

Wallace packaged his precious intellectual cargo and sent it by mail steamer to London, where it shortly thereafter appeared in the workaday scientific journal *Annals & Magazine of Natural History* in September 1855. He must have anguished over what the reaction might be (including no reaction at all—every author's worst nightmare), for he wrote to Darwin (on September 27, 1857, only a fragment of which exists), who wrote him back (on December 22, 1857): "You say that you have been somewhat surprised at no notice having been taken of your paper in the Annals. I cannot say that I am, for so very few naturalists care for anything beyond the mere description of species. But you must not suppose that your paper has not been attended to: two very good men, Sir C. Lyell, and Mr. E. Blyth at Calcutta, specially called my attention to it." But Darwin was careful to also add this qualifier: "Though agreeing with you on your conclusion in that paper, I believe I go much further than you; but it is too long a subject to enter on my speculative notions."[15]

Indeed, the Sarawak paper, as it came to be known, was read by Edward Blyth, the curator of the Asiatic Society's museum in Calcutta, India, who raved to Darwin in a letter dated December 8, 1855: "What think you of Wallace's paper in the Ann. N. Hist.? Good! Upon the whole! Wallace has, I think, put the matter well; and according to his theory, the various domestic races of animals have been fairly developed into species. A trump of a fact for friend Wallace to have hit upon!" Clearly Blyth was unaware of just how far Darwin had developed his theory, for he naively inquired: "What do you think of the paper in question? Has it at all unsettled your ideas regarding the persistence of species,—not perhaps so much from novelty of argument, as by the lucid collation of facts & phenomena."[16]

Lyell too read the Sarawak paper on November 26, 1855, and was so impressed that it stimulated him to open his own species notebook (the first of seven) to consider further the mutability of species and the mechanism of change. Earlier that year, in fact, he and his wife toured the Canary and Madeira island groups where he noticed that "the vegetation was so exclusively . . . unEuropean & so peculiar." In fact, he wrote, "it seems to me that many species have been created, as it were expressly for each island since they were disconnected & isolated in the sea."[17] But Lyell did much more than alert Darwin to Wallace's paper. He warned him that someone else was closing in on the species prize and that he had better get something—anything—into print. But Darwin was not ready. Five months later, on April 16, 1856, Lyell visited Darwin at Down, during which it appears that Darwin first disclosed the details of his theory, for Lyell told Darwin that Wallace's Sarawak law "seems explained by the Natural Selection Theory."[18] Recalling the beating Chambers took over his speculative treatise *Vestiges,* Darwin again indicated to Lyell that he was not prepared at this early date to go public. As he told the American botanist Asa Gray: "You will, perhaps, think it paltry in me, when I ask you not to mention my doctrine; the reason is, if any one, like the Author of the Vestiges, were to hear of them, he might easily work them in, and then I sh$^{d.}$ have to quote from a work perhaps despised by naturalists, & this would greatly injure any chance of my views being received by those alone whose opinion I value."[19] (Interestingly, despite his apparent enthusiasm for Darwin's theory, Lyell concluded that species were *not* mutable, and gripped firmly to that position long after the publication of Darwin's *Origin of Species.* In fact, Lyell did not "convert," as Darwin noted, until 1869.)

Darwin's reaction to the Sarawak paper is perplexing. On the one hand he told Wallace that he had gone "much further than you" in developing the theory, but six months earlier, on May 1, 1857 (when, perhaps, his theory was not so far along), he had written to Wallace that "I can plainly see that we have thought much alike & to a certain extent have come to similar conclusions. In regard to the Paper in Annals, I agree to the truth of almost every word of your paper; & I daresay that you will agree with me that it is very rare to find oneself agreeing pretty closely with any theoretical paper; for it is lamentable how each man draws his own different conclusions from the very same fact." On the other hand, on his copy of the article he scribbled "nothing very new" and "Uses my simile of tree" but that "it seems all creation with him." Then, in what can only be interpreted as a move toward establishing priority over his younger colleague, Darwin noted: "This summer will make the 20th year (!) since I opened my first-notebook, on the question how & in what way do species & varieties differ from each other.—I am now preparing my work for publication, but I find the subject so very large, that though I have written many chapters, I do not suppose I shall go to press for two years."[20]

At first blush it appears that Darwin missed Wallace's main point that, in fact, a creation model cannot account for the numerous varieties within species, and the gradual shifting of varieties into separate species in separate geographical locals, particularly islands (where a creator God, for example, apparently preferred old islands to new in terms of the number of endemic species). On further reflection, however, it seems hard to believe that Darwin could have missed what Wallace so clearly stated, in italicized emphasis no less—*"Every species has come into existence coincident both in space and time with a pre-existing closely allied species."* A more likely explanation is that Darwin was so far along in his own theorizing on the species question that, for him, there really *was* nothing new in Wallace's paper. In point of fact, Wallace had *not* solved the species problem because he still lacked a mechanism for *how* varieties could become separate species, and how geographically isolated areas, like islands, can produce *new* species. Darwin already knew that species are closely allied in both space and time. The *origin of species* remained a problem to be solved.

The Problem of the Species

It is with some irony to note that in his autobiography Wallace recalls that Stevens told him "he had heard several naturalists express regret that I was 'theorizing,' when what we had to do was to collect more facts";[21] ironic because at this time Wallace was at the height of his collecting career and, in fact, had gathered data cumulatively for over five years (counting the Amazon) before he presented his law. Wallace may have been heretical in his theorizing, and he may have operated from theory to data in his mind, but he knew what was expected and was careful to present his case with loads of empirical observations. But what was it, exactly, that Wallace and Darwin were theorizing about? What was the problem of the species to be solved?

The "species problem" may be addressed in a two-part question: Do species really exist in nature; and, if so, how can they change in order to account for the variety of life on earth? Pre-Darwinian creationists saw species as separate divine creations, individual thoughts of the creator, said Darwin's contemporary Louis Agassiz. The "problem" is that if there are such things as separate "kinds," or units of nature, then how can they change into other kinds, as Darwin and Wallace believed they had observed and inferred?

Ever since Darwin and Wallace, naturalists and evolutionary biologists have wrestled with how best organisms should be classified, as "lumpers" (those who see similarities) and "splitters" (those who see differences) debate whether a group of organisms should be one species or two, or not even categorized as species at all.[22] Interestingly, although he discovered the mechanism by which species change, Darwin demurred somewhat when he suggested "We shall have to treat species as . . . merely artificial combinations made for convenience. This may not be a cheering prospect; but we shall at least be freed from the vain search for the undiscovered and undiscoverable essence of the term species."[23] J.B.S. Haldane concurred with Darwin: "The concept of a species is a concession to our linguistic habits and neurological mechanisms."[24] Stephen Jay Gould's compromise is unique if unsatisfactory to some: "One can argue that our world of ceaseless flux alters so slowly that configurations of the moment may be treated as static."[25] Ernst Mayr is less conciliatory: "Species are the product of evolution and not of the human mind." In fact, Mayr's definition of a species is the one most commonly used today, memorized by generations of students: "Species are groups of actually or potentially interbreeding natural populations which are reproductively isolated from other such groups."[26]

The species problem, which still enjoys polemical debate, has a long and rich historical past. It dates as far back as Aristotle, who first proposed the problem in developing his taxonomic system, and reached a peak in the Enlightenment when such figures as John Ray, Carolus Linnaeus, and especially Georges Buffon, brought the problem into the scientific community for open examination. Between the publication of Ray's *Historia Plantarum* in 1686 and Buffon's magisterial *Historie Naturelle* (published between 1749 and 1785), the species problem moved from the casual musings of amateur naturalists to the austere arena of professional scholars and natural philosophers. In turn, it became one of the triggers that pushed the study of nature into a paradigmatic science of biology, affording Darwin and Wallace a serious problem to solve.

We may roughly divide the problem of the species into three time periods: (1) Ancient Greece to medieval naturalists; (2) the Enlightenment from 1686 to 1785 (from Ray to Buffon); and (3) Wallace and Darwin to the present. The period between (1) and (2) corresponds to the renaissance of Aristotle's taxonomic classification of organisms into kinds and groups, while the period between (2) and (3) is associated with the rise of evolutionary theory and the threat this posed to the concept of fixed and immutable species.

Discovering Nature: Ancient Greece to Medieval Naturalists

Alfred North Whitehead once opined that "the safest general characterization of the European philosophical tradition is that it consists in a series of footnotes to Plato."[27] This observation is true for the problem of the species. One of Plato's *idées fixes* was just that—the belief in fixed ideas that are independent of the phenomena of appearance, also known as *essentialism.* The essence of a cat is its "catness," a characteristic that exists independently of any particular cat observed in nature. The reason a cat is a cat is that it could not be anything else and still be a cat, a point rather meaningless until the consequences are considered—catness is a "kind" that cannot change. For those relegated to Plato's footnotes, their job was to slog through the confusing array of nature and sort organisms by type.

Ernst Mayr notes the importance of Plato's essentialism by demonstrating how Ionian philosophers before Plato had developed many evolutionary ideas, "such as unlimited time, spontaneous generation, changes in the environment, and an emphasis on ontogenetic change in the individual."[28] Thales, for example, said that water is the element out of which all things arise, including life, whereas Anaximander proposed that the first animal forms were created in water, then were surrounded by a type of "husk" so that they could survive on dry land where they subsequently gave rise to other creatures.[29] Anaximander's evolutionary model even had humans developing from other animals, which led historian of science George Sarton to call his ideas "a theory of organic evolution," and "a distant forerunner of Darwin."[30] Anaximenes, a pupil of Anaximander, introduced the idea of primordial terrestrial slime, a mixture of earth and water that, when stimulated by the sun's heat, gives rise to plants, animals, and even humans.[31] Heraclitus developed a theory of organic change where "all is flux, nothing is stationary. . . . There is nothing permanent except change."[32] Empedocles believed in abiogenesis, a type of spontaneous generation where plant life came first, then gave rise to basic animal forms, which later developed into more complex organisms, and so on up to humans.[33] The Ionian list is long, but Plato's subsequent essentialism put an end to such evolutionary thinking and natural history itself became a footnote to Plato.

Because Enlightenment thinkers adapted Platonic essentialism and thus could not relinquish the belief in the fixity of species, Mayr calls Plato "the great antihero of evolution." He goes so far as to say: "The rise of modern biological thought is, in part, the emancipation from Platonic thinking."[34] Plato's essentialism, however, created a worldview that set the stage for the development of the science of taxonomy and the invention of the species.

If Plato is the antihero of evolution, then Aristotle is the hero of the species. More than any other scientist before the Enlightenment, Aristotle opened the world of nature to quantification and analysis. Provided with thousands of specimens from round the world by his students, including Alexander the Great, Aristotle created a catalogue of nature that was not surpassed until the time of Linnaeus over two millennia later. Aristotle was the first to divide the natural world into separate fields of study, including (what we would call) zoology, botany, ethology, ecology, biogeography, comparative anatomy and physiology, embryology, and so forth. Through his zoologic observations in *Historia animalium,* for example, Aristotle laid the foundation of "taxonomy," or the study of classification. He recognized some 540 types of animals, identifying them as separate species through classifying them by physical characteristics such as blood versus bloodless, vertebrate versus invertebrate, egg-bearing versus live-bearing, weak-shelled versus hard-shelled, ecological habitat, and so on. He recognized, for instance, that the fishlike Cetacea were not fish at all, but marine mammals (whales, dolphins, and porpoises). His discovery that the bones of the flippers of Cetacea resembled those of land mammals, including humans, was the first identification of the modern principle of homology, one of the strongest lines of evidence in support of evolutionary theory.[35] Most important for the species problem, however, was Aristotle's introduction of the taxa *Genos,* or "genus," and *Eidos,* or "species." His *scala naturae* described a hierarchy of perfection, or a "great chain of being,"[36] in which species were classified from simple to complex. In the *Historia animalium* Aristotle observed: "Nature proceeds little by little from things lifeless to animal life in such a way that it is impossible to determine the exact line of demarcation, nor on which side thereof an intermediate form should lie. Thus, next after lifeless things in the upward scale comes the plant, and of plants one will differ from another as to its amount of apparent vitality; and, in a word, the whole genus of plants, whilst it is devoid of life as compared with an animal, is endowed with life as compared with other corporeal entities."[37]

This was no evolutionary model, however, as the scala ladder was static and species fixed at each rung. In this sense, Aristotle's essentialism was similar to Plato's in that the essence of a thing is permanent and unchanging. Since natural substances act according to their own properties (wood floats because of its inherent property of floatability), and these properties cannot change, this is a nonevolutionary worldview. In Aristotle's model, observed varieties of a species are actually mere degenerations from the species' underlying essence.

Between Aristotle and the Renaissance, two names stand out above all others as naturalists who treated species as natural entities. The first-century B.C.E. Roman poet Lucretius is sometimes incorrectly labeled as an early evolutionist who envisaged an ever-changing universe developing slowly to its present state. Excerpts from his great work *De Rerum Natura* are frequently cited as evidence for this "anticipation." For example, in Eli Minkoff's textbook *Evolutionary Biology,* the following passage from Lucretius is given under the subheading of "A Very Early Theory of Natural Selection": "In those days, again, many species must have died out altogether and failed to reproduce their kind. Every species that you now see drawing the breath of life has been protected and preserved from the beginning of the world either by cunning or by prowess or by speed. But those that were gifted with none of these natural assets, trapped in the tolls of their own destiny, were fair game and an easy prey for others, till nature brought their race to extinction."[38] Lucretius's process of preservation and extinction, however, was no anticipation of Darwinian natural selection. Rather, it was a system to preserve the purity of the essence of a species by eliminating the varieties on the extremes. Lucretius actually rejected the concept of species mutability: "But each thing has its own process of growth; All must preserve their mutual differences, Governed by Nature's irreversible law."[39]

The first-century C.E. Roman writer Dioscorides, whose work *De materia medica* was the foremost classical source of botanical terminology and the leading pharmacological text for the next sixteen hundred years, reified the species as a unit of nature through a thorough description of over 600 plants that he collected while traveling with the armies of Emperor Nero. *De materia medica* became the foundation of late medieval herbals when it was translated into seven languages and distributed throughout Europe. After his death, however, Dioscorides' disciples studied Dioscorides instead of nature. In time, copyists of the copyists of the copyists of Dioscorides created a whole new nature that had little correspondence to the real world. Leaves were drawn on branches for symmetry. Enlarged roots and stem systems were added to fill in oversized folio pages. Before the printing press, publishers used stock blocks of wood carved individually for roots, trunks, branches, and leaves, combining them into a conglomerate illustration of a tree that did not exist anywhere in the world. Copyists' fancy and imagination became the norm. The "Barnacle-tree," for example, actually grew barnacles; the "Tree-of-life" was enveloped by a serpent with a woman's head; and the Narcissus plant grew tiny human figures! So powerful was Dioscorides' influence over the centuries that late in the sixteenth century the chair of botany at the University of Bologna was conferred with the title "Reader of Dioscorides."[40] In other words, Dioscorides solidified the species, but his followers put an end to their study in nature.

What herbals were to botany, bestiaries were to zoology, and the effect was similar—a reification of separately created and autonomous species, each with its own peculiar set of characteristics. From Pliny's *Natural History* of the first century C.E. to Conrad Gesner's 1545 *Bibliotheca universalis* and Edward Topsell's 1607 *The Historie of Four-Footed Beastes,* the tradition of listing and classifying continued, and so long as no one bothered to check the original source—nature herself—the presumed truth was there to be seen in black and white. The contrast with nineteenth-century naturalists like Wallace and Darwin, who logged thousands of miles and tens of thousands of specimens before cautiously forwarding generalizations, could not be more dramatic. Empirical observation and verification were simply not part of the medieval mind. Plants and animals were God's creations, put on earth to serve as moral symbols for man. Ants were diligent, the lion was courageous, and all had their place in the great chain of being, from slime to plants to animals to man to angels to God. Emblematic of such hierarchical thinking is this excerpt from a fourteenth-century commentary on Genesis by Henry of Langenstein: "God, the creator, wishing there to be as many different types of things as were needed for the proper ordering of the universe and his glory . . . established in living creatures (eis) a marvelous array and diversity of parts and

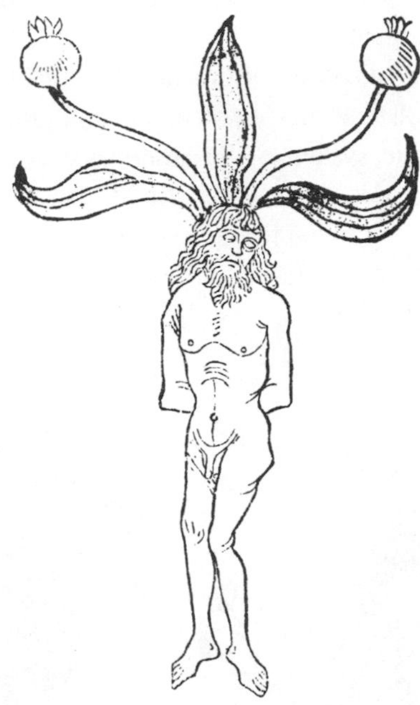

Figure 3-5 Solving the problem of the origin of species first required accurate representations of plants and animals, which did not come about until the sixteenth century. Before this time medieval scholars made copies of copies of copies, dating back centuries to the ancient texts considered canonical, without checking the original sources. Here we see "The true picture of the Lamia" (half-man half-beast) from Edward Topsell's 1607 *The Historie of Four-Footed Beastes,* and the plant mandragora (half-man half-plant) from the 1485 German *Herbarius.* (From Debus, 1978, 36, 44)

members, both interior and exterior, as well as a varied and wonderful placement of organs."[41]

Two major events, however, changed this picture and prepared the intellectual climate for the Enlightenment and the creation of the modern species: the invention of the printing press and the discovery of the New World. Among the many products of Gutenberg's printing press were copies in the vernacular of Dioscorides and the accompanying herbals and bestiaries. The publication of these works encouraged people to go out into nature and check the accuracy of the medieval drawings. To their surprise, the printed page had little in common with the natural world. Accuracy in plant and herb description became important because it could be checked, and this led to author bylines so that inaccuracies and mistakes could be linked to their source in an individual. Draftsmen, artists, and woodcarvers were employed to copy nature instead of Dioscorides, and these became the forerunners to the naturalists of the Enlightenment.

The great voyages of discovery from the fifteenth to nineteenth centuries produced a potpourri of new species of plants and animals that had no place in the ancient taxonomic system of Aristotle; nor were they to be found in the authority of Dioscorides or the Bible. Landlocked naturalists were excited by the new discoveries, but found themselves befuddled as to where to classify the new finds. Dioscorides's 500 plants had swelled to 6,000 by 1623 with the publication of a new herbal.

Figure 3-6 A new breed of artist-naturalist took hold in the sixteenth century in which authors checked with nature instead of classic texts before publishing their renditions of natural objects and living things. Here artists are preparing illustrations for Fuchs's 1542 *De historia stirpium.* (From Debus, 1978, 45)

The Enlightenment's Cartography of Nature

Just as the mariners and explorers of the New World needed a revised cartographical system to find their way through uncharted lands, the new naturalists needed a revised cartography of nature—a latitude and longitude of plants and animals—to make sense of the new world of species. The thousands of new specimens coming in from around the world, growing in tandem with the publications coming off printing presses all over the continent, left the 500-species taxonomic system of Aristotle wanting. The hundred years of Enlightenment concern with the problem of the species would see more progress than in the preceding millennium. Three major figures, John Ray, Carolus Linnaeus, and Georges Buffon, would define the species and in the process place the concept of evolution in the air, setting the stage for nineteenth-century thinkers to conceive of mechanisms of evolutionary change.

Where Carolus Linnaeus and Georges Buffon were contemporaries (both were born in 1707 and died in 1778), John Ray (1627?–1705) was their predecessor, and his work cut the initial path through nature's befuddling creations. Prior to Ray, there was no objective, unifying principle by which to separate species *eidos*. Differences were subjective, based on what the naturalist observed to be different. Judging inanimate objects such as minerals is not difficult because there is so little overlap in "kinds" of minerals. Complex organisms, however, show so much variation between kinds that distinctions are blurred. An objective criterion was needed, and John Ray was the first to proffer one. In his three-volume *Historia Plantarum* (1686–1704), Ray offered this unique definition of a species based on reproductive capability: "Thus, no matter what variations occur in the individuals or the species, if they spring from the seed of one and the same plant, they are accidental variations and not such as to distinguish a species. . . . Animals likewise that differ specifically preserve their distinct species permanently; one species never springs from the seed of another nor vice versa."[42]

This reproductive aspect of species was unique and important. It removed the subjective element from the naturalist's observations. If two organisms could produce viable offspring, then they were the same species and contained the same essence, regardless of how similar or dissimilar they may appear in a naturalist's subjective assessment. This definition, however, allowed the fixity of the species to be maintained. If variations became too extreme, there would be no reproduction and the variations would be eliminated. This fit well into the creationist doctrine in preserving the original kinds of God's creation, yet there was progress in Ray's formulation in that he eliminated the mythical creatures of the medieval imagination. In fact, the word "species"

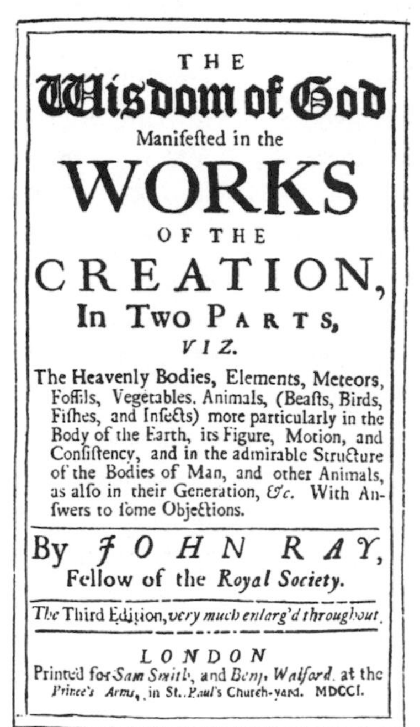

THE

Wisdom of God

Manifested in the

WORKS

OF THE

CREATION,

In Two Parts,

VIZ.

The Heavenly Bodies, Elements, Meteors, Fossils, Vegetables, Animals, (Beasts, Birds, Fishes, and Insects) more particularly in the Body of the Earth, its Figure, Motion, and Consistency, and in the admirable Structure of the Bodies of Man, and other Animals, as also in their Generation, *&c.* With Answers to some Objections.

By *JOHN RAY*,
Fellow of the *Royal Society*.

The Third Edition, *very much enlarg'd throughout.*

LONDON
Printed for *Sam Smith*, and *Benj. Walford*, at the *Prince's Arms*, in St. *Paul's* Church-yard. MDCCI.

Figure 3-7 The origin of species problem has a long and rich past. It dates as far back as Aristotle, who first proposed the problem in developing his taxonomic system, and reached a peak in the Enlightenment when John Ray, Carolus Linnaeus, and especially Georges Buffon brought the problem into the scientific community for open examination. Pictured are Carolus Linnaeus, Georges Buffon, and John Ray's *Wisdom of God.* The problem would not be solved for nearly a century, when Wallace and Darwin independently hit upon the mechanism of change—natural selection.

derives from the Latin *specere,* "to look at" or "to see," and this was precisely Ray's methodology.

Indeed, Ray's own desire to get a handle on nature's system was probably derived from his habit of taking walks in the countryside with his colleague at Cambridge, Francis Willughby. The friendship would prove a necessary one for Ray, who, on his refusal to take an oath accepting everything in the Book of Common Prayer (required by the 1662 Act of Uniformity passed by Charles II's Parliament), lost his fellowship at Trinity College, Cambridge. Willughby supported Ray's life as an independent scholar and their tours of the English countryside produced a *systema naturae,* a collaborative effort in which Ray catalogued the plants and Willughby the animals. Ray's reputation grew, leading to his election to the Royal Society, but his fierce independence and loyalty to the will left by Willughby (that stipulated an annual stipend for being an independent naturalist) caused Ray to reject an offer to take the prestigious position of Secretary of the Society. After Willughby's death, Ray published two more treatises under his colleague's name, and then began work on this magnum opus that would remove the species confusion. As he explained in his 1682 *Methodus Plantarum,*

> The number and variety of plants inevitably produce a sense of confusion in the mind of the student: but nothing is more helpful to clear understanding, prompt recognition and sound memory than a well-ordered arrangement into classes, primary and subordinate. A Method seemed to me useful to botanists, especially beginners; I promised long ago to produce and publish one, and have now done so at the request of some friends. But I would not have my readers expect something perfect or complete; something which would divide all plants so exactly as to include every species without leaving any in positions anomalous or peculiar, something which would so define each genus by its own characteristics that no species be left, so to speak, homeless or be found common to many genera.[43]

Continuing, Ray now laid the groundwork for what would later become Hutton's doctrine of uniformitarianism and Darwin's concept of gradualism by writing a polemic against catastrophism. As previously mentioned, Ray supported the ancient doctrine *natura non facit saltus,* "nature does not make leaps": "Nature, as the saying goes, makes no jumps and passes from extreme to extreme only through a mean. She always produces species intermediate between higher and lower types, species of doubtful classification linking one type with another and having something in common with both—as for example the so-called zoophytes between plants and animals."[44]

Once again, however, this was *not* a mechanism of evolutionary change, but a process for preserving the original type. As Ray noted, "Forms which are different in species always retain their specific natures, and one species does not grow from the seed of another species." What Ray did was fill in the gaps, showing how fossil species, now extinct, have a place in nature's chain of being. It was a *fixed* chain of being, but he allowed room for later evolutionists to show how species' immutability may become mutable: "Although this mark of unity of species is fairly constant . . . it is not invariable and infallible."[45]

What Gerardus Mercator and his Mercator projection was to the geographical world, the Swedish naturalist Carolus Linnaeus (1707–1778) and his taxonomic system of "binomial nomenclature" was to the botanical world. Linnaeus was trained for the ministry, but from the time of his youth, he enjoyed the outdoors and spent much of his time observing and cataloguing plants by their various characters. (A precocious child, by age eight he was called "the little botanist.") Linnaeus drew correspondences from the animal to the plant kingdom, particularly focusing on sexual organs, as evidenced in this passage from his 1760 *Sexes of Plants:* "Therefore the calyx is the bedchamber, the corolla the curtains, the filaments the spermatic vessels, the anthers the testes, the pollen the sperm, the stigma the vulva, the style the vagina, the germen the ovary, the pericarp the fecundating ovary, and the seed the ovum." Plant behavior, as well, had anthropormorphizing sexual overtones: "One husband in a marriage," or "Two husbands in a marriage," or "Twenty men in bed with one female," was typical rhetoric for a young man fascinated with sex.[46] A classic natural theologian of the Enlightenment period, Linnaeus studied the works of the Creator to better understand His words. Since God was not disorganized and confused, the seeming disarray of nature was only a manifestation of man's inability to discern God's meaningful pattern.

Linnaeus became a collector par excellence. Outgoing and gregarious, Linnaeus sent hundreds of disciples to gather specimens from all corners of the globe. Most seagoing voyages in the eighteenth century carried naturalists for the specific purpose of cataloguing new finds and bringing back unusual specimens (a practice that resulted a century later in the voyages of Wallace and Darwin). It was Linnaeus who made the discovery of new species valuable by appending the name of the discoverer to the species name. For example, in 1763 Captain Carl Gustaf Ekeberg of the Swedish East India Company brought back a large number of tea plants for Linnaeus's examination; Linnaeus rewarded him by assigning the Latinized name of *Ekebergia* to a tea plant. A glance through the index of a modern botanical text reveals the names of many an eighteenth-century naturalist, including and especially Linnaeus himself (e.g., *Linnaea borealis*). He and his students were prolific. In his first publication, Linnaeus gave binomial labels to 5,900 species of plants. Linnaeus's pupil Daniel Solander, the naturalist aboard Captain Cook's *Endeavour* voyage of 1768–1771, netted a remarkable catch of 1,200 new species and 100 new genera of plants. Another student named Peter Kalm personally sighted 90 new species, and Linnaeus awarded him taxonomic immortality by labeling an entire genus of plants *Kalmia.* Linnaeus was a prophet of God's creations and his naturalist students were his disciples. The metaphor is not exaggerated, as Linnaeus became known as "God's Registrar," and that *Deus creavit, Linnaeus disposuit*—"God created, Linnaeus classified."[47]

Linnaeus's taxonomic system, described in his *Species Plantarum* (1753) and his *Systema Naturae* (1758–1759), is essentially the modern one still in use today—the two-name "bionomial nomenclature." The main two categories—genus and species—were a continuance of the old system already in use. Modern humans, for example, are classified as *Homo sapiens* (wise man), whereas premodern humans are classified as *Homo habilis* (handy man), or *Homo erectus* (erect man), or *Homo neandertalensis* (man found in the valley of Neander). Creatures more apelike than human carried a different genus name. *Australopithicus africanus* is the southern ape-man from Africa. *Australopithicus afarensis* is the southern ape-man from the Afar region of Africa. And so on. Noting the almost infinite variety of organisms and the numerous ways to classify them by type from similarities to dissimilarities, Linnaeus added the categories kingdom, class, and order. For example, he had three kingdoms—Mineral, Plant, and Animal—that he distinguished by these determining features: "Stones grow; plants grow and live; animals grow, live, and feel."[48] For each kingdom there were numerous classes. For example, the animal kingdom had six: quadrupeds, birds, amphibia, fishes, insects, and worms. And, of course, each class had numerous orders, and so on down to the species level, of which there were hundreds of thousands.

The pre-Linnean nomenclature was bulky and inconsistent, and names were attempts at adequate descriptions, characteristics, and behaviors—resulting in cumbersome and unusable terminology. For example, a plant species identified by Clusius in 1576, *Concolvulus folio Altheae,* was labeled *Concolvulus argenteus Altheae folio* by Caspar Bauhin in 1623 and by Linnaeus in 1738 as *Convolvulus foliis ovatis divisis basi truncatis: laciniis intermediis duplo longioribus*![49] Linnaeus's bionomial nomenclature became an esperanto for biologists, a system for mapping the world of nature and adding borders to groups of organisms, from the most specific (species) to the most general (kingdom). "All groups of plants show relationships on all sides," Linnaeus wrote, "like countries on the map of the world."[50] Like Aristotle, Linnaeus saw the relationships between species as a ranking in the scale of nature, a manifestation of the great chain of being: "There is, as it were a certain chain of created beings, according to which they seem all to have been formed, and one thing differs so little from some other, that if we hit upon the right method we shall scarcely find any limits between them. Does not everyone perceive that there is a vast difference between a stone and a monkey? But if all the intermediate beings were set to view in order, it would be difficult to find the limits between them."[51]

Because of his theological preferences and his insistence on the static nature of species, Linnaeus is typically seen as an archfoe of evolutionism. Yet like Ray before him, his pioneering work in classification and the definition of species made the *origin* of species a legitimate scientific problem to be solved. "We must pursue the great chain of nature till we arrive at its origin," he wrote in the *Sexes of Plants*. "We should begin to contemplate her operations in the human frame and from thence continue our researches through the various tribes of quadrupeds, birds, reptiles, fishes, insects, and worms, till we arrive at the vegetable creation."[52] With so many thousands of species never before known, Linnaeus hinted at the possibility that maybe not *all* species now alive were created in the beginning. Hybridization was one possible way of accounting for this variability, but his musings on origins were beyond what his natural theology would allow, and he relinquished the problem to later thinkers, such as Buffon.

Georges-Louis Leclerc, Comte de Buffon (1707–1778), better known simply as Buffon, elevated natural history to a serious science through his 1779 volume *Epoques de la Nature,* itself part of his multivolume series *Histoire Naturelle* (1749–1785). Any history of evolutionary thought that ends with Darwin and Wallace must include these two themes: (1) the discovery of geological, or "deep" time; and (2) a definition of species that allows for change. Buffon's scientific calculation of the age of the earth (based on Newton's theory of planetary formation), and his objective definition of a species (following Ray's lead in arguing for a more "natural" definition based on reproductive capability) moved the study of evolutionary biology under the umbrella of the day's most advanced scientific analysis. Like Linnaeus before him, Buffon was an enthusiastic biophiliac whose unsullied love of the natural world took him away from his preselected profession of the law and into that of naturalist. Born into a wealthy family in Burgundy, Buffon was schooled in law at the University of Dijon. Then, at the University of Angers, he studied mathematics, botany, and medicine, rounding out his education with a membership in the French Academy, where he completed his development by publishing works on probability theory, forestry, chemistry, and biology. Originally born Georges-Louis Leclerc, his inheritance of the village of Buffon on his mother's death made him financially secure and provided him a life of leisure—which in Buffon's case was directed to the full-time study of the natural world.[53]

Buffon's contributions to a variety of sciences were without parallel in an age that was itself without parallel in intellectual enlightenment. He wrote on every subject, from minerals to man, and published them all in a single great work consisting of thirty-five volumes in numerous editions throughout his life. The *Histoire Naturelle* expanded as Buffon's interests widened, and the work ballooned well beyond its original, more modest goals. With each passing year and with each new edition of his work, Buffon's reputation grew publically and professionally. Writing for a general readership as well as for his colleagues, Buffon followed Linnaeus in his poetic descriptions of sex in the natural world: "There are few birds as ardent, as powerful in love as the sparrow; they have been seen to couple as many as twenty times in succession, always with the same eagerness, the same trepidation, the same expression of pleasure." Of pigeons he wrote of their "entire lifetime devoted to the service of love and to the care of its fruits." Such popularizing of science has played an important role in any age, and when it is done by professionals instead of amateurs the impact is even greater. His biographers note: "Buffon is the only important scientist ever to have achieved the rank of a major literary figure almost entirely through his original scientific publications. He was not simply a popularizer of scientific thought; he was concerned rather, with presenting directly to the general reading public scientific theories which he himself had devised, in a style distinguished by its power and beauty."[54]

Buffon's work on the discovery of deep time is of monumental importance. His extension for the age of the earth to 74,832 years was a giant leap beyond Bishop Ussher's biblical calculation of 4004 B.C. for the creation. In private Buffon estimated an age of three million years or more, possibly infinity, but did not want to shock his readers and so stuck with the more conservative estimate. Yet he confessed: "The more we extend the time, the closer to the truth."[55] Buffon also hinted at the possibility of slow, evolutionary change, supporting the dictum *natura non facit saltus:* "Nature's great workman is time. He marches ever with an even pace, and does nothing by leaps and bounds, but by degrees, gradations and successions he does all things; and the changes which he works—at first imperceptible—become little by little perceptible, and show themselves eventually in results about which there can be no mistake."[56]

Like Ray before him, Buffon argued for a more objective, unifying definition of a species. He felt Linnaeus's collection of similarities and dissimilarities to define each species was artificial, based on the naturalist's tendencies, not nature's. "One must not forget that these families are *our* creation, we have devised them only to comfort our minds."[57] In fact, in one passage, Buffon carries out his argument to show how artificial divisions can lead one to doubt completely the existence of species: "In general, the more one increases the number of one's divisions . . . the nearer one comes to the truth; since in reality, individuals alone exist in nature, while genera, orders, classes, exist only in our imagination."[58] A few pages later, Buffon shows how befuddling such arbitrary divisions can be: "Is it not better to make the dog, which is fissiped [many parts to the foot], follow the horse, which is soliped [one part to the foot], rather than have the horse followed by the zebra, which perhaps has nothing in common with the horse except that it is soliped? Does a lion, because it is fissiped, resemble a rat, which is also fissiped more closely than a horse resembles a dog?"[59]

In an attempt to remove the subjective element in defining the species, and following Ray's choice of reproductive isolation, Buffon's description is even closer to the modern definition of a species: "We should regard two animals as belonging to the same species if, by means of copulation, they can perpetuate themselves and preserve the likeness of the species; and we should regard them as belonging to different species if they are incapable of producing progeny by the same means." Buffon then offers this example, one that might be found in a modern text: "Thus the fox will be known to be a different species from the dog, if it proves to be the fact that from the mating of a male and a female of these two kinds of animals no offspring is born; and even if there should result a hybrid offspring, a sort of mule, this would suffice to prove that fox and dog are not of the same species—inasmuch as this mule would be sterile."[60]

Buffon also dabbled in biogeography, showing how climate can determine the makeup of species, and hinted at the future doctrine of uniformitarianism: "[I]n all places where the temperature is the same, one finds, not only the same species of reptiles, which have not been transported from other regions, but also the same species of fish, the same species of quadrupeds, the same species of birds, none of which have emigrated to the regions in which they are found. . . . The same temperature nourishes, and calls into being everywhere, the same species."[61] From here Buffon went one step further with this seemingly startling suggestion on human origins:

> If we once admit that there are families of plants and animals, so that the ass may be of the family of the horse, and that the one may only differ from the other through degeneration from a common ancestor, we might be driven to admit that the ape is of the family of man, that he is but degenerate man, and that he and man have had a common ancestor, even as the ass and horse have had. It would follow then that every family whether animal or vegetable, had sprung from a single stock, which after a succession of generations, had become higher in the case of some of its descendants and lower in that of others. . . . then there is no further limit to be set to the power of nature, and we should not be wrong in supposing that with sufficient time she could have evolved all other organized forms from one primordial type.[62]

Through the study of comparative anatomy Buffon also looked to the underlying unity of type by which one could infer a degree of relationship among species. These observations led to the principle of homology, where organisms have similar anatomy due to similar ancestry, such as the wing of a bat, the arm of a human, and the flipper of a whale. All show similar structure due to similar ancestry, though the environments to which they have adapted are radically different:

> If we choose the body of some animal or even that of man himself to serve as a model with which to compare the bodies of other organized beings, we shall find that . . . there exists a certain primitive and general design, which we can trace for a long way. . . . Even in the parts which contribute most to give variety to the external form of animals, there is a prodigious degree of resemblance, which irresistibly brings to our mind the idea of an original pattern after which all animals seem to have been conceived. The foot of the horse in appearance so different from the hand of man, is nevertheless composed of the same bones, and we have at the extremities of our fingers the same small hoofshaped bone which terminates the foot of that animal.[63]

From these many and diverse quotes, all taken from the *Histoire Naturelle,* one might be tempted to conclude that Buffon was an evolutionist, a genius out of time, but this is not the case. Instead of reaching the conclusion of descent with modification through an argument from homology, Buffon asks "whether this does not seem to show that the Creator in making all these used but a single main idea, though varying it in every conceivable manner?"[64] Further, the causative environmental agents that create variations do not lead to new species in Buffon's model, but instead are degenerations of originally created kinds that came into existence shortly after the cooling of the earth some 75,000 years ago. For Buffon there are species, and there are varieties *within* species, but he knew of no mechanism by which varieties could become new species and thus never developed an evolutionary theory. Even in the "ape to man" passage earlier, Buffon notes that the "missing links" between species prove the fixity of the same. He continued: "If one species had been produced by another, if, for example, the ass species came from the horse, the result could have been brought about only slowly and by gradations. There would therefore be between the horse and the ass a large number of intermediate animals. Why, then, do we not today see the representatives, the descendants of these intermediate species? Why is it that only the two extremes remain?" Buffon answers his own question: "Though it can not be demonstrated that the production of a species by degeneration from another species is an impossibility for nature, the number of probabilities against it is so enormous that even on philosophical grounds one can scarcely have any doubt upon the point."[65]

Buffon was no evolutionist. But he did set the stage for later generations through his many contributions that included his extension of the age of the earth, his recognition of near infinite organic variety, his rejection of *natura non facit saltus* (thus allowing in gradualism), his correlation of climate with species characteristics (implying adaptation, not special creation), his uniformitarianism of biogeographic phenomena such as climate and coloration (thus rejecting catastrophism), and his observations of homologous design between related species. Buffon was a great empiricist, but he was also a grand synthesizer who emphasized the need to search for general principles behind specific observations. Linnaeus's system of classification was important, but there is more to the study of natural history than the copious naming of organisms. Though the great synthesis would come in the next century, Buffon's demand for a more "philosophical" understanding of nature paved the way for the theory of evolution: "We ought to try to rise to something greater and still more worthy of occupying us—that is to say, to combine observations, to generalize the facts, to link them together by the force of analogy, and to endeavor to attain that high degree of knowledge in which particular effects are recognized as dependent upon more general effects. Nature is compared with herself in her larger processes, and thus ways are opened before us by which the different parts of physical science may be perfected."[66]

Science Non Facit Saltus

This long sequence in the development of evolutionary thought in relation to the problem of the species that brings us back to Wallace, may itself provide a historical example of a model of change in the history of science. Historian Robert Richards has outlined five models in the historiography of science that include: (1) *The Static Model* of the late Renaissance and early Enlightenment that saw science as the invariable passing on of knowledge from the ancients to the moderns relatively unaltered—new knowledge was merely a rediscovery of old; (2) *The Growth Model* of George Sarton that traces the development of science as a progressive cultural element climbing ever closer to the upper wall of truth and reality; (3) *The Revolutionary Model* of Alexandre Koyre, A. C. Crombie, Rupert Hall, and Charles Gillispie, who argue that "a revolution in thought, a decisive overthrow of distinctly ancient modes of conception, is necessary to set a discipline on the smooth course of modern science"; (4) *The Gestalt Model* of Thomas Kuhn and Michel Foucault in which a conglomerate whole of "social context, past experience, and familiar assumptions control our perceptual and conceptual experiences of things," and where "theories provide patterns within which data appear intelligible" and "constitute a 'conceptual Gestalt,' " instead of merely being the end result of bounteous data collection; and (5) *The Social-Psychological Model* of Robert Merton and Joseph Ben-David, more commonly known as the "sociology of science," that builds a case for social and psychological phenomena determining the development of a science that is embedded in a given culture, and "asserts that the structure of scientific knowledge is determined not by nature but by social patterns or psychological complexes."[67] Richards adds his own model, a modified version of Karl Popper and Stephen Toulmin's *Evolutionary Model,* where theories stand or fall like species that are more or less fit for survival in a jungle of competing hypotheses.[68]

The danger of this taxonomy of historiography is that, like the species problem, it does not allow for overlap, blending, and change. There may be "lumpers" and "splitters" among historians of science. The historical sequence presented here, from the ancient Greeks to Buffon, is a combination of the *Growth, Evolutionary, Gestalt,* and *Social-Psychological* models. These pre-Darwinian thinkers are considered "forerunners" because Darwin and Wallace synthesized elements from them all, then added the final element to the equation: a mechanism of change. As Mayr metaphorically described the historical significance of this achievement: "The fixed, essentialistic species was the fortress to be stormed and destroyed; once this had been accomplished, evolutionary thinking rushed through the breach like a flood through a break in a dike."[69]

To draw a biological analogy, to the extent that nature does not proceed by leaps alone, neither does science. On close examination, most great scientific *revolutions* are more like gradual *evolutions.* If there is a revolutionary element, it is rare and usually occurs only after painstaking developments that set the stage for the revolutionist. This pattern might be called an *Interrupted Gradualism* model in which the slow, gradual changes in theories are periodically interrupted by an extra element that turns them into something completely different—"reproductively isolated" from other theories. The developments in the solution to the problem of the species, leading to the discovery of a mechanism of change, are well represented by this model. To paraphrase the old axiom, most of the time *science non facit saltus.*

Alfred Russel Wallace was about to test that model on a tiny island in the Malay Archipelago, where he discovered an entomological *terra incognita* that would lead him to trigger a revolution in evolution.

4

The Mystery of Mysteries Solved

During this period of ambitious collecting, cataloguing, and synthesizing in the Malay Archipelago, Wallace barely survived; he was often weak, sick, and starving, not to mention poor. His youthful assistant Allen resigned after a year and a half, moving to Singapore. Wallace replaced him with a Malaysian youth named Ali, who remained "my faithful companion" through the end of the expedition. Samuel Stevens continued his success in agenting Wallace's collections throughout England, but the transaction process was slow and Wallace often had to wait weeks or months for the arrival of a remuneration in order to make his next travel connection. A typical package sent to England is noted in a letter to Stevens on August 21, 1856, in which Wallace includes: "Birds for sale about 300. Butterflies in papers 150. Mammalia 9. Beetles 250. Land & Fresh Water Shells 100. Miscellaneous 65." For this Wallace expected to get "£60 which I think is the lowest sum it will fetch as no collection of birds has ever been made here before and I am sure there are many new and rare things among them."[1]

Run Amok

New and unusual finds were cherished both for their contribution to the understanding of the natural history and biogeography of an area, and for what they might fetch back home that would allow Wallace to continue the expedition. Stevens must have been amused at his observations of the indigenous peoples, interspersed throughout letters that contained mostly business transactions. From Macassar, on September 27, 1856, for example, Wallace told Stephens:

> The people here have some peculiar practices. "Amok" is as we say "running a muck" is common here. There was one last week, a debt of a few dollars was claimed of a man who could not pay it so he murdered his creditor, and then knowing he could be found out and punished he "run a muck" killed four people, wounded four more and died what the natives call an honorable death! A friend here seeing I had my mattress on the floor of a bamboo house which is open beneath, told me it was very dangerous as there were many bad people about who might come at night and push their spears up through me from below.[2]

Wallace finished out 1855 in Singapore and Borneo; moved through Bali, Lombok, Celebes, and Ké in 1856; and Aru, New Guinea, Timor, Banda, and Ambonin in 1857. Although the going was hard, he still managed a respectable outpouring of scientific papers, including three on orangutans in the *Annals & Magazine of Natural History* (later combined into a chapter in his book on *The Malay Archipelago*), a general paper on "Observations on the Zoology of Borneo" in *Zoologist,* "Attempts at a Natural Arrangement of Birds" in the *Annals & Magazine of Natural History,* and "Notes of a Journey up the Sadong River, in North-west Borneo" in the *Proceedings of the Royal Geographical Society of London.*

The Inconsistent Creator

In a paper he published in the *Transactions of the Entomological Society of London,* entitled "On the Habits and Transformations of a Species of *Ornithoptera,* Allied to *O. priamus,* Inhabiting the Aru Islands, near New Guinea," Wallace struggled with the species problem in an Aru butterfly that featured black and iridescent green wings, a golden body and crimson breast, and three black spots on each hind wing. Its genus was clearly *Ornithoptera,* but what was its species? It was similar to *Ornithoptera priamus* from the island of Amboyna, which had four black spots on each hind wing. But it was also like *Ornithoptera poseidon* from New Guinea, which had two black spots. Was the Aru specimen with its three black spots an intermediate species between *O. priamus* and *O. poseidon,* or was it a variety of one of the two? If the former, could a simple dot denote an entirely new species? If the latter, to which of these similar species could this variety belong? In a creationist model of static species this was like being a little bit pregnant. In an evolutionist model of nonstatic species this was possibly an example of a transitional form.[3]

In a subsequent and short (two-page) "Note on the Theory of Permanent and Geographical Varieties," he explained the inconsistences that come with a creationist model:

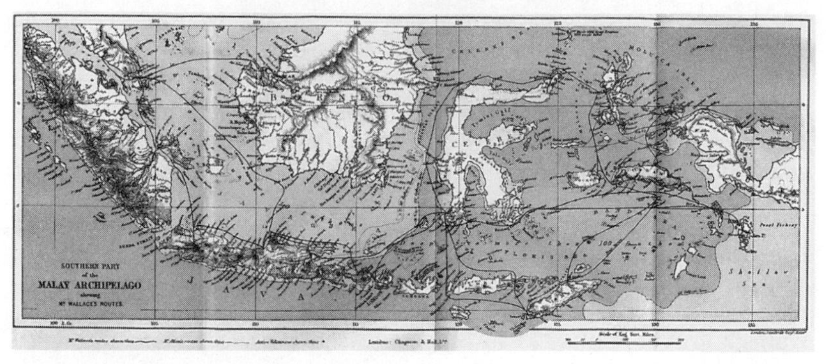

Figure 4-1 A map of the southern part of the Malay Archipelago showing (darker lines) Wallace's routes throughout the various islands, including Matra, Java, Borneo, Celebe, Timor, New Guinea, and Jilolo (Gilolo). Off the west coast of Jilolo is the tiny island of Ternate, where Wallace discovered and described the primary mechanism of evolutionary change—natural selection—in an essay he promptly sent to Charles Darwin in March 1858. This map demonstrates the vast diversity of ecosystems, biogeographies, species, and peoples that Wallace encountered in the eight years of his second major expedition. (From *My Life,* 1905, v. I, 384)

> Now the generally adopted opinion is that species are absolute independent creations, which during their whole existence never vary from one to another, while varieties are not independent creations, but are or have been produced by ordinary generation from a parent species. There does, therefore (if this definition is true), exist such an absolute and essential difference in the nature of these two things that we are warranted in looking for some other character to distinguish them than one of mere degree, which is necessarily undefinable. If there is no other character, that fact is one of the strongest arguments against the independent creation of species, for why should a special act of creation be required to call into existence an organism differing only in degree from another which has been produced by existing laws? If an amount of permanent difference, represented by any number up to 10, may be produced by the ordinary course of nature, it is surely most illogical to suppose, and very hard to believe, that an amount of difference represented by 11 required a special act of creation to call it into existence.[4]

In case the reader missed it, Wallace provided one additional analogy that brings him right to the brink of a solution to the species problem:

> Let A and B be two species having the smallest amount of difference a species can have. These you say are certainly distinct; where a smaller amount of difference exists we will call it a variety. You afterwards discover a group of individuals C, which differ from A less than B does, but in an opposite direction; the amount of difference between A and C is only half that between A and B: you therefore say C is a variety of A. Again you discover another group D,

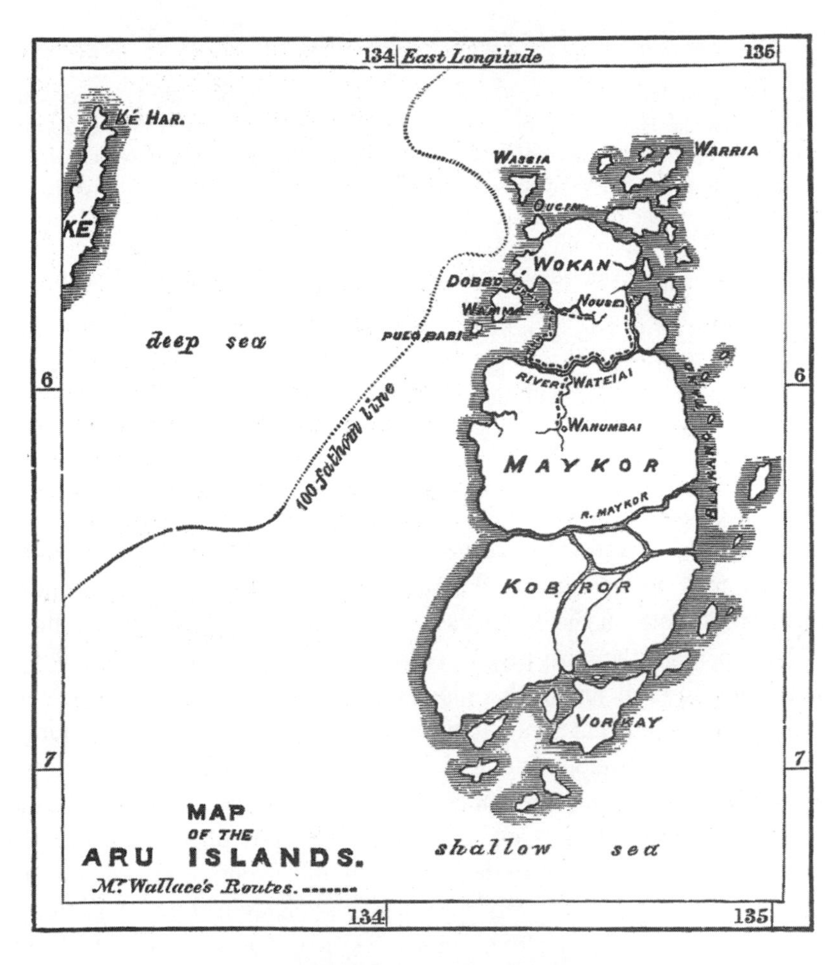

Figure 4-2 Wallace's map of the Aru Islands. Wallace struggled to solve the species problem for the butterfly species *Ornithoptera poseidon* (top) and *Ornithoptera priamus* (bottom) that featured black and iridescent green wings and two and three black spots respectively on each hind wing. (Map from *The Malay Archipelago,* 1869)

> exactly intermediate between A and B. If you keep to your rule you are now forced to make B a variety, or if you are positive B is a species, then C and D must also become species, as well as all other permanent varieties which differ as much as these do: yet you say some of these groups are special creations, others not. Strange that such widely different origins should produce such identical results. To escape this difficulty there is but one way: you must consider every group of individuals presenting permanent characters, however slight, to constitute a species; while those only which are subject to such variation as to make us believe they have descended from a parent species, or that we know have so descended, are to be classed as varieties. These two doctrines, of "permanent varieties" and of "specially created unvarying species," are inconsistent with each other.[5]

He was so close to the solution! But here the paper ends without the denouement. It was published in *Zoologist* in January 1858, the annus mirabilis of evolutionary theory, just as Wallace caught a steamer to the islands of Ternate and Gilolo, in the Moluccas, where there was, as he told Bates on the 25th, "perhaps the most perfect entomological *terra incognita* now to be found. I think I shall stay in this place two or three years, as it is the centre of a most interesting and almost unknown region."[6] Here he was working on a more complete development of the 1855 Sarawak paper that was, as he explained to Bates, "only the announcement of the theory, not its development. I have prepared the plan and written portions of an extensive work embracing the subject in all its bearings and endeavouring to prove what in the paper I have only indicated."[7] Though the Sarawak Law explained the geographical alliance of species in time and space, it did not account for either descent or divergence of species.[8] "I had no conception of how or why each new form had come into existence with all its beautiful adaptations to its special mode of life," he later recalled.[9] That would all change during one of the greatest moments of serendipity in the history of science.

On the Tendency

Wallace's extensive stay would be cut short by illness, but the development of his theory would not. Once again stricken with malaria, trembling, delirious, and fighting for his life, it occurred to Wallace that death would befall individuals unequally throughout a species. The stronger, healthier, and faster would be more likely to survive, while the less favored would die. Thus, there was a selection for and against certain individuals, depending on the variety of their characteristics.

Throughout his career Wallace was frequently asked—and seemed always to oblige his inquirers—to recount the events that led him to solving the

Figure 4-3 Wallace's temporary home in the Aru Islands, showing the rugged nature of such expeditions and the quiet solitude that afforded him the time for thoughtful inquiry into the great questions of science and philosophy. (From *My Life,* 1905, v. I, 357)

problem. Forty years after the event, for example, in a retrospective volume entitled *The Wonderful Century,* Wallace recalled the insightful experience that led him to the discovery and description of what would become known as natural selection (and using terms coined later):

> During one of these fits, while again considering the problem of the origin of species, something led me to think of Malthus' *Essay on Population* . . . and the "positive checks" . . . which he adduced as keeping all savage populations nearly stationary. It then occurred to me that these checks must also act upon animals, and keep down their numbers; and as they increase so much faster than man does, while their numbers are always very nearly or quite stationary, it was clear that these checks in their case must be far more powerful, since a number equal to the whole increase must be cut off by them every year. While vaguely thinking how this would affect any species, there suddenly flashed upon me the idea of *the survival of the fittest*—that the individuals removed by these checks must be, on the whole, *inferior* to those that survived. Then, considering the *variations* continually occurring in every fresh generation of animals or plants, and the changes of climate, of food, of enemies always in progress, the whole method of specific modification became clear to me, and in the two hours of my fit I had thought the main points of the theory.[10]

That evening Wallace "sketched out the draft of a paper," and in two nights penned his complete theory (twelve typeset pages), "On the Tendency of Varieties to Depart Indefinitely from the Original Type,"[11] a title that reflected his outline of the problem in the Aru butterfly paper. Three years to the month after he penned the Sarawak Law, Wallace had at last deduced the mechanism to explain *how* species had become more or less closely allied in space and time. And, like Darwin does in the *Origin,* Wallace begins his analysis by comparing species instability in domestic animals versus those in the wild, but to a different end: "One of the strongest arguments which have been adduced to prove the original and permanent distinctness of species is, that *varieties* produced in a state of domesticity are more or less unstable, and often have a tendency, if left to themselves, to return to the normal form of the parent species; and this instability is considered to be a distinctive peculiarity of all varieties, even of those occurring among wild animals in a state of nature, and to constitute a provision for preserving unchanged the originally created distinct species." But Wallace sets out in this paper "to show that this assumption is altogether false, that there is a general principle in nature which will cause many *varieties* to survive the parent species, and to give rise to successive variations departing further and further from the original type, and which also produces, in domesticated animals, the tendency of varieties to return to the parent form."[12] Already Wallace has departed from Darwin in this contrast, as the latter compared the artificial selection of domesticated animals to the natural selection of wild animals.

The starting point for Wallace's theory is the observation that "the life of wild animals is a struggle for existence." But why, he wonders, are some species so abundant "while others closely allied to them are very rare"? The answer begins with a Malthusian-style analysis of populations that, if left unchecked, would increase dramatically. Yet it is evident that this does not happen; indeed, it cannot happen because in a short span of time "the population must have reached its limits, and have become stationary, in a very few years after the origin of each species." The reason is that life is Hobbesian—nasty, brutish, and short—as borne out in this ornithological analysis: "It is evident, therefore, that each year an immense numbers of birds must perish—as many in fact as are born; and as on the lowest calculation the progeny are each year twice as numerous as their parents, it follows that, whatever be the average number of individuals existing in any given country, *twice that number must perish annually,*—a striking result, but one which seems at least highly probable, and is perhaps under rather than over the truth. It would therefore appear that, as far as the continuance of the species and the keeping up the average number of individuals are concerned, large broods are superfluous. On the average all above *one* become food for hawks and kites, wild cats and weasels, or perish of cold and hunger as winter comes on. This is strikingly proved by the case of particular species; for we find that their abundance in individuals bears no relation whatever to their fertility in producing offspring."[13] (Wallace's exception is, ironically, the passenger pigeon, whose fecundity was legion in his day with flocks so dense that they blackened the noonday sky but, tragically, succumbed to the predations of human hunters and habitat destruction and is now extinct.)

Wallace's first deduction is then drawn from these observations. Death, he reasons, will not befall individuals equally. "The numbers that die annually must be immense; and as the individual existence of each animal depends upon itself, those that die must be the weakest—the very young, the aged, and the diseased,—while those that prolong their existence can only be the most perfect in health and vigour—those who are best able to obtain food regularly, and avoid their numerous enemies. It is, as we commenced by remarking, 'a struggle for existence,' in which the weakest and least perfectly organized must always succumb."[14] Note that Wallace's selection mechanism is negative in orientation: the weakest are being selected against. But he re-orients the direction of the selection process when he draws his second deduction to groups and species: "Now it is clear that what takes place among the individuals of a species must also occur among the several allied species of a group,—viz. that those which are best adapted to obtain a regular supply of food, and to defend themselves against the attacks of their enemies and the vicissitudes of the seasons, must necessarily obtain and preserve a superiority in population; while those species which from some defect of power or organization are the least capable of counteracting the vicissitudes of food, supply, &c., must diminish in numbers, and, in extreme cases, become altogether extinct. Between these extremes the species will present various degrees of capacity for ensuring the means of preserving life; and it is thus we account for the abundance or rarity of species."[15]

The key to the selection process in either direction—for the stronger, against the weaker—is the fact that species vary. Variation—the remarkable diversity of individuals and species that Wallace had been observing for nearly eight years—is what drives the selective process directionally, as he clearly stated in the subtitle of the next section of the paper: "Useful Variations will tend to Increase; useless or hurtful Variations to Diminish." Examples provided include the selection process going in both directions: "An antelope with shorter or weaker legs must necessarily suffer more from the attacks of the feline carnivora; the passenger pigeon with less powerful wings would sooner or later be affected in its powers of procuring a regular supply of food; and in both cases the result must necessarily be a diminution of the population of the modified species. If, on the other hand, any species should produce a variety having slightly increased powers of preserving existence, that variety must inevitably in time acquire a superiority in numbers."[16]

But we still do not have a mechanism for the origin of new varieties, and thus new species. This requires a third deduction on Wallace's part. Those variations most likely to survive will displace those that are not, and over time "the *variety* would now have replaced the *species,* of which it would be a more perfectly developed and more highly organized form. It would be in all respects better adapted to secure its safety, and to prolong its individual existence and that of the race. Such a variety *could not* return to the original form; for that form is an inferior one, and could never compete with it for existence." Thus, we see where Wallace derived his title of "The Tendency of Varieties to Depart Indefinitely from the Original Type." The successful variations win out over the unsuccessful ones, and "this new, improved, and populous race might itself, in course of time, give rise to new varieties, exhibiting several diverging modifications of form, any of which, tending to increase the facilities for preserving existence, must, by the same general law, in their turn become predominant. Here, then, we have *progression and continued divergence* deduced from the general laws which regulate the existence of animals in a state of nature, and from the undisputed fact that varieties do frequently occur."[17]

In the next section of the paper Wallace returns to the question of domesticated animals but, again unlike Darwin, he uses these examples to a different end—he is more interested in the process of variation, Darwin more in selection. Wild animals, Wallace explains, "depend upon the full exercise and healthy condition of all their senses and physical powers, whereas, among the latter [domesticated], these are only partially exercised, and in some cases are absolutely unused. . . . Now when a variety of such an animal occurs, having increased power or capacity in any organ or sense, such increase is totally useless, is never called into action, and may even exist without the animal ever becoming aware of it. In the wild animal, on the contrary, all its faculties and powers being brought into full action for the necessities of existence, any increase becomes immediately available, is strengthened by exercise, and must even slightly modify the food, the habits, and the whole economy of the race. It creates as it were a new animal, one of superior powers, and which will necessarily increase in numbers and outlive those inferior to it." Without variation and, more important, those variations that matter most to the survival of the organism, new species cannot arise. And whereas Darwin uses domesticates as a centerpiece in his opening arguments in the *Origin,* Wallace says "that no inferences as to varieties in a state of nature can be deduced from the observation of those occurring among domestic animals. The two are so much opposed to each other in every circumstance of their existence, that what applies to the one is almost sure not to apply to the other."[18]

Wallace could not have known Darwin's reasoning in the *Origin* since it was yet to be published, and since he was on the other side of the planet he could not have shared in conversation with the now retiring naturalist. So Wallace's critique of the domesticate analogue was directed at no one in particular. But Wallace did specifically target the other great evolutionist of the age, Lamarck. The textbook version of our age has Darwin's theory of natural selection slaying the egregious dragon of Lamarck's theory of acquired characteristics, which, thanks to modern genetics, we know cannot happen. (One need only observe the Jewish rite of circumcision, ongoing now for several thousand years, to see that an alteration in characters in the parents is not passed down to the offspring.) But by the 1850s most naturalists believed that Lamarck was wrong. "The hypothesis of Lamarck," Wallace writes, "has been repeatedly and easily refuted by all writers on the subject of varieties and species, and it seems to have been considered that when this was done the whole question has been finally settled." Settled, but not explained. Variation and selection, not acquired characteristics, were the explanation, and, as always, Wallace offers specific case studies, including the now infamous giraffe example: "The powerful retractile talons of the falcon- and the cat-tribes have not been produced or increased by the volition of those animals; but among the different varieties which occurred in the earlier and less highly organized forms of these groups, *those always survived longest which had the greatest facilities for seizing their prey.* Neither did the giraffe acquire its long neck by desiring to reach the foliage of the more lofty shrubs, and constantly stretching its neck for the purpose, but because any varieties which occurred among its antitypes with a longer neck than usual *at once secured a fresh range of pasture over the same ground as their shorter-necked companions, and on the first scarcity of food were thereby enabled to outlive them.*" Accenting his points through italicized clauses, Wallace turns to what he was studying the most when he penned the essay—insects. In fact, in the three-and-a-half years since his arrival in the Malay Archipelago, Wallace had collected no fewer than 8,540 species of insects.[19] With that quantity of varieties he was able to distinguish between varieties and species. "Even the peculiar colours of many animals, especially insects, so closely resembling the soil or the leaves or the trunks on which they habitually reside, are explained on the same principle; for though in the course of ages varieties of many tints may have occurred, *yet those races having colours best adapted to concealment from their enemies would inevitably survive the longest.*"[20]

Wallace then concludes his analysis by restating his thesis: "We believe we have now shown that there is a tendency in nature to the continued progression of certain classes of *varieties* further and further from the original type—a progression to which there appears no reason to assign any definite limits." Given enough time this simple process leads to "all the extraordinary modifications of form, instinct, and habits which they exhibit."[21] With that he signed off, "Ternate, February, 1858," slipped the essay and a cover letter into a mail pouch, and sent it to the one man he knew would be most interested in reading it.

"I Never Saw a More Striking Coincidence"

Mail ships were not a daily occurrence in the Malay Archipelago, but sometime in early March, perhaps on the ninth, Alfred Wallace posted his manuscript to Charles Darwin. As far as we know, Darwin received Wallace's paper on June 18 (more on this and the priority question in the next chapter). We do not know which passages in the Ternate paper caught Darwin's eye first, but one can imagine that he must have been taken aback by Wallace's conclusion that "there is a general principle in nature which will cause many varieties to survive the parent species, and to give rise to successive variations, departing further and further from the original type." Whatever it was, Darwin was stunned by what he read. He immediately put pen to paper and wrote his friend and colleague Charles Lyell, in a letter dated simply "18th" (it was Darwin's custom not to include the year on his letters): "Some year or so ago, you recommended me to read a paper by Wallace in the Annals, which had interested you & as I was writing to him, I knew this would please him much, so I told him. He has to day sent me the enclosed & asked me to forward it to you. It seems to me well worth reading. Your words have come true with a vengeance that I sh$^{d.}$ be forestalled." Then, seemingly for posterity (and priority), Darwin nudged Lyell's memory: "You said this when I explained to you here very briefly my views of 'Natural Selection' depending on the Struggle for existence." Seeing the similarities of their theories rather than the differences, Darwin continued his grief: "I never saw a more striking coincidence. If Wallace had my M.S. sketch written out in 1842 he could not have made a better short extract! Even his terms now stand as Heads of my Chapters." It was potentially one of the greatest disasters that could befall the scientist who had been so patient in the careful construction and proper defense of his views. "So all my originality, whatever it may amount to, will be smashed."[22]

Although the two theories were not as closely parallel as Darwin initially thought, it does not matter because Darwin *thought* they were virtually identical. On the opening pages of the Introduction to the *Origin of Species,* after he had much time to reflect on Wallace's ideas presented in the essay, Darwin still made note that he had "been induced" to publish "this Abstract" before he could complete his originally planned larger work, "as Mr. Wallace, who is now studying the natural history of the Malay archipelago, has arrived at almost exactly the same general conclusions that I have on the origin of species."[23] Even in his autobiography, written two decades after the event, Darwin still recalled that Wallace's "essay contained exactly the same theory as mine."[24] This is surprising since, as we have just read in Wallace's original paper, there are distinct differences that Darwin surely could not have missed. Some historians, for example, point out that Wallace emphasized *environmental selection,* or the elimination of the unfit, whereas Darwin tended to focus on *competitive selection,* or the success of the fit that secondarily causes the elimination of the unfit.[25] But we have just seen that Wallace only *began* with the elimination of the unfit; he quickly noted the flip side of the equation where the fit are *selected* for survival. Other historians suggest that Wallace thought natural selection operated on varieties already formed, whereas Darwin saw it as creating varieties out of individual differences.[26]

Regardless of how later commentators may parse the theories, in 1858 and 1859 it is clear that Darwin perceived them to be quite similar; certainly close enough to challenge his sense of priority and cause considerable consternation. As he told Lyell on the 25th, after a week of hand wringing: "I would far rather burn my whole book, than that he or any other man should think that I have behaved in a paltry spirit." Darwin's concern was understandable. He feared complete loss of priority of the theory on which he had worked so long. Beginning in 1838 and 1839 with the opening of the "M" and "N" notebooks, and further developed in his essay sketches of 1842 and 1844, Darwin's tactical delay in publishing his theory was about to backfire. He was building his reputation as a keen observer and first-rate zoologist in order to lessen the shock of what he knew would be his controversial theory of transmutation. Now someone else might beat him to the punch. He inquired of Lyell: "Do you not think his having sent me this sketch ties my hands? If I could honorably publish, I would state that I was induced now to publish a sketch . . . from Wallace having sent me an outline of my general conclusions." Darwin then requested that his colleague and confidant "send this and your answer to Hooker . . . for then I shall have the opinion of my two best and kindest friends."[27]

The solution derived by Lyell and Hooker was to read both Wallace's paper and Darwin's sketch of 1844, along with a letter Darwin had written to Asa Gray on September 5, 1857, outlining his ideas (and thus establishing priority under the rules of that time), at the July 1, 1858, meeting of the Linnean Society of London under the title:

> "On the Tendency of Species to Form Varieties;
> and on the Perpetuation of Varieties and Species by Natural Means of Selection."
> By Charles Darwin, Esq., F.R.S., F.L.S., & F.G.S., and Alfred Wallace, Esq.
> Communicated by Sir Charles Lyell, F.R.S., F.L.S.,
> and J. D. Hooker, Esq., M.D., V.P.R.S., F.L.S., &c.

The composite set included, in exact order and wording from the table of contents, the following sequence that clearly established a chronology that put Darwin's name first in priority:

A. Letter from Charles Lyell and Jos. D. Hooker. London, June 20th, 1858.
B. Extract from an unpublished Work on Species, by C. Darwin, Esq., consisting of a portion of a Chapter entitled, "On the Variation of Organic Beings in a state of Nature; on the Natural Means of Selection; on the Comparison of Domestic Races and true Species." (1844)
C. Abstract of a Letter from C. Darwin, Esp., to Prof. Asa Gray, Boston, U.S., dated Down, September 5th, 1857.
D. "On the Tendency of Varieties to depart indefinitely from the Original Type." By Alfred Russel Wallace.[28]

Although some believe a conspiracy was afoot to exclude or attenuate Wallace's priority, Lyell and Hooker made it clear in the opening lines of their letter (addressed to J. J. Bennett, Esq., Secretary of the Linnean Society, dated "London, June 30th, 1858") that both men deserve ample recognition: "The accompanying papers, which we have the honour of communicating to the Linnean Society, and which all relate to the same subject, viz. the Laws which affect the production of varieties, races, and species, contain the results of the investigations of two indefatigable naturalists, Mr. Charles Darwin and Mr. Alfred Wallace. These gentlemen having, independently and unknown to one another, conceived the same very ingenious theory to account for the appearance and perpetuation of varieties and of specific forms on our planet, may both fairly claim the merit of being original thinkers in this important line of inquiry; but neither of them having published his views, though Mr. Darwin has for many years past been repeatedly urged by us to do so, and both authors having now unreservedly placed their papers in our hands, we think it would best promote the interests of science that a selection from them should be laid before the Linnean Society."[29]

And so it was, but not with Wallace's permission (unreservedly or not), because all this took place before the letter to him explaining the arrangement arrived in Ternate. Although Lyell and Hooker eschewed any judgment on the question of priority, they left no doubt about how the reader should perceive the arrangement: "We are not solely considering the relative claims to priority of himself and his friend, but the interests of science generally; for we feel it to be desirable that views founded on a wide deduction from facts, and matured by years of reflection, should constitute at once a goal from which others may start, and that, while the scientific world is waiting for the appearance of Mr. Darwin's complete work, some of the leading results of his labours, as well as those of his able correspondent, should together be laid before the public."[30]

From this point forward, Wallace's correspondence with Darwin and other leading naturalists became more frequent. For a man on the outside (in every sense of the word), it was more than he could have expected. After he found out about the publication he wrote to his boyhood friend, George Silk, urging him to pick up a copy of the Linnean *Proceedings* with comments by Lyell and Hooker, especially since "as I know neither of them I am a little proud."[31] Perhaps most significant, a letter on January 25, 1859, from the sage of Down himself, finds Darwin equating Wallace with Lyell and Hooker as intellectual stimulants to the publication of the *Origin of Species,* now only ten months away from release: "I owe indirectly much to you and them for I almost think that Lyell would have proved right and I should never have completed my larger work, for I have found my abstract hard enough with more poor health; but, now, thank God, I am in my last chapter but one."[32] Whether this was a simple acknowledgment by Darwin of those who had helped, or a reminder from a very competitive scientist hell-bent on not being beaten out for the prize this close to the the finish line, or both, is difficult to say. But when Wallace returned to London in 1862, he soon came to meet and befriend a number of scientific and intellectual luminaries such as Herbert Spencer, Charles Lyell, Joseph Hooker, Francis Galton, Thomas Huxley, William Crookes, E. B. Poulton, Karl Pearson, Raphael Meldola, and even John Stuart Mill. This was only possible because of his new and stellar reputation in science afforded him by this arrangement. Given his station in life, Wallace could not help but be pleased that his essay elevated him into the ranks of world-class scientists.

The Line

The remainder of the Malay expedition was one of wholesale collecting, biogeographical observing, and verification of the theory of natural selection. At one point Wallace reported averaging a remarkable forty-nine new species finds a day, with a high of seventy-eight in one particularly good catch. He was an enthusiast for collecting, as he explained to his friend (and later brother-in-law) Thomas Sims, who was encouraging Wallace to return home before illness could overcome him. "I have much to do yet before I can return with satisfaction of mind . . . I feel that my work is here as well as my pleasure; and why should I not follow out my vocation? . . . So far from being angry at being called an enthusiast (as you seem to suppose), it is my pride and glory to be worthy to be so called. Who ever did anything good or great who was not an enthusiast?" He was especially enthusiastic to further understand "the relations of animals to space and time, or, in other words, their geographical and geological distribution and its causes."[33]

One day in 1858, upon ruminating on the closely allied species of the

Malay Archipelago, Wallace noticed that some are more allied than others as a function of the geography in which they are found. One train of thought, of course, led to the Ternate paper. But a related strain led him to think more on a sharp division he noticed that existed between species allied to Asia and species allied to Australia. He remembered his earlier work on the differential distribution of monkeys along the Lower Amazon and Rio Negro, where he had noticed that different species inhabited different sides of the river, and that the river was acting as a reproductive isolating mechanism. Then, in the February 1858 issue of the *Proceedings* of the Linnean Society, Wallace read a paper by the British ornithologist Philip Lutley Sclater, on the "Geographical Distribution of Birds," in which the author noted a break in bird distribution between the western and eastern islands of the Malay Archipelago. This caught Wallace's attention, because he too had noticed a similar division. As he wrote Bates in 1859: "In this archipelago there are two distinct faunas rigidly circumscribed, which differ as much as do those of Africa and South America, and more than those of Europe and North America: yet there is nothing on the map or on the face of the islands to mark their limits. The boundary line often passes between islands closer than others in the same group."[34] This faunal split between Asia and Australia led Wallace to conclude "that the same division will hold good in every branch of Zoology," and he promptly penned a paper on the subject, "On the Zoological Geography of the Malay Archipelago," published in the *Zoological Proceedings* of the Linnean Society in November 1859, the same month that Darwin's *Origin of Species* was published.

This paper is, arguably, the second or third most important paper on natural history that Wallace ever wrote, for it led to his identification of a line that can be drawn "among the islands, which shall so divide them that one-half shall truly belong to Asia, while the other shall no less certainly be allied to Australia." This biogeographical line that would eventually (and still) bear his name ("The Wallace Line," initially labeled as such by Thomas Huxley) was further evidence for Wallace of the geographical isolation and biological transformation that species may experience over space and time. "The Australian and Indian regions of Zoology are very strongly contrasted. In one the Marsupial order constitutes the great mass of the mammalia,—in the other not a solitary marsupial animal exists. Marsupials of at least two genera (Cuscus and Belideus) are found all over the Moluccas and in Celebes; but none have been detected in the adjacent islands of Java and Borneo."[35]

As far as Wallace knew—and few in science had more time in the field than he to make such observations—this distribution was "the most anomalous yet known, and in fact altogether unique. I am aware of no other spot upon the earth which contains a number of species, in several distinct classes

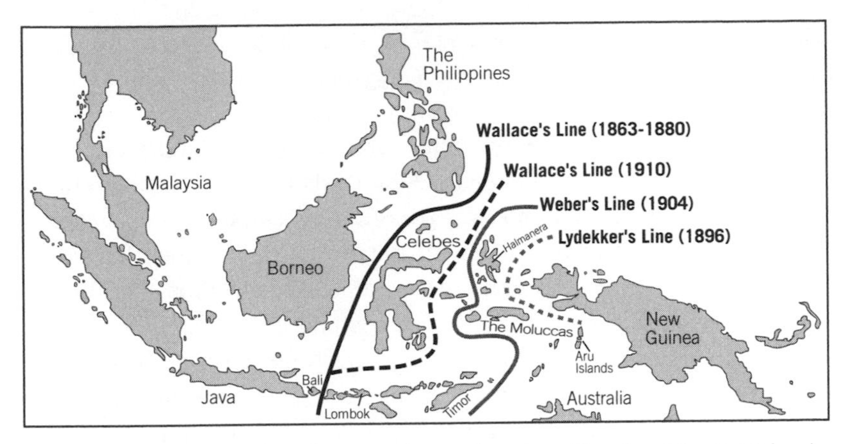

Figure 4-4 Wallace's Line. See text for explanation. (Rendered by Pat Linse)

of animals, the nearest allies to which do not exist in any of the countries which on every side surround it, but which are to be found only in another primary division of the globe, separated from them all by a vast expanse of ocean." Once again he turned to theory to tie together his observations, this time suggesting that this unique division may be in violation of a "law" of nature. "In no other case are the species of a genus or the genera of a family distributed in two distinct areas separated by countries in which they do not exist; so that it has come to be considered a law in geographical distribution, 'that both species and groups inhabit continuous areas.' "[36]

The cause of this break, not fully understood by Wallace at the time, has to do with the depth of sea levels around the islands and the ice-age cycles effecting those levels. And the line varies depending on the species under question. Wallace tweaked the line himself in 1910, shifting it eastward between Celebes and the Moluccas, and others have proposed lines as well, such as Lydekker's line drawn on the edge of the Australian shelf in 1896, and Weber's line involving the distribution of freshwater fish in 1904. Always the empiricist, Wallace admitted that "it may be said: 'The separation between these two regions is not so absolute. There is some transition. There are species and genera common to the eastern and western islands.' This is true, yet (in my opinion) proves no transition in the proper sense of the word; and the nature and amount of the resemblance only shows more strongly the absolute and original distinctness of the two divisions. The exception here clearly proves the rule."[37] Now that he had the mechanism for how such transformation occurred, his observations took on a whole new meaning.

Wallace was not only a prodigious collector, he was also a prolific writer, publishing hundreds of scientific articles, letters, notes, and monographs, all in respected scientific and scholarly journals, on a wide variety of topics, such as "Geographical Distribution of Birds" in *Ibis* (1859), "Notes of a Voyage to New Guinea" in the *Journal of the Geographical Society* (1860), "Notes on *Semioptera wallacii*" in *Proceedings of the Zoological Society of London* (1860), "Zoological Geography of Malay Archipelago" in the *Proceedings of the Linnean Society* (1860), "On some New Birds from the Northern Moluccas" in *Ibis* (1862), and on and on. This literary corpus is even more impressive considering he wrote them all out by hand in camps, on boats—catch-as-catch-can in the middle of a tropical expedition—mailing the manuscripts by steamship and hoping they would arrive safely. Living austerely with nothing to do but collect and write from dawn to dusk, Wallace was a veritable scientific and literary engine. It is evident from reading these early works that he could write clearly, concisely, and rapidly. He often refers to the penning of an article in a day or two, frequently between bouts of malaria or under the harshest of living conditions.

A Strange and Terrible Cannibal Monster

In addition to his published works, Wallace kept a detailed journal filling four handwritten (and still unpublished) volumes, now residing at the Linnean Society of London. His observations range from humorous to sensitive to dramatic, as the following three passages (respectively) reflect. In the first we see Wallace's description of the reactions of native peoples to his long beard, white skin, and, especially, his towering height (just over six feet):

> One of the most disagreeable features of travelling or residing in this country is the excessive terror I invariably excite. Wherever I go dogs bark, children scream, women run & men stare with astonishment as though I were some strange and terrible cannibal monster. Even the pack horses on the roads & paths start aside when I appear & rush into the jungle. . . . If I come suddenly upon a well where women are drawing water or children bathing, a sudden flight is the certain result, which things occurring day after day are to say the least of them very unpleasant & annoying, more particularly to a person who likes not to be disliked, & who has never been accustomed to consider himself an ogre or any other monster.[38]

Wallace recounted his experience with a Mias, or orangutan, which was an infant "probably not above a month old when I obtained it." He bathed the primate every day, "which appeared to have a good effect," though "it winced a little and made ridiculously wry faces when the cold water was poured over its head but enjoyed the rubbing dry amazingly, and especially having the hair of its back & head brushed afterwards. When first I obtained it it clung desperately tight with its four hands to whatever it could lay hold of, and having once seized my whiskers & beard I could not get it off for some time, as it doubtless felt quite at home being accustomed to cling almost from birth to the long hair of its mother."[39]

Finally, the following unemotional rendering of a sanctioned murder reveals a culture that must have been as curious to the English naturalist as the diversity of nature itself:

> Some years ago one of the English residents here had one of the native Balinese women for his temporary wife. The girl however offended against the law by receiving a flower or a sirih leaf or some such trifle from another man. This was reported to the Rajah, (to some of whose wives the girl was related) & he instantly sent to the Englishman's house ordering him to give the woman up as she must be krissed. In vain he begged & prayed, & offered to pay any fine the Rajah might impose, & refused to give her up without he was compelled by force. This the Rajah did not wish to resort to so he let the matter drop, and a short time after sent one of his followers to the house, who beckoned the girl to the door, & then saying "the Rajah sends you this" stabbed her to the heart.[40]

As often as not observations such as these were used by Wallace to support some scientific hypothesis or bolster an argument. Such is the case with his

Figure 4-5 The python incident. In *The Malay Archipelago,* Wallace recounts yet another adventure in the tropics when he recalled awakening one morning to discover a giant python curled up three feet from his head that had kept him up much of the night with a rustling noise. The account was illustrated with this engraving of his Malaysian assistants trying to remove the intruder from his hut. (From *The Malay Archipelago,* 1869)

Figure 4-6 Artifacts from Wallace's travels in the Malay Archipelago, including his wax seal for letters and a portable sextant used for navigating and recording precise locations for his countless biological and geological observations. (Courtesy of Alfred John Russel Wallace and Richard Russel Wallace)

observations on "man and nature in all its aspects," made in his many thousands of hours of solitude in the jungle. For example, he told his brother-in-law Thomas Sims from Delli, Timor, on March 15, 1861: "I have since wandered among men of many races and many religions. In my solitude I have pondered much on the incomprehensible subjects of space, eternity, life and death. I think I have fairly heard and fairly weighed the evidence on both sides, and I remain an *utter disbeliever* in almost all that you consider the most sacred truths." And though he called himself a "sceptic" who knows the "falsehood [of Christianity] as a general rule," he could subsequently claim "I am thankful I can see much to admire in all religions. To the mass of mankind religion of some kind is a necessity. But whether there be a God or whatever be His nature; whether we have an immortal soul or not, or whatever may be our state after death, I can have no fear of having to suffer for the study of nature and the search for truth." Such convictions, however, could go only so far for a public figure like Wallace, and he instructed Sims that this postscript on religion is "for yourself; show the *letter only* to my mother."[41]

For his eight-year stay in the Malay Archipelago, Wallace was a one-man collecting machine. But in time various illnesses and injuries, coupled with the fatigue of travel and hauling equipment, wore down the seemingly implacable naturalist. In the spring of 1862 he headed for home after compiling an almost unbelievable collection of 125,660 total specimens, including 310 mammals, 100 reptiles, 8,050 birds, 7,500 shells, 13,100 butterflies, 83,200 beetles, and 13,400 "other insects." This time he and his collections made it back to England safely.

Unlike most of his fellow naturalists, Wallace also had a flare and passion for something deeper—to put the nearly infinite variety of nature's pieces together into a puzzle so that as a historical scientist and research naturalist he could solve the riddle of the mystery of mysteries. But he was not the only one working on this problem.

5

A Gentlemanly Arrangement

It was Wallace's combination of broad observational scope and penetrating theoretical depth that set him apart from most of his contemporaries and led him to his discovery about the mutable nature of species and the interdependency of organisms on their geographical locales. Wallace was demonstrating the practice of science at its best—the blending of process and product into an art form described over a century later by Nobel laureate biologist Sir Peter Medawar as "the art of making difficult problems soluble by devising means of getting at them."[1]

The art of the soluble. Our historical understanding of Wallace's discovery of the immutable nature of species helps us see that the fitful and sometimes quirky progress of science is more explicable as an interaction of steady historical trends punctuated by serendipitous flashes of insight. Science is not an asymptotic curve of stately progress toward Truth, or the unfolding of the shroud covering Reality. Rather, it consists of long periods of paradigmatic status quo, occasionally interrupted by shifts in the shared paradigm, resulting in a new and different way of interpreting nature. The *particulars* of a specific historical event, however, do not always fit the philosophers' *universal* concept of how science is suppose to change.[2] Each is unique unto itself. Because of the contingent nature of history, no two paradigms or paradigm shifts are ever the same.

The independent discovery and description of natural selection by Charles Darwin and Alfred Russel Wallace, which became the driving engine behind the larger paradigm shift in evolutionary thought, provides a case study in the interactive nature of contingency and necessity in science. This particular episode, however, is especially revealing about how science works because so much is at stake and, for some, the question of priority has not been resolved.

A Delicate Arrangement or a Gentlemanly One?

The matter of who was first in the discovery and description of natural selection has been the subject of much confusion for three reasons: (1) the letter and essay from Wallace to Darwin in the spring of 1858 is missing, making direct and tangible resolution impossible; (2) a misunderstanding of intellectual property and how priority disputes were settled at that time; and (3) the pugnacious zero-sum game (win–lose) model of priority held by some scientific communities does not recognize the cumulative, interactive, and social nature of the scientific enterprise.

Wallace's co-discoverer status with Darwin is generally accepted by all biologists and historians. The question some raise, however, is this: Should Wallace be given even more credit? In his maximally tendentious 1980 work, *A Delicate Arrangement*, journalist Arnold Brackman makes an emotional appeal for Wallace's case, suggesting that Charles Lyell and Joseph Hooker, with Darwin's knowledge (but not his direction), conspired to negate Wallace's credit, while simultaneously boosting Darwin's.[3] Specifically, Brackman claims that Darwin received Wallace's letter and essay earlier than the announced June 18, 1858, date, and that he probably spent that time fleshing out the missing pieces of his theory from Wallace's essay, then feigned surprise and distress over Wallace's parallel ideas.

The strongest associative evidence we have on the question of the date of arrival of Wallace's paper is another letter sent by Wallace to Frederick Bates, the younger brother of his naturalist colleague and Amazon companion Henry Walter Bates. The letter is dated March 2 and is assumed to have been mailed on the same steamship as the letter to Darwin on March 9 (mail ships were not a daily occurrence). The Bates letter appeared in London on June 3. The clearly dated, postmarked letter (no envelope—the letter itself was addressed and postmarked) is in possession of Wallace's grandson, Alfred John Russel Wallace, and is shown in Figure 5-1. The cover of the letter bears Wallace's direction "via Southampton" and was postmarked "Singapore Apr 21 58' " and "London Ju 3 58'." In the letter Wallace tells Bates of the seemingly incoherent diversity of insect coloration in the Malay, and notes that "such facts as these puzzled me for a long time, but I have lately worked out a theory which accounts for them naturally."[4] That theory, "lately worked out," was obviously in reference to the essay sent to Darwin, the original autograph manuscript and cover letter of which is, tragically, nowhere to be found.

Thomas Huxley's son Leonard called the Wallace situation "a delicate arrangement," and Arnold Brackman focuses his attention on the noun, not the adjective. His argument is that since Darwin had been working on his theory for twenty years, and that because he was an established scientist with a

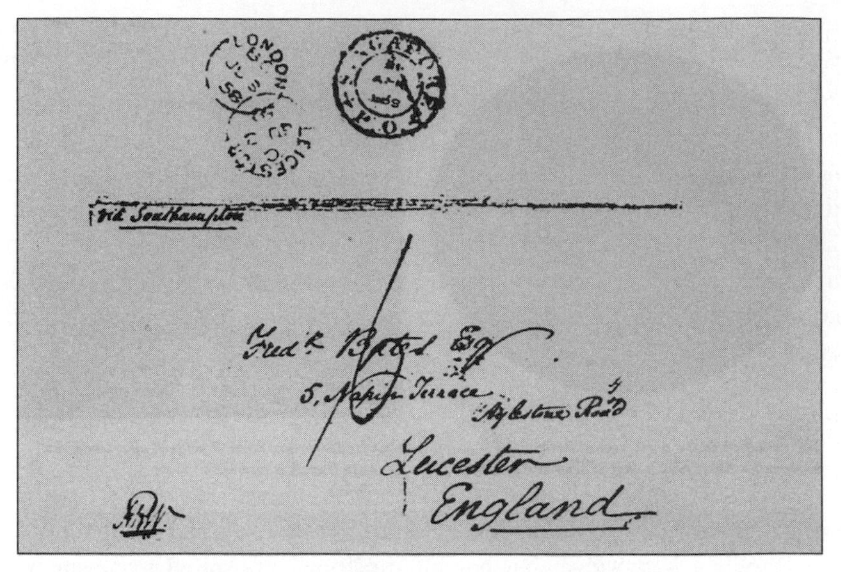

Figure 5-1 The envelope of Wallace's letter to Frederick Bates, stamped arrival in London on June 3, 1858, believed to have been sent on the same steamship from the Malay Archipelago as his letter and paper to Darwin outlining his theory of natural selection. (Courtesy of Alfred John Russel Wallace and Richard Russel Wallace)

recognized role within the scientific community, when this young amateur naturalist appeared on the verge of scooping his senior, Lyell and Hooker determined that Darwin should be given the lion's share of the credit or else no one would accept the theory of evolution. Wallace was not part of the traditional scientific community in England (owing to his working-class background and lack of formal university education), and since he spent most of his professional life outside England, it was necessary for there to be an organized conspiracy by the intellectual elite surrounding Darwin to lessen the value of Wallace's contribution. Wallace, with a working-class mentality, deferred to his superior. Brackman outlines his thesis:

> No matter how heinous is a conspiracy, the participants—especially if it is successful—are apt to develop a plausible rationale for gilding it. "I do not think that Wallace can think my conduct unfair in allowing you and Hooker to do whatever you thought fair," Darwin wrote to Lyell. The message was clear: Lyell and Hooker bore historical responsibility for the cover-up. Darwin did not "allow" Lyell and Hooker to act independently. In this instance, he appeared helpless, informed powerful friends of his impending doom, pointed subtly in the direction of a solution, let his friends solve the problem by dubious means, and went along with the solution—claiming it, of course, as theirs.[5]

There is no doubt that the Darwin–Wallace situation was a "delicate" one. Any time there is a question of scientific priority—and in this case the priority

Figure 5-2 Joseph Hooker and Charles Lyell, friends and colleagues of both Charles Darwin and Alfred Wallace. (From Burkhardt, 1991, 90, 122)

Figure 5-3 The Linnean Society of London meeting room as it presently looks with the original furniture from the 1858 meeting. The room itself, in another location, is no longer used by the Linnean Society. (Author photo)

is over one of the half-dozen most important ideas in the history of Western civilization—the situation could be nothing but delicate. But with the primary evidence missing in this historical mystery, we have to piece together from circumstantial evidence what really happened at Down. The extreme interpretation of a conspiratorial cover-up is not supported by the evidence. If Darwin were going to rig (or allow to be rigged) the editorial presentation of the papers to award him priority or, worse, plagiarize from Wallace certain key ideas (such as the principle of the divergence of species), why announce the arrival of Wallace's paper and submit it for publication in the first place? Why not either just take what was needed, or, if Wallace's paper added nothing new to the theory, destroy it and the cover letter and blame the loss on an inefficient postal service, or the mishandling of his mail at Down, or whatever? If one is going to accuse Darwin of such devious finagling and delicate arrangements, or even worse, plagiarization, then would not the same guileful and scheming personality think of complete elimination of Wallace's essay as a successful strategy?

There is no question that much confusion surrounds the critical period of the spring and summer of 1858, and biologist John Langdon Brooks presents an "alternative reconstruction"[6] in which he suggests that Darwin's letter to Lyell, dated "18th" and assumed by most to be June, was actually May 18, 1858. Darwin held the letter and essay for a month then, "after much soul-searching, he restudied Wallace's Ternate manuscript and, with recourse again to Wallace's 1855 paper, wrote the material on [divergence] and inserted it into the text of his chapter on 'Natural Selection.' "[7]

Brooks's subsequent analysis of various manuscripts and letters after that incident, then, are all based on the assumption that Darwin received Wallace's letter and essay on May 18, which he derived by studying the schedules of the Dutch East Indies mail service and of the Peninsular & Oriental Company. But his analysis is inconsistent. Earlier in his book Brooks says that "the evidence indicates that Darwin must have received Wallace's manuscript on either of two dates in May. Receipt on May 18 would leave 25 days for completion of those folios [on divergence] by June 12 [the date Darwin wrote down his ideas on divergence in his manuscript]; May 28–29 would leave scarcely two weeks. But it must be conceded that desperation will make the pen move quickly."[8]

Conspiracy-mongering by historians also makes the pen move quickly . . . too quickly. First, Brooks suggests the dates of May 18 or May 28–29 for the arrival of Wallace's letter and essay, then he tells us he thinks the "Down 18th" letter to Lyell announcing the arrival of Wallace's letter and essay was actually written on May 18, thus completely negating the May 28–29 option. Worse, Brooks assumes the Wallace–Bates letter that arrived in London (and postmarked) June 3 was in the same batch as the Wallace–Darwin letter and essay. This is not a historical fact but an inference, but even if true, it negates both May dates and, assuming Darwin did not lie in the letter to Lyell about the arrival of the Wallace material on the same day ("18th"), then the arrival date must be June, not May. Finally, Brooks fails to mention that the Dutch East Indies mail service schedules show that another batch of mail from the East Indies arrived in London on June 17. The most logical conclusion is that the Wallace letter that Darwin references as arriving "to day" on the "18th" was in this batch of mail.

Wallace biographer H. L. McKinney has consistency problems as well. He first concludes that the mail from Malaya to London averaged ten weeks in transition, and thus "ten weeks from 9 March, when the communication was mailed, is precisely 18 May, one month before Darwin acknowledged receiving it." McKinney then points to the Wallace–Bates June 3 letter and concludes: "It is only reasonable to assume that Wallace's communication to Darwin arrived at the same time and was delivered to Darwin at Down House on 3 June 1858, the same day Bates's letter arrived in Leicester." To account for the delay from May 18 to June 3, McKinney explains: "Knowing the numerous delays in such matters, we should perhaps allow some leeway, although one month appears to be an excessive allowance."[9] Fine, but then why no "delays" and "leeway" for the Bates letter? Or, in the other direction, why no leeway for Wallace's manuscript from June 3 to June 18? And what was Darwin doing with Wallace's manuscript in that time? McKinney wisely ends his discussion "with a series of question marks," but then hints that Darwin might have been filling in the gaps "on divergence in his long version of the Origin; he finished that section on 12 June."[10]

So which is it? Either the Bates letter is damning evidence, or it is not. Brooks and McKinney cannot have it both ways. They cannot use the Wallace–Bates letter as evidence that the Wallace–Darwin materials arrived on June 3, and then have Darwin writing Lyell announcing same on May 18 (as Brooks does); or that the Darwin letter was delayed while the Bates letter was not (in McKinney's case). Either way, to accuse one of the greatest scientists in history of committing one of the most heinous crimes in science on one of the most important aspects of his theory, one better have compelling evidence.[11] Modern skeptics are fond of saying that extraordinary claims require extraordinary evidence. These claims against Darwin are truly extraordinary, but the evidence is not.

In a considerably more modest revisionist vain, historian Charles Smith suggests that Wallace, while overtly accepting of the publication arrangement (because it was the Victorian polite reaction, and it did help his career), might later have been covertly distressed when he realized how significantly his thoughts and Darwin's differed on a number of matters. Smith notes, for example, that on at least five occasions during his life Wallace emphasized the fact that he did not give his approval for publication, that he was not given the opportunity to see the page proofs before the printing of the journal, and that he was not made aware of what the order of names and papers would be in the journal before publication.[12] All of this is true, but on none of these occasions did he complain about the arrangements. He simply stated the facts of what happened. Smith also doubts "that Wallace wanted to publish the 1858 essay at all, possibly because he was still debating how to fit humankind's development into the scheme of things,"[13] and "it appears to me that before 1858 Wallace's position on the nature of adaptation had steered him quite away from a natural selection-like interpretation of development, and instead in a more teleological, even Newtonian, direction. . . . When in 1858 the principle of natural selection finally occurred to him, it provided a solution to the first problem [the evolution of many animal traits], but not to the second [the evolution of certain human characteristics]." Spiritualism, says Smith, provided him with a solution to the second problem of the evolution of human consciousness.[14]

It is true that in the *Origin of Species* Darwin only briefly mentioned possible applications of the theory to humans: "In the distant future I see open fields for far more important researches. Psychology will be based on a new foundation, that of the necessary acquirement of each mental power and capacity by gradation. Light will be thrown on the origin of man and his history."[15] (Later editions included the modifier "much" before "light.") It is also true that although Darwin published a book on human evolution, *The Descent of Man,* in 1871, Wallace beat him to the punch in 1864 with an article entitled "The Origin of Human-Races and the Antiquity of Man Deduced from the Theory of 'Natural Selection,' " a paper he considered one of the most important works he ever wrote. But Wallace's response to both the arrangement and Darwin's work is anything but reticent. In early 1860 Wallace received a copy of the *Origin of Species,* and his praise could not have been higher. He promptly read through it "five or six times," then told Bates on Christmas Eve that year that "Mr. Darwin has created a new science and a new philosophy; and I believe that never has such a complete illustration of a new branch of human knowledge been due to the labours and researches of a single man. Never have such vast masses of widely scattered and hitherto quite unconnected facts been combined into a system and brought to bear upon the establishment of such a grand and new and simple philosophy."[16] These are hardly the words of a man in the least bit distressed over his affiliation with Darwin. Wallace even dedicated his most commercially successful book, *The Malay Archipelago,* to Darwin.

Solving the Priority Mystery

Even without a complete set of original documents from which to piece together every detail of what happened in this historical mystery, we can solve the case in the same manner that all historical scientists resolve disputes when direct observation and empirical evidence is scant or nonexistent—a preponderance of evidence and a consilience of inductions that converge to a single conclusion. There are several lines of evidence to explore.

On the matter of what happened to the original Wallace essay and letter it should be said that Darwin's near obsession at keeping all correspondence does make one a bit suspicious. Yet even here further reflection shows that most likely nothing was afoot. Darwin's son Francis, for example, when compiling his father's letters for publication five years after his death, noted that before 1862 Darwin was not the paper pack rat of his later years: "It was his custom to file all letters received, and when his slender stock of files ('spits' as he called them) was exhausted, he would burn the letters of several years, in order that he might make use of the liberated 'spits.' This process, carried on for years, destroyed nearly all letters received before 1862. After that date he was persuaded to keep the more interesting letters, and these are preserved in an accessible form."[17] Surely Darwin would not have destroyed one of the most important letters and manuscripts he ever received. On this we can be certain, unless Darwin never got the materials back after publication. Recall that the last person to handle Wallace's manuscript, so far as we can trace it, was not Darwin but someone at the Linnean Society of London who typeset it for publication in the *Proceedings* in August. And before that Lyell had the manuscript, which Darwin sent him along with his June 18 letter of despair. If there is a culprit in the caper of the missing manuscript, Darwin is not at the top of the list.

As for the charge that Darwin took materials from Wallace's 1858 manuscript, such as the principle of divergence, Darwin's contribution to the joint Linnean Society papers did not include materials developed in 1858; rather, he included a letter to the American botanist Asa Gray, written in September 1857, almost a year before the Wallace essay. If Darwin had cribbed the concept of divergence from Wallace, why submit this older version? In any case, the principle of divergence was listed in the table of contents of his *Natural Selection* manuscript in March 1857. Specifically, Darwin offered this account to Gray of the principle of divergence:

> Another principle, which may be called the principle of divergence, plays, I believe, an important part in the origin of species. The same spot will support more life if occupied by very diverse forms. We see this in the many generic

> forms in a square yard of turf, and in the plants or insects on any little uniform islet, belonging almost invariably to as many genera and families as species. We can understand the meaning of this fact amongst the higher animals, whose habits we understand. We know that it has been experimentally shown that a plot of land will yield a greater weight if sown with several species and genera of grasses, than if sown with only two or three species. Now, every organic being, by propagating so rapidly, may be said to be striving its utmost to increase in numbers. So it will be with the offspring of any species after it has become diversified into varieties, or subspecies, or true species. And it follows, I think, from the foregoing facts, that the varying offspring of each species will try (only few will succeed) to seize on as many and as diverse places in the economy of nature as possible. Each new variety or species, when formed, will generally take the place of, and thus exterminate its less well-fitted parent. This I believe to be the origin of the classification and affinities of organic beings at all times; for organic beings always *seem* to branch and sub-branch like the limbs of a tree from a common trunk, the flourishing and diverging twigs destroying the less vigorous—the dead and lost branches rudely representing extinct genera and families.[18]

A content analysis of Darwin's letters also points to a June 18 date. Consider what Darwin was doing before the arrival of the Wallace manuscript. In 1858 Darwin was knee-deep in writing a massive, multivolume work entitled *Natural Selection*. He planned on taking several more years to complete it, and without outside pressure to publish he was in no hurry. He had seen the fallout of other theorists who had published prematurely (Chambers's *Vestiges* being the most obvious), and he was not about to be subjected to that kind of criticism. But Wallace's 1858 letter and essay changed all that, and the change is most notable in Darwin's correspondence immediately after June 18. Before that date there was no hint of rushing to complete his book. On May 16 he told Joseph Hooker "I presume you will have done the horrid job of considering my M.S.—I am in no sort of hurry, more especially as I know full well you will be dreadfully severe." On May 18 he reiterated to Hooker: "There is not least hurry in world about my M.S. I would rather you leave it till you feel disengaged, if that time ever comes to you." On the same day Darwin wrote to Syms Covington, his former assistant on the *Beagle*, that his big work was going to take some time to complete: "This work will be my biggest; it treats on the origin of varieties of our domestic animals and plants, and on the origin of species in a state of nature. I have to discuss every branch of natural history, and the work is beyond my strength and tries me sorely. I have just returned from staying a fortnight at a water-cure establishment, where I bathe thrice a day, and loiter about all day long doing nothing, and for the time it does me wonderful good."[19] These are hardly the words of a man about to see his life's work forestalled by another, and his letters in the days leading up to June 18 are filled with trivial chitchat and musings on the minutiae of natural history.

After June 18 the Darwin correspondence changes dramatically, not only because of the arrival of Wallace's letter, but because two of his children took seriously ill, one with diphtheria and the other, Charles Waring Darwin, with scarlet fever, bringing on his death on the 28th. (The funeral was held on July 1, preventing Darwin from attending the Linnean Society meeting at which his and Wallace's papers were read.) In a June 25 letter to Lyell, for example, after again berating himself for caring about priority, Darwin calls it a "trumpery affair" and a "trumpery letter influenced by trumpery feelings."[20] After the July 1 meeting and the resolution of the priority question, his anxiety waned and by July 13 he could reflect to Hooker that he "always thought it very possible that I might be forestalled, but I fancied that I had grand enough soul not to care; but I found myself mistaken & punished; I had, however, quite resigned myself & had written half a letter to Wallace to give up all priority to him & shd. certainly not have changed had it not been for Lyell's & yours quite extraordinary kindness." Any lingering doubts about whether Darwin had a hand in arranging the sequence of the Linnean Society papers are put to rest in his acknowledgment to Hooker that "I am much more than satisfied at what took place at Linn. Socy—I had thought that your letter & mine to Asa Gray were to be only an appendix to Wallace's paper."[21]

As for Wallace's knowledge of and response to the so-called "delicate arrangement," a letter he wrote to Joseph Hooker on October 6, 1858, from Ternate, reveals that Wallace not only knew about the arrangement but that he was, in fact, pleased with how it was handled (Figure 5-4):[22]

My dear Sir

I beg leave to acknowledge the receipt of your letter of July last, sent me by Mr. Darwin, & informing me of the steps you had taken with reference to a paper I had communicated to that gentleman. Allow me in the first place sincerely to thank yourself & Sir Charles Lyell for your kind offices on this occasion, & to assure you of the gratification afforded me both by the course you have pursued & the favourable opinions of my essay which you have so kindly expressed. I cannot but consider myself a favoured party in this matter, because it has hitherto been too much the practice in cases of this sort to impute *all* the merit to the first discoverer of a new fact or a new theory, & little or none to any other party who may, quite independently, have arrived at the same result a few years or a few hours later.

I also look upon it as a most fortunate circumstance that I had a short time ago commenced a correspondence with Mr. Darwin on the subject of "Varieties," since it has led to the earlier publication of a portion of his researches & has secured to him a claim of priority which an independent publication either

Jos. D. Hooker, M.D. F.R.S.

Ternate, Moluccas, Oct. 6. 1858.

My dear Sir

I beg leave to acknowledge the receipt of your letter of July last, sent me by Mr. Darwin, & informing me of the steps you had taken with reference to a paper I had communicated to that gentleman. Allow me in the first place sincerely to thank yourself & Sir Charles Lyell for your kind offices on this occasion, & to assure you of the gratification afforded me both by the course you have pursued & the favourable opinions of my essay which you have so kindly expressed. I cannot but consider myself a favoured party in this matter, because it has hitherto been too much the practice in cases of this sort to impute all the merit to the first discoverer of a new fact or a new theory, & little or none to any other party who may, quite independently, have arrived at the same result a few years or a few hours later.

I also look upon it as a most fortunate circumstance that I had a short time ago commenced a correspondence with Mr. Darwin on the subject of "Varieties", since it has led to the earlier publication of a portion of his researches & has secured to him a claim of priority which an independent publication either by myself or some other party might have injuriously affected,— for it is evident that the time has now arrived when these & similar views will be promulgated & must be fairly discussed.

It would have caused me much pain & regret had Mr. Darwin's excess of generosity led him to make public my paper unaccompanied by his own much earlier & I doubt not much more complete views on the same subject, & I must again thank you for the course you have adopted, which while strictly just to both parties, is so favourable to myself.

Being on the eve of a fresh journey I can now add no more than to thank you for your kind advice as to a speedy return to England;— but I dare say you well know & feel, that to induce a Naturalist to quit his researches at their most interesting point requires some more cogent argument than the prospective loss of health.

I remain

My dear Sir

Yours very sincerely

Alfred R. Wallace

J. D. Hooker, M.D.

Figure 5-4 Alfred Wallace's letter to Joseph Hooker, October 6, 1858, from Ternate, that puts the lie to the claim that he did not know about the arrangement of his and Darwin's papers at the Linnean Society of London, or that he was displeased with how it was handled. (Courtesy of Quentin Keynes and the Henry E. Huntington Library and Art Gallery, San Marino, CA)

by myself or some other party might have injuriously affected,—for it is evident that the time has now arrived when these & similar views will be promulgated & *must* be fairly discussed.

It would have caused me such pain & regret had Mr. Darwin's excess of generosity led him to make public my paper unaccompanied by his own much earlier & I doubt not much more complete views on the same subject, & I must again thank you for the course you have adopted, which while strictly just to both parties, is so favourable to myself.

Being on the eve of a fresh journey I can now add no more than to thank you for your kind advice as to a speedy return to England;—but I dare say you well know & feel, that to induce a Naturalist to quit his researches at their most interesting point requires some more cogent argument than the prospective loss of health.

I remain / My dear Sir / Yours very sincerely / Alfred R. Wallace

This is a very revealing letter on several levels. First, it is polite and thoughtful, but not excessively so by the standards of the day, where how one expressed oneself was nearly as important as what was being communicated. Second, by stating it five times in this one letter Wallace has gone out of his way to show that he is gratified with how the situation was handled. The message could not be clearer, thereby silencing the voices of the Darwin revisionists who would have us believe that the actions of Darwin and his colleagues were nefarious. Third, the final paragraph displays his deep and longing passion for natural history, and he expresses it with such style and panache! In fact, when he penned this letter he was preparing to depart Ternate for the nearby island of Batchian, near Gilolo, where he subsequently spent six months collecting. Fourth, the letter offers a glimpse into the rules of intellectual property and priority determination in previous centuries. Today, priority is established largely by publication—first in print or to the patent office wins the prize. But in earlier centuries it was enough to have communicated an idea to someone else as long as there was a paper trail. The lengths that scientists like Galileo and Newton went to in order to establish priority without actually publishing an idea were almost amusing, constructing elaborate and obfuscating anagrams and puzzles with the idea embedded within, posted to friends and colleagues through traceable documentation. In this fashion priority could be established without the discovery being revealed before it was fully developed. Note Wallace's comment that he considers himself a "favoured party" because "it has hitherto been too much the practice in cases of this sort to impute all the merit to the first discoverer of a new fact or a new theory, & little or none to any other party who may, quite independently, have arrived at the same result a few years or a few hours later." In other words, he feels fortunate that he received any credit at all because by the priority rules of the day Darwin was clearly the winner, having established priority with his 1844 essay shared with his colleagues Lyell and Hooker, and his letter to Asa Gray on September 5, 1857.[23]

Since we have already examined the powerful role that birth order and personality play in the receptivity and openness to heretical ideas in the history of science, it is worth noting here that Frank Sulloway analyzed over a hundred episodes of priority dispute and found that of the more than 200 individuals in his database, firstborns, with a stronger drive to always finish ahead of their laterborn siblings, "were 3.2 times more likely than laterborns to initiate priority disputes or to pursue them in an uncompromising manner. Galileo, Newton, Leibniz, and Freud are some of the firstborns who vigorously asserted their priority and sought to tarnish the careers of any scientists who got in their way. Many of the laterborns involved in these disputes were dragged into them by firstborns, whose charges of plagiarism forced the accused to defend a valid priority."[24] Two of the most notable laterborns who refused to be pulled into a priority dispute were Charles Darwin and Alfred Wallace.

Finally, consider Darwin's response to Wallace's letter, on January 25, 1859:

> I was extremely much pleased at receiving three days ago your letter to me & that to Dr. Hooker. Permit me to say how heartily I admire the spirit in which they are written. Though I had absolutely nothing whatever to do in leading Lyell & Hooker to what they thought a fair course of action, yet I naturally could not but feel anxious to hear what your impression would be. I owe indirectly much to you & them [Lyell and Hooker] for I almost think that Lyell would have proved right and I should never have completed my larger work, for I have found my abstract hard enough with my poor health. . . . Everyone whom I have seen has thought your paper very well written & interesting. It puts my extracts, (written in 1839, now just 20 years ago!) which I must say in apology were never for an instant for publication, in the shade.[25]

Eight months later Darwin received another paper from Wallace (this one on the geographical distribution of species in the Malay Archipelago) that he also forwarded to the Linnean Society for presentation and publication. The originals of this letter and paper are also missing, but no one has concocted a conspiracy about that fact.[26] Though the arrangement may have been delicate, it was worked out between the two men in a gentlemanly way.

Darwin's Surprise or Chagrin?

What is surprising in this whole matter, if anything, is Darwin's apparent astonishment at the receipt of Wallace's essay. A clipping of a letter in the Darwin archives at Cambridge from Wallace to Darwin, dated (in Darwin's hand) September 27, 1857, clearly shows that Wallace was continuing work on the problem of the origin of species that he had begun with the publication of his 1855 Sarawak paper, for which he voices to Darwin his disappointment in a lack of response. The fragment begins in mid-sentence: ". . . of May last, that my views on the order of succession of species were in accordance with your own, for I had begun to be a little disappointed that my paper had neither excited discussion nor even elicited opposition. The mere statement and illustration of the theory in that paper is of course but preliminary to an attempt at a detailed proof of it, the plan of which I have arranged, and in part written, but which of course requires much [research in English libraries] & collections, a labor which I look. . . . " (The letter extract is truncated because of deliberate cutting by Darwin, who clipped sections from letters and articles for later use in constructing manuscripts.[27])

It seems clear from this passage that the only thing Darwin could have been surprised about was how quickly Wallace completed a promised "detailed proof" of the theory that did, in fact, loosely parallel his and result in the 1858 essay sent to Down in the spring. But it leaves one to wonder what plan Wallace was working on that he had already written part of, since, by his own account, the 1858 essay was composed in the course of two nights in late February, a full five months after this letter to Darwin. Did his feverish discovery overturn the ideas he was developing in this 1857 plan? If not, what happened to this manuscript? If so, then why did Wallace not expand the 1858 essay into a longer book-length manuscript? One possible answer may be found in a letter written to Bates between these two dates, on January 4, 1858, in which Wallace discusses what appears to be this same "plan" or "work."

> To persons who have not thought much on the subject I fear my paper on the succession of species [the Sarawak Law of 1855] will not appear so clear as it does to you. That paper is, of course, only the announcement of the theory, not its development. I have prepared the plan & written portions of an extensive work embracing the subject in all its bearings & endeavouring to prove what in the paper I have only indicated. I have been much gratified by a letter from Darwin, in which he says that he agrees with "almost every word" of my paper. He is now preparing for publication his great work on Species & Varieties, for which he has been collecting information 20 years. He may save me the trouble of writing the 2nd part of my hypothesis, by proving that there is no difference in nature between the origin of species & varieties, or he may give me trouble by arriving at another conclusion, but at all events his facts will be given for me to work upon. Your collections and my own will furnish most valuable material to illustrate & prove the universal applicability of the hypothesis.[28]

Here a plausible scenario presents itself. Wallace, after years of collecting and observing, formed a hypothesis—"On the Law which has regulated the Introduction of New Species" (the 1855 "Sarawak Law"). Lacking further supportive evidence for a mechanism to drive evolutionary change, coupled to the fact that he perceived his paper to be largely ignored by the scientific community, Wallace continued about his business of collecting in relative anonymity, but never abandoned his ultimate quest to understand the origin of species. He knew that Darwin had been working on the problem for twenty years and was currently writing his "big species book" (originally entitled *Natural Selection,* later changed to *On the Origin of Species*). Wallace, in no position (either logistically in his travels, or scientifically in his research) to complete a work thorough enough to be well received, decided to sit back and wait to see what Darwin would produce. If Darwin was successful (i.e., if Wallace agreed with his arguments), then he would have no need to repeat what had already been done ("He may save me the trouble of writing the second part of my hypothesis"). If Darwin was not successful ("he may give me trouble by arriving at another conclusion"), then Wallace could respond accordingly with his own theory and data. It seems clear that Darwin's *Origin* satisfied the first set of criteria, and Wallace never did write his own "big species book" until he published *Darwinism* in 1889, the very title of which indicates his own leanings on the priority question.

The September 27, 1857, clipping also indicates that, if anything, instead of surprised Darwin should have been a little chagrined at the arrival of Wallace's Ternate paper, having been forewarned by Lyell that he should publish. Darwin's response to this portent indicates his dislike of publishing solely for the sake of priority, yet stating his own fear of being forestalled. On May 3, 1856, Darwin wrote to Lyell: "I rather hate the idea of writing for priority, yet I certainly should be vexed if anyone were to publish my doctrine before me."[29] His hand forced by Wallace in 1858, Darwin found a solution to his apparent dilemma (i.e., publish for priority sake only, or be completely scooped) by writing a book that was midway between a brief sketch and a magnum opus—*On the Origin of Species.*

The Zero-Sum Game of Science

This priority incident is emblematic of a larger problem in the history of science in which the cumulative, interactive, and social nature of the enterprise is not always recognized. Science is often seen as a zero-sum game. To the extent that science may be modeled as a game with rules, modern game theory may grant us a deeper understanding of the tension between competitiveness and cooperation by distinguishing between zero-sum and plus-sum models. In zero-sum games the gain of one participant means the loss of the other, and the more one gains the more the other loses.[30] Winning a game by six points means that one's opponent must lose by six points, and thus they sum to zero (6+−6 = 0). This antagonistic win–lose model, however, misses the interdependent, sometimes cooperative, and always social nature of the scientific process. Wallace's priority credit and recognition for scientific achievement can and should be significantly enhanced without taking anything away from Darwin. It is a relationship modern evolutionary biologists would describe as "reciprocal altruism," where "I'll scratch your back if you'll scratch mine."[31] A game theory task that demonstrates the effectiveness of reciprocal relationships is called the Prisoner's Dilemma, in which two prisoners have several options: (1) they can cooperate with each other and get light sentence terms; (2) if one defects while the other cooperates, the defector is freed while the cooperator gets an even longer jail sentence; or (3) both can defect, in which case both receive longer jail stays. When this game is iterated, or repeated, the majority of responses produced are cooperative, as this strategy leads in the long run to "the greatest good for the greatest number." In the short run—that is, in a noniterated or one-trial game—defection is the rule. Over time, however, consistent defectors lose out.[32]

The zero-sum model is at the heart of most disputes of scientific priority because it assumes that the only way one scientist can profit is through the loss of another. Clearly, Newton and Newtonian supporters saw Newton's gain in the priority of the invention of the calculus to be Leibniz's loss, and vice versa, leading to centuries of contentious debate and bitter disagreement. Likewise, some perceive Darwin's gain as Wallace's loss, and Wallace's gain as Darwin's loss. Because of this it becomes difficult in most of these debates to tease out the facts from the emotion, the information from the rhetoric. This disputatious posturing on both sides wedges historians into a defensive stance that compels an attack-or-be-attacked response. Thus, the antagonism between scholars and historians in both camps could be attenuated by the rejection of the zero-sum model. Darwin, and especially Wallace, clearly did, as they recognized the gain to be had through cooperative interaction. Consider this exchange of letters between the two men. The first, an April 6, 1859, letter from Darwin to Wallace, reveals a man paying the highest respect for a fellow winner in this game of scientific cooperation:

> You cannot tell how I admire your spirit, in the manner in which you have taken all that was done about establishing our papers. I had actually written a letter to you, stating that I could not publish anything before you had published. I had not sent that letter to the post when I received one from Lyell and Hooker, urging me to send some ms. to them, and allow them to act as they thought fair and honourably to both of us. I did so.[33]

Wallace always responded to Darwin with an equally generous dose of recognition, as in this passage from a May 29, 1864, letter:

> As to the theory of Natural Selection itself, I shall always maintain it to be actually yours and yours only. You had worked it out in details I had never thought of, years before I had a ray of light on the subject, and my paper would never have convinced anybody or been noticed as more than an ingenious speculation, whereas your book has revolutionized the study of Natural History, and carried away captive the best men of the present age.[34]

This is no attempt to whitewash Darwin, nor is it a naive and unrealistic portrayal of what in reality was the highly competitive world of nineteenth-century science in which Darwin was fully ensconced. For several weeks Darwin did a lot of ethical hand-wringing. On June 25, 1858, just a week after the letter announcing his dismay over receiving the Wallace manuscript, he wrote Lyell again, expressing his moral dilemma over the priority question. It is a most revealing passage that shows not the anxiety of a man who has done something unethical, but a just and fair individual caught in the tension between win–lose and win–win scenarios:

> There is nothing in Wallace's sketch which is not written out much fuller in my sketch copied in 1844, & read by Hooker some dozen years ago. About a year ago I sent a short sketch of which I have copy of my views (owing to correspondence on several points) to Asa Gray, so that I could most truly say & prove that I take nothing from Wallace. I sh^d. be extremely glad now to publish a sketch of my general views in about a dozen pages or so. But I cannot persuade myself that I can do so honourably. Wallace says nothing about publication, & I enclose his letter.—But as I had not intended to publish any sketch, can I do so honourably because Wallace has sent me an outline of his doctrine?—I would far rather burn my whole book, than that he or any other man sh^d. think that I had behaved in such a paltry spirit. Do you not think that his having sent me this sketch ties my hands?[35]

As we saw, Lyell certainly did not think it did, and the outcome was win–win for both scientists. (It is interesting to note that not only Alfred Wallace, but his grandson John, were and are satisfied with the historical priority outcome. After a lengthy conversation on this question, John Wallace noted: "I can't understand what all the fuss is about. Grandfather was satisfied with the arrangement, none of us desire to call it 'Wallace's theory of natural selection,' but many of the Darwin people seem defensive about it."[36] There is no doubt about the latter, but it is understandable because the aforementioned Wallace defenders have embraced the zero-sum model, causing them to give more credit to Wallace while simultaneously taking credit away from Darwin. Darwin scholars, in turn, adopt the zero-sum model in defense, as they feel Wallace's gain is Darwin's loss.)

Zero-Sum Pratfalls: Credit Where It's Due

Some writers have approached this priority question from yet another angle—that Darwin's overall contribution, irrespective of Wallace, has been overrated and that he was merely reiterating what others had already brought to the forefront of knowledge. An example of this sort of ax-grinding literature is Francis Hitching's *Neck of the Giraffe,* where the author states: "It is one of the less pleasing sides of Darwin's otherwise affable and scholarly nature that he could never bring himself to acknowledge a debt to the many predecessors in his field who were puzzling about the origin of species."[37] This statement follows a lengthy discussion by Hitching of Lyell's influence on Darwin and of Darwin's *willingness* to credit him, with such acknowledgments as "I feel as if my books came half out of Sir Charles Lyell's brain" and "I saw more of Lyell than of any other man, both before and after my marriage."[38]

First of all, Darwin's reference to the influence of Lyell on his books pertained to his geological knowledge, not natural selection or evolution. Also, is this not an example of Darwin's giving credit where it was due? (He was greatly influenced by Lyell's work on the *Beagle* voyage and on his return, but only on geological matters, not biological.) Following this inconsistency, Hitching cites an example in Loren Eiseley's book, *Darwin and the Mysterious Mr. X,* to establish Darwin's unwillingness to give appropriate credit to the English naturalist Edward Blyth, who in 1835 and 1837 published in the *British Magazine of Natural History* his theories about competition among species. Eiseley chronicles passages "that are almost word-for-word identical between Darwin and Blyth," with nothing more than "a cryptic reference in a letter" to the fact that Darwin had read these articles. After castigating Darwin for this specific violation, Hitching now "forgives" him because "it was not the Victorian fashion to acknowledge predecessors and give references in the way it is mostly done in science today."[39]

This may be true, in part, but there is a deeper misunderstanding of Darwin that only a cursory review of the literature can produce. Hitching, for example, quotes Blyth in his observation that "among animals which procure their food by means of their agility, strength, or delicacy of sense, the one best organized must always obtain the greatest quantity; and must, therefore, become physically the strongest and be thus enabled, by routing its opponents, to transmit its superior qualities to a greater number of offspring. The same law therefore, which was intended by providence to keep up the typical qualities of a species can be easily converted by man, into a means of raising different varieties."[40] Similarly, in Charles Lyell's *Principles of Geology* one finds the following passage that also sounds anticipatory: "Nature is constantly at war with herself and thus there will always be individuals who perish by disease or by the actions of predation. In a variable species it would not be the typical individual who succumbed to death, but the deviant ones—too great, too small, too thin legged, too thick legged—that would most often be the victims of predation or accident."[41]

On first blush this does seem to resemble the process of eliminating the unfit, but in this case it is not for the creation of new species. Lyell and Blyth (and many others) were promulgating the accepted dogma of the day—that the forces of natural selection were acting to preserve created kinds by *extinguishing* the extreme varieties, not the process of changing them into new and different kinds. This *essentialistic* belief in the fixity of species was a vital part of the understanding of nature since the time of Aristotle. As a matter of fact, Lyell's *Principles of Geology* was, according to Ernst Mayr, a "superbly argued case against the modification of species." For example, Lyell claimed that "varieties have strict limits, and can never vary more than a small amount away from the original type." And: "There are fixed limits beyond which the descendants from common parents can never deviate from a certain type. It is idle . . . to dispute about the abstract possibility of the conversion of one species into another, when there are known causes, so much more active in their nature, which must always intervene and prevent the actual accomplishment of such conversions."[42] Blyth's quote above thus makes sense, as he notes that this mechanism "intended by providence to keep up the typical qualities of a species" can also be used artificially by breeders "as a means of raising different varieties." In the opening statement of his 1858 Ternate paper, Wallace summarizes this argument to show how he is using the same mechanism for completely opposite ends—species mutability, not stability:

> One of the strongest arguments which have been adduced to prove the original and permanent distinctness of species is, that *varieties* produced in a state of domesticity are more or less unstable, and often have a tendency, if left to themselves, to return to the normal form of the parent species; and this instability is considered to be a distinctive peculiarity of all varieties, even of those occurring among wild animals in a state of nature, and to constitute a provision for preserving unchanged the originally created distinct species.[43]

It seems highly probable that to the extent Darwin (or, for that matter, Wallace) did not refer to certain writers, it was *not* because they had "anticipated" him, or that they plagiarized his ideas (or that acknowledgments were not given in Victorian England, therefore we can forgive them), but because these ideas were in the air and well known to most and, in fact, were in complete opposition to what Darwin and Wallace were proposing. It should be mentioned, too, that the "anticipation" of Darwin and Wallace might have gone unnoticed by the two naturalists. A perfect example of this was the Scottish botanist Patrick Matthew, who, in the appendix of his 1831 *On Naval Timber and Arboriculture,* proposed a mechanism of organic change similar to natural selection, and after the *Gardeners' Chronicle* had run a review by Huxley of the *Origin,* Matthew wrote in claiming priority. Quoting himself from his appendix, Matthew summarized his version of natural selection:

> There is a law universal in Nature, tending to render every reproductive being the best possibly suited to its condition that its kind, or that organised matter, is susceptible of, which appears intended to model the physical and mental or instinctive powers, to their highest perfection, and to continue them so. This law sustains the lion in his strength, the hare in her swiftness, and the fox in his wiles. As Nature, in all her modifications of life, has a power of increase far beyond what is needed to supply the place of what falls by Time's decay, those individuals who possess not the requisite strength, swiftness, hardihood, or cunning, fall prematurely without reproducing—either a prey to their natural devourers, or sinking under disease, generally induced by want of nourishment, their place being occupied by the more perfect of their own kind, who are pressing on the means of subsistence.[44]

Matthew's description does sound similar to Darwin's, but the fact is Darwin never saw it because it was buried away in an appendix of a book on Naval timber. Darwin penned the following explanation in a response that is hardly the stance of an ideological plagiarizer or scientific schemer:

> I have been much interested by Mr. Patrick Matthew's communication in the Number of your Paper, dated April 7. I freely acknowledge that Mr. Matthew has anticipated by many years the explanation which I have offered of the origin of species, under the name of natural selection. I think that no one will feel surprised that neither I, nor apparently any other naturalist, has heard of Mr. Matthew's views, considering how briefly they are given, and that they appeared in the Appendix to a work on Naval Timber and Arboriculture. I can do no more than offer my apologies to Mr. Matthew for my entire ignorance of his publication.[45]

Another fallacy in the ideological manifestation of the zero-sum model is the idea that some sizable dollop of thought is completely original to a thinker. Very few of our thoughts are unsullied original creations that spontaneously generate from neural impulses. We think (mostly) with language, the vast majority of which is not original to us. We are all synthesizers, of a sort, some better at it than others. Darwin and Wallace, among their peers, synthesized a vast quantity of biological and geological phenomena in a parallel fashion different from what anyone else had done. But most of the bits and pieces were already there. What they did with those intellectual parcels is what is original to them. Matthew, Blyth, or others (e.g., William Charles Wells, discussed as another "precursor")[46] may have predated Darwin and Wallace with a similar idea, but they did nothing with it. Wallace, and especially Darwin, took this mechanism of species preservation and changed it into one of species transmutation, then constructed a research program to test the theory, and in the process took a giant leap forward in our understanding of the origin of species.

The Plus-Sum Game of Science

A plus-sum model—the gain of one is the gain of another—recognizes the contingent, cooperative, and interdependent nature of scientific discovery. Both Darwin and Wallace profited by the profit of the other. Both were winners in the game to understand the origin of species. An 1870 letter of "reflection" from Darwin to Wallace shows the special win–win nature of their relationship: "I hope it is a satisfaction to you to reflect—and very few things in my life have been more satisfactory to me—that we have never felt any jealousy towards each other, though in one sense rivals." In the most gentlemanly fashion Wallace always politely addressed Darwin in all their correspondence, and Darwin always responded in kind. "I was much pleased to receive your note this morning," reads a typical letter opening from Wallace to Darwin. "Hoping your health is now quite restored," "I sincerely trust that your little boy is by this time convalescent," and so on.[47] Darwin and Wallace used each other and each other's ideas to their mutual benefit, and the world of science is better off for it, as Wallace explained in an 1886 interview: "I arrived at the theory independently of Darwin, no doubt, and communicated it to him before he had published anything on the subject." But, Wallace continued in response to a question about how this affected his relations with Darwin, "we have been on the most friendly terms throughout up to the very time of his death; we were always exceedingly friendly."[48]

Another shortcoming of the zero-sum model is the assumption that the ideas under priority dispute are identical, leading to the conclusion that only one individual can be first in discovery. But a law of nature is the product of both the *discovery* and *description* of a phenomenon. Two individuals may make the same discovery, but they may not make the same description. This is the case with Darwin and Wallace, where their theories of evolution by means of natural selection are similar and complementary, but not identical. In an 1898 interview Wallace explained how he thought his theory differed from Darwin's: "Sexual selection resulting from the fighting of males is indisputable, but, differing from Darwin, I do not believe there is any selection through the choice of the females, and the drift of scientific opinion is towards my view. Again, I do not believe in the transmission of acquired characters, the evidence seeming to me to be against it, and this is the chief point on which there is a growth of scientific opinion against Darwin. The discussion is still proceeding, naturalists now being about equally divided. Herbert Spencer takes the same view as Darwin, but Mr. Francis Galton and Weismann between them have almost certainly proved the non-heredity of acquired variations. But neither of these questions affects Darwin's fundamental principles."[49]

In fact, Wallace and Darwin differed on a number of important theoretical points, including (as enumerated by Wallace): "1. The Origin of Man as an Intellectual and Moral Being. 2. Sexual Selection through Female Choice. 3. Arctic Plants in the Southern Hemisphere, and on Isolated Mountain-tops within the Tropics. 4. Pangenesis, and the Heredity of Acquired Characters." Nevertheless, Wallace emphasized that "none of my differences of opinion from Darwin imply any real divergence as to the overwhelming importance of the great principle of natural selection, while in several directions I believe that I have extended and strengthened it. The principle of 'utility,' which is one of its chief foundation-stones, I have always advocated unreservedly; while in extending this principle to almost every kind and degree of coloration, and in maintaining the power of natural selection to increase the infertility of hybrid unions, I have considerably extended its range."[50]

Where Wallace and Darwin really parted company was on numerous social issues. As we will see, Wallace attempted a much more thoroughgoing and all-embracing worldview than Darwin, and as such he found himself at odds with his more conservative colleagues who were more cautious in their philosophical speculations from the scientific data.

Through their numerous intellectual exchanges in letters, papers, and books, Darwin and Wallace stimulated each other in both knowledge and theory, with a net gain profit for both, making them genuinely co-discoverers, but not co-describers of the same theory. The historical record, however, has been read differently, beginning with the ranking of the joint papers presented at the July 1, 1858, Linnean Society meeting that placed Darwin's 1844 extract and his 1857 Asa Gray letter ahead of Wallace's 1858 essay. If considered by dates of ideas alone, then the ranking is chronologically correct. (It is also alphabetically correct, which was how the names were listed.) But, in fact, what has happened is that Darwin has become a household name and Wallace all but forgotten. This historical reality, of course, was not caused by the ranking of their names at this meeting. In actual fact, according to the Linnean Society president, Thomas Bell, in a reflection of the year's activities, nothing of significance happened in 1858: "The year which has passed . . . has not, indeed, been marked by any of those striking discoveries which at once revolutionize, so to speak, the department of science on which they bear."[51] Obviously Bell and his colleagues did not grasp the significance of the theory of natural selection at its time of presentation. (Although Hooker obviously did, commenting that "the interest excited was intense, but the subject was too novel and too ominous for the old school to enter the lists, before armouring. After the meeting it was talked over with bated breath: Lyell's approval, and perhaps in a small way mine, as his lieutenant in the affair, rather overawed the Fellows, who would otherwise have flown out against the doctrine. We had, too, the vantage ground of being familiar with the authors and their theme."[52]) Darwin's fame and importance accrued over many decades of sound scientific work and a thorough research program, not through a "delicate arrangement" and clandestine priority ranking of his name over Wallace's. Besides, other than later noting that his paper "was printed without my knowledge, and of course without any correction of proofs," Wallace was most delighted to finally gain the recognition of the scientific community he had desired for so many years, as he indicated to his mother on October 6, 1858, the same day he wrote Darwin: "I have received letters from Mr. Darwin and Dr. Hooker, two of the most eminent naturalists in England, which has highly gratified me. I sent Mr. Darwin an essay on a subject on which he is now writing a great work. He showed it to Dr. Hooker and Sir C. Lyell, who thought so highly of it that they immediately read it before the Linnean Society. This assures me the acquaintance and assistance of these eminent men on my return home."[53]

Consider Wallace's position at this time. He was a relatively unknown thirty-five-year-old amateur naturalist whose only major theoretical work—the 1855 Sarawak Law paper—was largely ignored (or, at least, so he thought). He had been away from England and the center of scientific activity already four years, and was, by all rights, still cutting his teeth on such weighty theoretical matters. Darwin, by contrast, was forty-nine years old, fairly well known in scientific circles, had already published numerous important scientific books and articles, and had shared his theoretical ideas with the most important scientists in England and America. Wallace did not feel the loser, because he was not. An essay written in two nights, sent to the right place at the right time, put him in the scientific inner circle and into the historical record—his name next to Darwin's—forever. Anyone who thinks that this was Wallace's loss should reconsider the circumstances in light of the plus-sum model of scientific priority. The gain of Darwin was the gain of Wallace.

6

Scientific Heresy and Ideological Murder

Throughout his long working career Alfred Russel Wallace was consistently and persistently a heretic scientist, not only for championing the then radical theory of evolution by means of natural selection, but for the many eccentric and fringe causes he supported. He was also a heretic scientist doing good science (at least in his own mind), whether he was in the tropical rain forest of the Amazon and Malay Archipelago, or in a séance with a medium levitating tables and calling forth spirits from the "other side." Wallace attempted to construct a scientific worldview that was consistent with the evidence as well as his deepest beliefs about human nature and the laws of the universe. Whether he succeeded in this construction or not is irrelevant (though most of his contemporaries think he did not), because Wallace believed he did. Never was this more apparent than in the near falling-out he had with Darwin over the evolution of the human mind, an extrapolation of the theory of natural selection that Darwin could not endorse.

Advancing Science

On his arrival home in the British Isles from the Malay Archipelago, on April 1, 1862, Wallace moved in with his mother, his sister Fanny, and brother-in-law Thomas Sims. Atop the house he wedged himself into a large room filled with crates of his specimens sent from the East. He then procured "the largest and most comfortable easy-chair I could find in the neighbourhood" and for the next month began to go through them, sorting the specimens by island and locality.[1] Wallace's initial concerns included recovering from malaria (the effects of which were to plague him intermittently throughout his life), obtaining employment (he was never successful in securing permanent full-time work), finding a life-companion, and, most important, culling the scientific leads accumulated during the previous eight years. He was still relatively young at thirty-nine, but the tropics had taken their toll. His physical recovery delayed his personal introduction to Darwin at Down until late summer 1862, by which time Darwin had already published the third edition of the *Origin,* containing an "Historical Sketch" from Aristotle through Lamarck "on the origin of species . . . propounded by Mr. Wallace and myself," an acknowledgment that must have aided him in his recovery. Before long Wallace met with and was befriended by Herbert Spencer and Charles Lyell. These associations came easy, as Wallace was frequently asked to speak to various societies where he had the opportunity to meet and mingle with both the luminaries and the rank and file of British science.

Figure 6-1 Alfred Russel Wallace in 1862 at age thirty-nine, shortly after his return home after eight years in the Malay Archipelago during which he collected a remarkable 125,660 specimens. (From Marchant, 1916, frontispiece)

Not long after his return, for example, Wallace delivered a series of lectures at the Zoological Society of London. On May 27 he gave a "Narrative of Search after Birds of Paradise," in which he began: "Having visited most of the islands inhabited by the paradise birds, in the hope of obtaining good specimens of many of the species, and some knowledge of their habits and distribution, I have thought that an outline of my several voyages, with the causes that have led to their only partial success, might not prove uninteresting." Indeed, apparently they were not uninteresting, for he was invited back on June 10 for a talk "On some New and Rare Birds from New Guinea"; and on June 24 he discussed "Descriptions of Three New Species of Pitta from the Moluccas." These were supplemented by a lecture on "On the Physical Geography of the Malay Archipelago" on June 8 to the Royal Geographical Society and, the following year, with such lectures as: "On the Geographical Distribution of Animal Life" at the August 31, 1863, meeting of Section C on Zoology and Botany of the British Association for the Advancement of Science, and on September 1 at the same meeting, but to Section E on Geography and Ethnology, he lectured on "On the Varieties of Men in the Malay Archipelago." Later that year he spoke again to the Zoological Society of London at their November 10 meeting, this time about "On the Identification of the Hirundo esculenta of Linnaeus, with a Synopsis of the Described Species of *Collocalia,*" and again to the same group on November 24 he discussed "A List of the Birds Inhabiting the Islands of Timor, Flores, and Lombock, with Descriptions of the New Species." His expertise in entomology led to his presidency of the Entomological Society, for which he gave the Presidential Address in January 1871. All these lectures became papers published in the journals of these prestigious scientific societies, and later were edited into chapters for his numerous books on natural history.[2]

Wallace was at the height of his literary power and was making the most of his time back in civilization. He even hosted a public display of his most impressive specimens in Thomas Sims's photographic gallery. "The entire series of my parrots, pigeons, and paradise birds, when laid out on long tables covered with white paper, formed a display of brilliant colours, strange forms, and exquisite texture that could hardly be surpassed," he recalled in a rare moment of immodesty, "and when to these were added the most curious and beautiful among the warblers, flycatchers, drongos, starlings, gapers, ground thrushes, woodpeckers, barbets, cuckoos, trogons, kingfishers, hornbills, and pheasants, the general effect of the whole, and the impression it gave of the inexhaustible variety and beauty of nature in her richest treasure houses, was far superior to that of any collection of stuffed and mounted birds I have ever seen."[3] Throughout 1867 and 1868, between his lectures, papers, and collections Wallace wove together a volume on the Malay Archipelago that was

Figure 6-2 Alfred Russel Wallace in 1869 at age forty-six, holding a copy of his bestselling book *The Malay Archipelago,* which quickly became a classic in travel literature and natural history. (From *My Life,* 1905, v. I, 385)

part natural history and part travelogue, and included "accounts of the manners and character of the people." He shopped the manuscript around and found a publisher in Macmillan & Co., London, and since they "wished the book to be well illustrated, I had to spend a good deal of time in deciding on the plates and getting them drawn, either from my own sketches, from photographs, or from actual specimens, and having obtained the services of the best artists and wood engravers then in London, the result was, on the whole, satisfactory."[4] Indeed it was. *The Malay Archipelago; The Land of the Orang-utan and the Bird of Paradise; A Narrative of Travel with Studies of Man and Nature,* published in two volumes in February 1869, became Wallace's most commercially successful book and the second most-cited work in his literary corpus.[5] It is considered a classic and remains in print nearly a century and a half after its release.

Wallace was also enmeshed in the mainstream of British science and, as a glance through his remarkable bibliography of 747 published works shows, he never let up. From his first voyage to the Amazon in 1848 to the end of his life in 1913, Wallace averaged 11.92 publications per year, including books, articles, reviews, and letters, a remarkable intellectual outpouring. Since he was less productive during his travels than he was after his return to England, the average from 1862 to 1913 rises to 13.5 publications per year, a stunning productivity value of over one publication per month throughout a very long career, during which there were far fewer journals and magazines in which to publish than there are today. (By comparison, between 1923 and 1999 evolutionary biologist Ernst Mayr totaled 704 publications for an average of 9.3 per year; between 1965 and 1999 paleontologist Stephen Jay Gould totaled 593 publications for an average of 17.4 per year; between 1958 and 1999 evolutionary biologist Jared Diamond totaled 549 publications for an average of 13.4 per year, and evolutionary biologist Edward O. Wilson totaled 380 publications for an average of 7.6 per year.) By any age or standard, Wallace was among the most productive scientists, in both quantity and quality of publications.

Restless Domesticity

As single-minded as he could be, Wallace did have a personal life that he occasionally attended to, most notably on his return to England his first deep love. She was the eldest daughter of his chess-playing partner Lewis Leslie, twenty-eight years old and identified by Wallace only as "Miss L——." She was Marion Leslie, described by her pursuer as "very agreeable though quiet, pleasant looking, well educated, and fond of art and literature, and I soon began to feel an affection for her, and to hope that she would become my wife." Wallace's social skills, however, appeared to be in their infancy, and after about a year of courtship that was evidently unidirectional, Wallace "wrote to her, describing my feelings and asking if she could in any way respond to my affection." The answer was negative. "Evidently my undemonstrative manner had given her no intimation of my intentions." Nevertheless, Miss L—— begged "that I would not allow her refusal to break off my visits to her father."[6]

The chess games with Lewis continued, as Wallace strategized on how best to checkmate his daughter. At the urging of his sister and mother, Wallace invested another year in trying to bring around Marion's affections. For a while it looked good. The couple met two or three times a week and, as was customary, Miss Leslie's father "told me that his daughter had a small income of her own, and asked that I should settle an equal amount on her. This was satisfactorily arranged, and at a subsequent meeting we were engaged."[7] Wallace was in love, but the bride had other intentions, despite her father's wishes. One day, on arriving at the Leslie home for his usual weekly visit, Wallace "was informed by the servant that Miss L—— was not at home, that she had gone away that morning, and would write." He was "staggered," and then informed by Lewis that the engagement was off and that his daughter would later write with an explanation. When it came Wallace "was hardly more enlightened." Apparently Miss L—— caught wind of another relationship that she thought Wallace was concealing from her—a widow friend of his mother. Wallace was befuddled. "The lady was the widow of an Indian officer, very pleasant and good-natured, and very gossipy, but as utterly remote in my mind from all ideas of marriage as would have been an aunt or a grandmother."[8] Wallace had not told Marion about the widow because there was nothing to tell. His feelings crushed, he wrote her "to explain my real feelings towards her, and assuring her that I had never had a moment's thought of any one but her, and hoping that this explanation would suffice." It did not. Wallace never again heard from or saw Marion Leslie, or her father.[9]

The broken engagement hurt Wallace deeply as he had "never in my life experienced such intensely painful emotion."[10] But nothing eases the pain of unrequited love better than to have it requited from another source. In the autumn of 1864 (by his own chronology it could not have been more than a few weeks, months at most, since his engagement was broken) he met the woman with whom he would spend the rest of his life. The initial meeting came through an acquaintance with William Mitten, a pharmacist and moss aficionado with whom Wallace spent many a long walk in the woods gathering specimens and sharing a common love of nature. But William Mitten had something else besides a passion for mosses that engaged Wallace's attention—to wit, an oldest daughter named Annie with whom he met and fell in love, though she was twenty-three years his junior. Within two years they were married, he forty-three, she twenty, a union that would last nearly half a century.

Though domesticity entered his life, Wallace remained the ever relentless naturalist—a monomaniac with a mission who never missed an opportunity to explore his environment. Even on their honeymoon the couple traveled to North Wales, where Alfred read Sir Andrew Ramsay's *The Old Glaciers of Switzerland and North Wales,* which enriched his personal observations of the countryside so that he "thoroughly enjoyed the fine examples of ice-groovings and striations, smoothed rock-surfaces, *roches moutonnees,* perched blocks, and rock basins." Always the historical scientist, Wallace demonstrated how past processes may be inferred from present observations: "Every day revealed some fresh object of interest as we climbed among the higher *cwms* of Snowdon; and from what I saw during that first visit the Ice Age became

Figure 6-3 Annie Wallace (née Mitten), daughter of the botanist William Mitten, who married Alfred Russel Wallace in 1866. (Courtesy of Alfred John Russel Wallace and Richard Russel Wallace)

almost as much a reality to me as any fact of direct observation."[11] The honeymoon produced a publication, of course. "Ice-marks in North Wales" was published in the *Quarterly Journal of Science* in January 1867, in which he argued that lake basins were formed by the pressure of glaciers, not as previously thought by a scooping-out process by water.[12] Wallace's theory is still the accepted explanation.

Wallace's Heresy

Wallace's finances during these productive years were a source of concern to him. As a single man the proceeds from the sales of his collections from the East provided him with sufficient sustenance, but marriage and children required more resources, so he "was always on the lookout for some permanent congenial employment which would yet leave time for the study of my collections." That was always the rub for Wallace. Most jobs would not give him the time to pursue his passion for natural history and evolutionary theory, and what few jobs there were in these fields Wallace was not well suited to fill. "My deficient organ of language prevented me from ever becoming a good lecturer or having any taste for it," he complained, "while the experience of my first work on 'The Amazon' did not encourage me to think that I could write anything that would much more than pay expenses." When an opening for assistant secretary of the Royal Geographical Society came up, Wallace was beaten out for the job by his own Amazon expedition partner, Henry Walter Bates, who "had just published his 'Naturalist on the Amazon,' and was, besides, much better qualified than myself by his business experience and his knowledege of German."[13]

The job loss did not disappoint, however, as Wallace was fixated on his newly emerging heresy that would find him at odds with his more conservative colleagues, including and especially Charles Darwin. Wallace's many and powerful connections in the scientific community would now come in handy, and none wielded more power than the geologist Charles Lyell. Wallace was in regular correspondence with Lyell, whose caution helped keep his younger charge in check, as Wallace recalled: "He was by nature so exceedingly cautious and conservative, and always gave such great weight to difficulties that occurred to himself or that were put forth by others, that it was not easy to satisfy him on any novel view upon which two opinions existed or were possible."[14] Such discussions often went on during walks "across the park to St. Mark's Crescent for an hour's conversation; at other times he would ask me to lunch with him, either to meet some interesting visitor or for friendly talk," and included Darwin's controversial theory of pangenesis, theories of glaciation, the origin of Alpine lakes, and human evolution. It was on this latter point that Wallace broke with Darwin in what became the greatest heresy of this heretic scientist, and Lyell was the triggerman.

The first public announcement of Wallace's scientific heresy can be dated to the April 1869 issue of the *Quarterly Review,* in an article entitled "Sir Charles Lyell on Geological Climate and the Origin of Species."[15] In reviewing Charles Lyell's tenth edition of *Principles of Geology* and sixth edition of *Elements of Geology,* Wallace noted that Lyell had finally become a convert to the theory of evolution. But at this point Wallace himself was undergoing his own conversion, in a direction Lyell would see as intriguing but Darwin would find dismaying. For Wallace, the problem of evolution was the failure of natural selection to explain the enlarged human brain (compared to apes), as well as the organs of speech, the hand, and the external form of the body. "In the brain of the lowest savages and, as far as we know, of the prehistoric races, we have an organ . . . little inferior in size and complexity to that of the highest types. . . . But the mental requirements of the lowest savages, such as the Australians or the Andaman Islanders, are very little above those of many animals," he observed. "How then was an organ developed far beyond the needs of its possessor? Natural Selection could only have endowed the savage with a brain a little superior to that of an ape, whereas he actually possesses one but very little inferior to that of the average members of our learned societies."[16]

Since natural selection was the only law of nature Wallace knew of to explain the development of these structures, and since he decided that it could not adequately do so, he concluded, making a leap that would cost him dearly in the scientific community (but would gain him considerable support in other communities of knowledge), that "an Overruling Intelligence has watched over the action of those laws, so directing variations and so determining their accumulation, as finally to produce an organization sufficiently perfect to admit of, and even to aid in, the indefinite advancement of our mental and moral nature."[17] Wallace's reasoning was sound and consistent. Natural selection does not select for needs in the future. There is no such thing as preselection. (Although the term itself is still used, preselection means that a structure is initially selected for one purpose and later finds a different use that was not selected for previously. Bird wings, at least in some cases, were probably originally selected for thermal regulation and later became useful as aerodynamic structures. We would then say that the wings were preselected, even though nature, of course, had no preconception of the wing's future alternative use for flight.) Natural selection operates on the here-and-now level of the organism. The usefulness or uselessness (or even harmfulness) of a given structure or function can only matter to the organism, and thus to nature herself, *now,* not in the future. Nature did not know we would one day need a big brain in order to contemplate the heavens or compute complex mathematical problems; she merely selected among our ancestors those who were best able to survive and leave behind offspring through the ability to implement stone tool technology, organize group hunts, construct controlled fires, and whatever else it takes to survive in a natural environment. But since we *are* capable of such sublime and lofty mental functions, Wallace deduced, clearly natural selection could not have been the originator of a brain big enough to handle them. Only an "Overruling Intelligence" could have fashioned such a mind—a rational, albeit supranatural, leap of the imagination.

Lyell, always the cagey lawyer, queried Darwin first before committing himself to Wallace's heresy. In a letter to Darwin, Lyell quoted a passage from a letter Wallace had written him, in which Wallace outlines the argument in an even more dramatic fashion than he did in the *Quarterly* review. In fact, decades later in his autobiography, Wallace recalled this summary as "perhaps more simply and forcibly stated than in any of my published works":

> It seems to me that if we once admit the necessity of *any* action beyond "natural selection" in developing man, we have no reason whatever for confining that agency to his brain. On the mere doctrine of chances it seems to me in the highest degree improbable that so many points of structure, all tending to favour his mental development, should concur in man alone of all animals. If the erect posture, the freedom of the anterior limbs from purposes of locomotion, the powerful and opposable thumb, the naked skin, the great symmetry of form, the perfect organs of speech, and, in his mental faculties, calculation of numbers, ideas of symmetry, of justice, of abstract reasoning, of the infinite, of a future state, and many others, cannot be shown to be each and all *useful* to man in the very lowest state of civilization—how are we to explain their co-existence in him alone of the whole series of organized beings? Years ago I saw in London a bushman boy and girl, and the girl played very nicely on the piano. Blind Tom, the half-idiot negro slave, had a "musical ear" or brain, superior, perhaps, to that of the best living musicians. Unless Darwin can show me how this latent musical faculty in the lowest races can have been developed through *survival* of the fittest, can have been of *use* to the individual or the race, so as to cause those who possessed it in a fractionally greater degree than others to win in the struggle for life, I must believe that some other power (than natural selection) caused that development. It seems to me that the *onus probandi* will lie with those who maintain that man, body and mind, could have been developed from a quadrumanous animal by "natural selection."[18]

Notice that there is no mention of spiritualism or any other quasi-religious notion that some historians believe was the sole reason for Wallace's calling forth of an Overruling Intelligence. The entire argument hinges on the failure of natural selection to account for a *variety* of features, in which elsewhere he includes spiritualism, but not here. If spiritualism was the sole reason for the shift, then why is it not even mentioned in this letter that he called his most forcibly stated position of any of his published works?

Darwin's Dismay

Lyell supported Wallace's new stance, telling Darwin that "I rather hail Wallace's suggestion that there may be a Supreme Will and Power which may not abdicate its function of interference but may guide the forces and laws of Nature." Lyell was an important ally for Wallace who bolstered his confidence in his decision to break from the Darwinian camp. Darwin, unsurprisingly, was not so conciliatory. Anticipating his friend's reaction before the *Quarterly* paper came out, Wallace wrote Darwin on March 24 to warn him that "in my forthcoming article in the 'Quarterly' I venture for the *first time* on some limitations to the power of natural selection." Knowing how this new development would be received, Wallace continued: "I am afraid that Huxley and perhaps yourself will think them weak & unphilosophical. I merely wish you to know that they are in no way put in to please the Quarterly readers,—you will hardly suspect me of that,—but are the expression of a deep conviction founded on evidence which I have not alluded to in the article but which is to me absolutely unassailable." (Wallace's reference to the *Quarterly* readers is based on the fact that it was mostly read by Tories, conservatives who favored the preservation of the existing political and social order and supported the authority of the king over Parliament.)

Before he read the article, perhaps forecasting that the intellectually adventuresome Wallace had diverged down a track he would find a bit too skewed, Darwin wrote to his younger charge on March 27, "I shall be intensely curious to read the *Quarterly:* I hope you have not murdered too completely your own and my child."[19] After he read the article Darwin's response was predictably brusque. In the margin of Darwin's copy of the *Quarterly Review* article, next to the passage on the inadequacy of natural selection to endow man with a large brain, he wrote a firmly pressed "NO," underlined three times with numerous added exclamation points.[20] He then told Lyell that he was "dreadfully disappointed" in Wallace, and then wrote Wallace again, making the presumption "that your remarks on man are those to which you alluded in your note. If you had not told me, I should have thought that they had been added by someone else. As you expected, I differ grievously from you, and I am very sorry for it. I can see no necessity for calling in an additional and proximate cause in regard to Man. But the subject is too long for a letter."[21] Several months later, on January 26, 1870, with the not-so-subtle hint of a disappointed friend and mentor, after first praising his protégé, Darwin notes: "I am very glad you are going to publish all your papers on Natural Selection: I am sure you are right, and that they will do our cause much good. But I groan over Man—you write like a metamorphosed (in retrograde direction) naturalist, and you the author of the best paper that ever appeared in the *Anthropological Review!* Eheu! Eheu! Eheu!—Your miserable friend, C. Darwin."[22]

Wallace's reaction to Darwin's disappointment was immediate and understanding. "I can quite comprehend your feelings with regard to my 'unscientific' opinions as to Man, because a few years back I should myself have looked at them as equally wild and uncalled for." In addition to his skepticism of the ability of natural selection to account for the human mind and other features, Wallace was now diverging down a track Darwin would find quite impossible to follow—the investigation of spiritual phenomena that played an important role in this intellectual shift. "My opinions on the subject have been modified solely by the consideration of a series of remarkable phenomena, physical and mental, which I have now had every opportunity of fully testing, and which demonstrate the existence of forces and influences not yet recognised by science." Wallace, ever the heretic scientist willing to explore any and all aspects of the mysterious world around him, had become caught up in, and enthralled with the spiritualist renaissance that had become the rage of America and England over the past two decades. Knowing Darwin's reaction to this quirky interest, Wallace marshaled his allies who had corroborated these findings, and then requested that Darwin delay his pronouncement of insanity on him: "This will, I know, seem to you like some mental hallucination, but as I can assure you from personal communication with them, that Robert Chambers, Dr. Norris of Birmingham, the well-known physiologist, and C. F. Varley, the well-known electrician, who have all investigated the subject for years, agree with me both as to the facts and as to the main inferences to be drawn from them, I am in hopes that you will suspend your judgment for a time till we exhibit some corroborative symptoms of insanity."[23]

As a further indication of his self-perception as a heretic scientist, upon receiving a copy of Darwin's *Descent of Man,* Wallace wrote to him on January 27, 1871: "Many thanks for your first volume, which I have just finished reading through with the greatest pleasure and interest, and I have also to thank you for the great tenderness with which you have treated me and my heresies." Having paid his homage, however, two paragraphs later Wallace again refers to his "special heresy" in a different light: "Your chapters on Man are of intense interest, but as touching my special heresy not as yet altogether convincing, though of course I fully agree with every word and every argument which goes to prove the 'evolution' or 'development' of man out of a lower form. My only difficulties are as to whether you have accounted for *every* step of the development by ascertained laws."[24] Even late in his life, at age eighty-seven, Wallace still referred to his "heresies," as in a letter to W. T. Thiselton-Dyer on December 17, 1910, when he sent him a copy of a new edition of *Darwinism,* noting that "I therefore venture to hope that you will find something in my new book to interest you; while your frank opinion on any of my conclusions & heresies will be acceptable."[25]

How do we account for these heresies? Later we will explore numerous psychological explanations for Wallace's general heretical tendencies, but here we will focus on the cause of this particular heresy about natural selection. Despite his own assessment, noted previously, that "my opinions on the subject have been modified solely by the consideration of a series of remarkable phenomena"—by which he meant spiritualism—this was a *proximate* reason. A deeper cause (but not an ultimate cause, to which we must turn to psychology to assess) is, paradoxically, an overemphasis on his own theory of natural selection.

Heretical Explanations

Why did Alfred Wallace retreat from his own naturalistic interpretations in favor of supranatural intervention when it came to the origin and evolution of the human mind? He was, after all, co-founder of the theory of evolution by means of natural selection and the self-styled defender of Darwinism who once confessed that "I am more Darwinian than Darwin himself."[26] Since the time of his break from Darwin, scientists, scholars, and historians have speculated as to the cause, ranging from spiritual inclinations to religious predilections. But something deeper is afoot in Wallace's heresy, as paleontologist and historian of science Stephen Jay Gould hints at when he observes that "Wallace did not abandon natural selection at the human threshold. Rather, it was his peculiarly rigid view of natural selection that led him, quite consistently, to reject it for the human mind. His position never varied—natural selection is the only cause of major evolutionary change."[27] Although Gould's is a monocausal explanation (confessing that "I cannot analyze Wallace's psyche, and will not comment on his deeper motives for holding fast to the unbridgeable gap between human intellect and the behavior of mere animals"), his assessment that "the traditional account of it is not only incorrect, but precisely backwards" is essentially correct. But that traditional account must be considered seriously.

Historian of science Malcolm Jay Kottler claims that "as early as the 1870s, Anton Dohrn, in his short paper 'Englische Kritiker und Anti-kritiker uber den Darwinismus,' felt that the intense religiosity dominant among the English had ultimately been behind Wallace's divergence from Darwin. In Wallace's case this national religious conviction had been expressed through a belief in spiritualism." Kottler rejects this theory, then baldly states his own monocausal interpretation that "Wallace's spiritualist beliefs were the origin of his doubts about the ability of natural selection to account for all of man."[28] Indeed, Kottler has good reason and historical evidence for such a conclusion, since Wallace states so himself in the letter to Darwin in which he talks about the "series of remarkable phenomena" that "modified solely" his beliefs about the evolution of man and mind. Therefore, Kottler concludes, "Something happened between 1864 and 1869 to change his mind: the crucial event was Wallace's conversion to spiritualism."[29]

Kottler's position, though compelling and powerfully argued, is founded on the single statement from Wallace. But a single quote does not an ideological shift make, even if it is from the ideologist himself. When we examine the whole man, that is, analyze *and* synthesize him, we see that this argument is too narrow in attributing Wallace's shift entirely to spiritualistic beliefs. Kottler, of course, is aware of Wallace's hyper-selectionism, but dismisses it nonetheless:

> It is clear that Wallace was able to present a case against natural selection, which his contemporaries considered formidable, without reference to psychical phenomena or spiritualism. It is tempting therefore to conclude that perhaps, contrary to my thesis, Wallace had two independent grounds for his divergence—scientific and spiritual. According to this alternative viewpoint, Wallace *originally* concluded that natural selection was inadequate in the origin of man on the basis of his utilitarian analysis of various human features. Thus Wallace's simultaneous discovery of spiritualism was not the origin of his doubts about natural selection's sufficiency in man's development, but spiritualist phenomena provided further evidence, and in spiritualism Wallace found an explanation for those human features inexplicable by natural selection on purely utilitarian grounds.[30]

This is a reasonable argument, but one that Kottler finds "unlikely" because "it fails to explain what prompted Wallace's new analysis of man" between his 1864 paper and his 1869 paper.[31] According to Kottler, the belief in spiritualism came first, but since "his colleagues rejected spiritualism, Wallace attempted to convince them of the validity of his new view with a strict utilitarian analysis of man." So it was not Wallace's doubts in the power of natural selection to account for all human structures that led him to his new view on man, but "spiritualism stimulated Wallace to reconsider the utility of various human features, and the results of this new analysis (a foregone conclusion?) reinforced his earlier doubts which had been created by spiritualism."[32]

It is obvious that these spiritual phenomena played an important role in Wallace's thinking, but so too did many other factors in his thoughts and culture. In the context of his entire corpus of writings, especially his correspondence, it is obvious that Wallace's hyper-selectionism was *far* more than a mere justification for a belief in spiritualism. Wallace's hyper-selectionism was potent, sustaining, and pervasive in his entire worldview, and led him to a number of scientific controversies.[33]

Historian Joel Schwartz disagrees with Kottler and argues that "Wallace's initial departure from the Darwinian view of human evolution in 1864 cannot be attributed to his belief in spiritualism, which commenced in 1865. After 1865 . . . Wallace's religious views were responsible for widening the gulf between Darwin and himself."[34] Schwartz presents evidence that Wallace's interest in spiritualism predated by several years his papers on human evolution. In an interview with W. B. Northrop, published in the *Outlook* in 1913, Wallace clarified the confusion over the origins of his interest in spiritualism:

> When I returned from abroad [the Malay Archipelago] I had read a good deal about Spiritualism, and, like most people, believed it to be a fraud and a delusion. This was in 1862. At that time I met a Mrs. Marshall, who was a celebrated medium in London, and after attending a number of her meetings,

> and examining the whole question with an open mind and with all the scientific application I could bring to bear upon it, I came to the conclusion that Spiritualism was genuine. However, I did not allow myself to be carried away, but I waited for three years and undertook a most rigorous examination of the whole subject, and was then convinced of the evidence and genuineness of Spiritualism.[35]

Curiously, Schwartz concludes from this that "Wallace was receptive to spiritualism because it filled a religious void in his life. He belonged to no organized church and, prior to his conversion in 1865, probably considered himself an agnostic. After 1865 his attitude changed: spiritualism was no longer a phenomenon that required investigation, it was his religion."[36] Unfortunately Schwartz never defines what he means by "religion," or what he thinks Wallace might have meant by it, thereby vitiating his argument entirely. (Words such as "supernatural," "spiritual," or "higher intelligence" do not necessarily mean divine providence in the traditional religious sense. Using the *OED*'s broadest definition of *religion* as "devotion to some principle; strict fidelity or faithfulness," if anything was Wallace's religion, it was not spiritualism, but scientism. And, as we will see, Wallace's belief in spiritualism was based on a rational, scientific analysis of the phenomena, not on blind faith, typically associated with religious devotion.) Schwartz then ventures an explanation of *why* Wallace departed from Darwin, that being "his inability to bridge his scientific and moral beliefs." According to Schwartz, this "arose from his disenchantment with life in Victorian England and with the answers that the scientific community offered as an explanation of that world." Thus, in the end, Wallace's "split with Darwin also expressed his desire for a new and better world, which his evolutionary scheme could provide and the Darwinian mechanism could not."[37]

It is true that Wallace's evolutionary worldview was much broader in scope and more open to supernatural intervention than Darwin's, but the reason is far more complex that just a disillusionment with his culture. Historian John Durant broadens the horizon of Wallace's complex ideological development in noting that "this deviation can, and clearly should be interpreted without resorting to the drastic measure of splitting him in two—the 'man of science' on one side, and the 'man of nonsense' on the other."[38] Durant's is an apt description, for many of Wallace's contemporaries, and historians since, created an intellectually schizophrenic Wallace. McKinney, for example, wrote off this part of Wallace's life as a Jekyll and Hyde metamorphosis. In claiming that Wallace invented as "a colorful story" his location on Ternate where he made his breakthrough discovery of natural selection, and that he was actually on the island of Gilolo, McKinney dismisses this as that inexplicable "other" Wallace:

> Perhaps this incident also helps us to understand better the later "aberrations" of Wallace the English naturalist: he was keenly interested in spiritualism, socialism, and the campaign against vaccination; he supported land nationalization and engaged in other activities which have done much to vitiate his reputation as a scientist. The Jekyll side of his character has very deep roots beginning with his early naive acceptance in 1845 of Robert Chambers' heretical theory of evolution, a theory rejected by most other scientists. His subsequent alteration of the account of his discovery of natural selection on Gilolo is simply another illuminating incident in a fascinating career.[39]

Durant does not fall into the trap of dismissing what is not understood. His interpretation is heavily slanted toward sociological factors. For example, Durant looks at Wallace's "attitude towards the scientific community," his "unconventional background and his ambiguous position within the scientific community," his "status as a self-appointed heretic," and, Durant suggests, "many aspects of Wallace's unusual outlook may be seen as the products of an enduring tension between his philosophy of nature and his hopes for the future of man and society."[40] For Durant, Wallace's naturalism "was moulded in a radical, working-class environment in which an optimistic faith in progress through the uninterrupted operation of beneficent natural laws was wedded uncomfortably to the ideals of social equality and political reform."[41] Wallace's spiritualism, in Durant's analysis, is not interpreted as the *single* cause in the shift away from Darwinism, so much as it reaffirmed Wallace's "faith in the possibility of human progress in a number of ways. First, it provided the evolutionary process with an assured goal. . . . Secondly, spiritualism offered an explanation of how the moral and intellectual faculties required for this progress had come into being. . . . Finally spiritualism provided the incentive for altruistic social conduct which Wallace had tried with so little success to derive from Darwinism."[42] In Durant's closing statement one might replace the word "naturalistic" with "scientific," which encompasses Wallace's methodology as well as his worldview: "The truly surprising feature of Wallace's long career is not that he became involved with so many cranky or pseudo-scientific causes, but rather that, through it all, he clung to a view of man and society which was still, essentially, naturalistic."[43] Durant has made an excellent start toward an integrative understanding of Wallace's thoughts and culture, and his interpretation is closer to the realities of this very complex man than that of others. But there is no substitute for a thorough content analysis of Wallace's own words on the subject of the limits of natural selection, and what those limits imply for evolutionary theory and for Wallace's scientism. As we will see, Wallace reasoned his way to his position and, for him at least, his theory was the natural outcome of a series of discoveries and the consequent logical deductions.

The Limits of Natural Selection

Wallace's intellectual style is a classic case of what the philosopher of science Karl Popper called "conjecture and refutation,"[44] except that Wallace was much better at the former and less willing to admit the latter when it was obvious to his more conservative colleagues. Although too many scientists cling to the quaint notion of science operating on "true Baconian methods," it does not and never has. Scientists freely and routinely conjecture, throwing guesses and hypotheses out to see what sticks and waiting to see which conjectures are subsequently refuted. The personality of a scientist enters the formula in how conservative or liberal they are with pushing conjectures beyond what the data may support. Wallace was at the extreme end of the liberal side of the equation, and his "heresy" on man is a case study.

This scientistic worldview is most clearly and powerfully presented in Wallace's 1870 paper "The Limits of Natural Selection as Applied to Man," incorporating the 1869 review article of Lyell's books,[45] and an 1869 letter in the journal *Scientific Opinion,* entitled "The Origin of Moral Intuitions."[46] In the 1870 paper (published as a chapter in his book *Contributions to the Theory of Natural Selection*), Wallace warns his readers up front of what to expect: "It will, therefore, probably excite some surprise among my readers to find that I do not consider that all nature can be explained on the principles of which I am so ardent an advocate; and that I am now myself going to state objections, and to place limits, to the power of natural selection. I believe, however, that there are such limits; and that just as surely as we can trace the action of natural laws in the development of organic forms, and can clearly conceive that fuller knowledge would enable us to follow step by step the whole process of that development, so surely can we trace the action of some unknown higher law, beyond and independent of all those laws of which we have any knowledge." He also admits the heretical nature of his theory in proferring a force that is beyond those known to science: "I must confess that this theory has the disadvantage of requiring the intervention of some distinct individual intelligence. . . . It therefore implies that the great laws which govern the material universe were insufficient for this production, unless we consider . . . that the controlling action of such higher intelligences is a necessary part of those laws."[47] Aware that his is not the first of such theories, Wallace notes the law of "unconscious intelligence . . . put forth by Dr. Laycock and adopted by Mr. Murphy" has "the double disadvantage of being both unintelligible and incapable of any kind of proof." Wallace, on the other hand, even while admitting that it has yet to be proven, boldly conjectures that his theory "may or may not have a foundation, but it is an intelligible theory, and is not, in its nature, incapable of proof; and it rests on facts and arguments of an exactly similar kind to those which would enable a sufficiently powerful intellectual to deduce."[48]

Wallace begins with an extensive analysis of brain size among humans and other primates, to make the point that despite the wide range of brain size within a species, the differences between human and nonhuman primates are still significant (for which he uses, of course, the most up-to-date data on cranial capacity in the relatively new science of physical anthropology):

> We have seen that the average cranial capacity of the lowest savages is probably not less than *five-sixths* of that of the highest civilised races, while the brain of the anthropoid apes scarcely amounts to *one-third* of that of man, in both cases taking the average; or the proportions may be more clearly represented by the following figures: Anthropoid apes,10; savages, 26; civilised man, 32. But do these figures at all approximately represent the relative intellect of the three groups? Is the savage really no further removed from the philosopher, and so much removed from the ape, as these figures would indicate? In considering this question, we must not forget that the heads of savages vary in size almost as much as those of civilised Europeans. Thus, while the largest Teutonic skull in Dr. Davis's collection is 112.4 cubic inches, there is an Araucanian of 115.5, an Esquimaux of 113.1, a Marquesan of 110.6, a Negro of 105.8, and even an Australian of 104.5 cubic inches. We may, therefore, fairly compare the savage with the highest European on the one side, and with the orang, chimpanzee, or gorilla, on the other, and see whether there is any relative proportion between brain and intellect.[49]

Citing studies published by Francis Galton in *Hereditary Genius,* in which "he remarks on the enormous difference between the intellectual power and grasp of the well-trained mathematician or man of science, and the average Englishman," Wallace proceeds with his comparison of differential abilities in brains of roughly equal size: "The number of marks obtained by high wranglers is often more than thirty times as great as that of the men at the bottom of the honour list, who are still of fair mathematical ability; and it is the opinion of skilled examiners that even this does not represent the full difference of intellectual power. If, now, we descend to those savage tribes who only count to three or five, and who find it impossible to comprehend the addition of two and three without having the objects actually before them, we feel that the chasm between them and the good mathematician is so vast that a thousand to one will probably not fully express it. Yet we know that the mass of brain might be nearly the same in both, or might not differ in a greater proportion than as 5 to 6; whence we may fairly infer that the savage possesses a brain capable, if cultivated and developed, of performing work of a kind and degree far beyond what he ever requires it to do."

Wallace then considers such abstractions as law, government, science, and even such games as chess (a favorite pastime of his), noting that "savages" lack all such cognitive expressions and would, in fact, in an evolutionary environment have no need for such matters. Even more, not only would nature have no reason to select for such cognitive abilities, "any considerable development of these would, in fact, be useless or even hurtful to him, since they would to some extent interfere with the supremacy of those perceptive and animal faculties on which his very existence often depends, in the severe struggle he has to carry on against nature and his fellow-man. Yet the rudiments of all these powers and feelings undoubtedly exist in him, since one or other of them frequently manifest themselves in exceptional cases, or when some special circumstances call them forth." Therefore, he concludes, "the general, moral, and intellectual development of the savage is not less removed from that of civilised man than has been shown to be the case in the one department of mathematics; and from the fact that all the moral and intellectual faculties do occasionally manifest themselves, we may fairly conclude that they are always latent, and that the large brain of the savage man is much beyond his actual requirements in the savage state." And, Wallace continues, compared to animals whose development is entirely accounted for by natural selection, even savages contain a brain far beyond what could possibly be required in nature:

> A brain one-half larger than that of the gorilla would, according to the evidence before us, fully have sufficed for the limited mental development of the savage; and we must therefore admit that the large brain he actually possesses could never have been solely developed by any of those laws of evolution, whose essence is, that they lead to a degree of organisation exactly proportionate to the wants of each species, never beyond those wants—that no preparation can be made for the future development of the race—that one part of the body can never increase in size or complexity, except in strict co-ordination to the pressing wants of the whole. The brain of prehistoric and of savage man seems to me to prove the existence of some power distinct from that which has guided the development of the lower animals through their ever-varying forms of being.

The middle sections of this lengthy paper review additional human features Wallace believes are inexplicable through natural selection, thereby corroborating the first part of the argument: the distribution of body hair, naked skin, feet and hands, the voice box and speech, the ability to sing, artistic notions of form, color, and composition, mathematical reasoning and geometrical spatial abilities, morality and ethical systems, and especially such concepts as space and time, eternity and infinity. "How were all or any of these faculties first developed, when they could have been of no possible use to man in his early stages of barbarism? How could natural selection, or survival of the fittest in the struggle for existence, at all favour the development of mental powers so entirely removed from the material necessities of savage men, and which even now, with our comparatively high civilisation, are, in their farthest developments, in advance of the age, and appear to have relation rather to the future of the race than to its actual status?"

Wallace then sets out to argue the logical necessity for the existence of this higher intelligence. He cites Professor Tyndall's presidential address to the Physical Section of the British Association at Norwich, delivered in 1868, in which Tyndall poses the age-old mind–brain problem: "How are these physical processes connected with the facts of consciousness? The chasm between the two classes of phenomena would still remain intellectually impassable." Wallace then offers Huxley's answer to the problem: our "thoughts are the expression of molecular changes in that matter of life which is the source of our other vital phenomena." But Wallace then claims he is unable "to find any clue in Professor Huxley's writings" to bridge the gap from molecules to thought, and feels strongly that he must do so because Huxley's "expression of opinion . . . will have great weight with many persons." Wallace begins with the materialist position that molecules, even if structured into levels of "greater and greater complexity, even if carried to an infinite extent, cannot, of itself, have the slightest tendency to originate consciousness." Consciousness, he argues, is a qualitative phenomenon, not quantitative. It cannot be spontaneously generated with just more molecules, as if there were some critical mass that when reached produces consciousness. "If a material element, or a combination of a thousand material elements in a molecule, are alike unconscious, it is impossible for us to believe that the mere addition of one, two, or a thousand other material elements to form a more complex molecule, could in any way tend to produce a self-conscious existence. There is no escape from this dilemma,—either all matter is conscious, or consciousness is, or pertains to, something distinct from matter, and in the latter case its presence in material forms is a proof of the existence of conscious beings, outside of, and independent of, what we term matter."

In a lengthy footnote in this paper, Wallace demonstrates that a successive series of more and more complex inorganic compounds could conceivably lead to simple life, and then more and more complex life, and so on up the ladder. But not so for consciousness. Consciousness either exists or it does not. The problem is in how to get from zero to one—from no consciousness to even a little consciousness. "We cannot *conceive* a gradual transition from absolute unconsciousness to consciousness," Wallace argues, because "the mere rudiment of sensation or self-consciousness is infinitely removed from absolutely . . . unconscious matter." Once again calling on Darwin's bulldog, Wallace agrees "with Professor Huxley that *protoplasm* is the 'matter of life' and the cause of organisation, but we cannot . . . conceive that *protoplasm* is the primary source of sensation and consciousness, or that it can ever of itself become conscious in the same way as we may perhaps conceive that it may become *alive*."

Wallace thus concludes that we cannot prove the existence of matter, just the force it gives off. "When we touch matter we only really experience sensations of resistance, implying repulsive force." Furthermore, our own free will cannot be explained by any known natural force ("gravitation, cohesion, repulsion, heat, electricity, etc."); therefore, there must be another force that accounts for our free will. Without this supernatural force, that is, if there were only the known natural forces, "a certain amount of freedom in willing is annihilated, and it is inconceivable how or why there should have arisen any consciousness or any apparent will, in such purely automatic organisms." Thus, Wallace deduces, finishing in a poetic flourish:

> If, therefore, we have traced one force, however minute, to an origin in our own WILL, while we have no knowledge of any other primary cause of force, it does not seem an improbable conclusion that all force may be will-force; and thus, that the whole universe is not merely dependent on, but actually is, the WILL of higher intelligences or of one Supreme Intelligence. It has been often said that the true poet is a seer and in the noble verse of an American poetess we find expressed what may prove to be the highest fact of science, the noblest truth of philosophy:
>
> God of the Granite and the Rose!
> Soul of the Sparrow and the Bee!
> The mighty tide of Being flows
> Through countless channels, Lord, from Thee.
> It leaps to life in grass and flowers,
> Through every grade of being runs,
> While from Creation's radiant towers
> Its glory flames in Stars and Suns.

Nowhere in this paper does Wallace argue for or even hint at spiritualism, God, or religion. In an intriguing footnote in the second edition, however, he does broach the subject only to squelch his critics who accused him of invoking God as a first-cause argument:

> Some of my critics seem quite to have misunderstood my meaning in this part of the argument. They have accused me of unnecessarily and unphilosophically appealing to "first causes" in order to get over a difficulty—of believing that "our brains are made by God and our lungs by natural selection;" and that, in point of fact, "man is God's domestic animal." An eminent French critic, M. Claparede, makes me continually call in the aid of—*"une Force superieure,"* the capital F meaning, I imagine, that this "higher Force" is the Deity. I can

> only explain this misconception by the incapacity of the modern cultivated mind to realise the existence of any higher intelligence between itself and Deity. Angels and archangels, spirits and demons, have been so long banished from our belief as to have become actually unthinkable as actual existences, and nothing in modern philosophy takes their place. Yet the grand law of "continuity," the last outcome of modern science, which seems absolute throughout the realms of matter, force, and mind, so far as we can explore them, cannot surely fail to be true beyond the narrow sphere of our vision, and leave an infinite chasm between man and the Great Mind of the universe. Such a supposition seems to me in the highest degree improbable.

Wallace continues in the footnote to clarify his multiple descriptions of this directing power, in order to show that he only meant whatever this force was, and however it acted, it did so through nature and nature's laws:

> Now, in referring to the origin of man, and its possible determining causes, I have used the words "some other power"—"some intelligent power"—"a superior intelligence"—"a controlling intelligence," and only in reference to the origin of universal forces and laws have I spoken of the will or power of "one Supreme Intelligence." These are the only expressions I have used in alluding to the power which I believe has acted in the case of man, and they were purposely chosen to show that I reject the hypothesis of "first causes" for any and every special effect in the universe, except in the same sense that the action of man or of any other intelligent being is a first cause. In using such terms I wished to show plainly that I contemplated the possibility that the development of the essentially human portions of man's structure and intellect may have been determined by the directing influence of some higher intelligent beings, acting through natural and universal laws.

Clearly, Wallace's heresy had nothing to do with God or spiritualism in any supernatural sense, as these natural and universal laws, he believed, could be fully incorporated into the type of empirical science he had practiced all his life. It was not spiritualism, but scientism, at work in Wallace's worldview: "These speculations are usually held to be far beyond the bounds of science; but they appear to me to be more legitimate deductions from the facts of science than those which consist in reducing the whole universe . . . to matter conceived and defined so as to be philosophically inconceivable," he explained in reaching the apex of his argument in which he concluded that the ancient philosophical doctrine of the existence of a uniquely human spirituality has finally been proven through the enterprise of modern science: "Philosophy had long demonstrated our incapacity to prove the existence of matter, as usually conceived; while it admitted the demonstration to each of us of our own self-conscious, spiritual existence. Science has now worked its way up to the same result, and this agreement between them should give us some confidence in their combined teaching. The view we have now arrived at seems to me more grand and sublime, as well as far simpler, than any other. It exhibits the universe as a universe of intelligence and willpower; and by enabling us to rid ourselves of the impossibility of thinking of mind, but as connected with our old notions of matter, opens up infinite possibilities of existence, connected with infinitely varied manifestations of force, totally distinct from, yet as real as, what we term matter." In Wallace's worldview, then, there is no supernatural. There is only the natural and unexplained phenomenon yet to be incorporated into the natural. It was one of Wallace's career goals to be the scientist who brings more of the apparent supernatural into the natural.

Every Fresh Truth Is Received Unwillingly

Over the course of the next two decades Wallace's marriage of the philosophical and the scientific grew ever stronger. In his greatest synthetic book-length work, *Darwinism,* in the final chapter on "Darwinism Applied to Man," Wallace includes mathematical reasoning and artistic skills among those "outgrowths of the human intellect which have no immediate influence on the survival of individuals or of tribes, or on the success of nations in their struggles for supremacy or for existence." He then gives an evolutionary/developmental sequence of three stages that cannot be accounted for by natural selection, including (1) "the change from inorganic to organic, when the earliest vegetable cell, or the living protoplasm out of which it arose, first appeared," (2) the "introduction of sensation or consciousness," and (3) "the existence in man of a number of his most characteristic and noblest faculties" such as mathematical reasoning, aesthetic appreciation, and abstract thinking.[50]

As in the 1870 paper, Wallace never makes direct reference to spiritualism, phrenology, or any other supernatural phenomena, or to God or religion. The reason is that these were only a part of a much grander scientific worldview that was derived through logical reasoning. Wallace's various experiences in and experiments on spiritualism were simply incorporated into this grander worldview. Natural selection, and the entire Darwinian paradigm, fit snugly into his scientistic vision of man evolving into a higher state of physical, intellectual, and spiritual development:

> The Darwinian theory . . . not only does not oppose, but lends a decided support to, a belief in the spiritual nature of man. It shows us how man's body may have been developed from that of a lower animal from under the law of natural selection; but it also teaches us that we possess intellectual and moral faculties

> which could not have been so developed, but must have had another origin; and for this origin we can only find an adequate cause in the unseen universe of Spirit.[51]

The remainder of Wallace's life was devoted to fleshing out the details of a scientism that encompassed so many different issues and controversies. Despite a lifelong interest in spiritualism, Wallace called himself a "scientific skeptic," but clearly had a broader view of science, and skepticism, than most of his contemporaries who saw physics as the queen of the sciences. To Wallace this was far too limiting a view because "there are whole regions of science in which there is no such regular sequence of cause and effect and no power of prediction," he wrote in an 1885 on the "Harmony of Spiritualism and Science" in response to criticism of his work. "Even within the domain of physics we have the science of meteorology in which there is no precise sequences of effects; and when we come to the more complex phenomena of life we can rarely predict results and are continually face to face with insoluble problems; yet no one maintains that meteorology and biology are not sciences—still less that they are out of harmony with or opposed to science." If such accepted sciences as meteorology and biology lack "uniformity" and cannot predict "what will happen under all circumstances," then the study of spiritualism should be treated with no less respect.[52]

For over half a century Wallace tried to reconcile his vision of science, his conviction to natural law, his theory of evolution, and his belief in spiritualism. In the context of his personality, thoughts, and culture, this is not so ill conceived. In the final year of his life Wallace observed that "truth is born into this world only with pangs and tribulations, and every fresh truth is received unwillingly. To expect the world to receive a new truth, or even an old truth, without challenging it, is to look for one of those miracles which do not occur."[53] Whether he was right or wrong, a grand synthesizer and thematic thinker like Wallace could not resist the challenge.

7

A Scientist Among the Spiritualists

Historians have a most unusual task among seekers of truth. In order to think ourselves into the minds of our predecessors to understand how they thought, we must *forget* what we know because we might unfairly judge them by our standards—they did not know what we know. On the other hand, in order to glean lessons from the past to understand which ideas were dead ends and which led to the modern worldview, we must *remember* what we know and compare their ideas with ours in order to make history meaningful and of service to us—the application of Darwin's dictum and Wallace's wisdom to the study of history. It is a tricky balance to maintain, especially when traveling along the borderlands of science where what we might today call pseudoscience a different age would call science. A case study in exploring the boundary issues in the nature of science, pseudoscience, and nonscience can be found in the investigations by Alfred Russel Wallace of a number of different subjects, most intriguing his intense interest in all matters spiritual. Wallace merits our attention in this regard not only because he was honest and passionate in his researches (lots of people are, but that does not make them good investigators), but because he was considered one of the greatest scientists of his age. How does an eminent scientist, through a series of investigations (as opposed to compartmentalized religious or spiritual beliefs), come to accept suprascientific or supernatural ideas?

The New Science of the Mind: Spiritualism and Phrenology

The rebirth of interest in spiritualism and phrenology in the mid- to late nineteenth century, by both the general public and the scientific community, added to Wallace's heretical polemics on the shortcomings of natural selection when applied to cognitive domains. Commingling his teleological thinking about the directional nature of evolution with the spiritual phenomena he was observing, Wallace understood the ultimate purpose of nature to be the development of the spirit—the final end of an immeasurably long evolutionary process.

The arrival of phrenology on the European continent preceded that of spiritualism by two decades. It was introduced in the 1790s by the Viennese physician Franz Joseph Gall, and picked up momentum in the 1820s. Phrenology is based on a few basic tenets: the mind is an aggregate of mental processes localized in specific brain areas (for Gall, it was a composite of thirty-seven independent faculties, propensities, and sentiments, each with its own brain area); the larger the localized area the more powerful that specific mental process. Since the skull in infant development is plastic and malleable it ossifies over the brain, forming external "bumps" or "valleys," indicating an individual's internal mental faculties. Gall's first protégé, Johann Gaspar Spurzheim, added the notion that certain personality characteristics and moral propensities, such as an evil disposition, were a result of an imbalance between the faculties. Where Gall sought to build a science of the mind through phrenology, Spurzheim hoped to expand the field's horizon beyond the individual and into the realm of social and political action. This approach attracted a Scottish lawyer named George Combe to the phrenological movement, which would, in time, result in a seminal work read by Wallace.

The social history of this movement is convoluted. Before 1820 phrenology was widely criticized by both the general public and the intelligentsia. But it experienced a boom in the decades from 1820 to 1840, supported at first by radicals in opposition to any form of established authority. During these decades science was being redefined by experimentalists, who portrayed the process as one that belonged in the laboratory.[1] Science was not just an organized body of knowledge, it was a set of methods designed to answer questions about the world. This new approach to science in general and phrenology in particular attracted advocates within the bourgeois class as its proponents worked to emphasize the empirical and quantifiable quality of its claims (through a wide range of mechanical devices placed over the head of the client that gave an air of "hard science" at work). Phrenology was also grafted onto the medical profession, adding additional credence to its claims and bolstering its credibility within the general public.[2]

From 1840 on, however, phrenology declined in credibility within the scientific community, though it remained popular in the working classes, especially among the most radical, which well describes Wallace. In 1844, in fact, the working-class naturalist first read about phrenology in a book entitled

Constitution of Man Considered in Relation to External Objects, authored by George Combe, an ideological disciple of Spurzheim. Combe was not only a skilled lawyer, he threw himself into the phrenological movement full force and founded the Edinburgh Phrenological Society. Combe turned Spurzheim's phrenology into a natural philosophy of the mind, attempting to explain human emotion and suffering in the context of natural laws governing thought.

Wallace, always the social and political speculator in search of grand underlying causes, immediately took to Combe's philosophy. Linking phrenology to mesmerism ("phreno-mesmerism" it was called), Wallace followed the new trend of defining science by its experimental protocols and began his lifelong quest for a scientific basis of such phenomena in the mid-1840s with a number of what he considered to be rigorous tests. In 1844, two decades before his public conversion to spiritualism, after hearing a lecture on mesmerism by Spencer Hall Wallace set out to try it on a suggestible youngster:

> Giving him a glass of water and telling him it was wine or brandy, he would drink it, and soon show all the signs of intoxication, while if I told him his shirt was on fire he would instantly strip himself naked to get it off. I also found that he had community of sensation with myself when in the trance. If I held his hand he tasted whatever I put in my mouth, and the same thing occurred if one or two persons intervened between him and myself; and if another person put substances at random into my mouth, or pinched or pricked me in various parts of the body, however secretly, he instantly felt the same sensation, and would describe it, and put his hand to the spot where he felt the pain.
>
> In like manner any sense could be temporarily paralyzed so that a light could be flashed on his eyes or a pistol fired behind his head without his showing the slightest sign of having seen or heard anything. More curious still was the taking away the memory so completely that he could not tell his own name, and would adopt any name that was suggested to him, and perhaps remark how stupid he was to have forgotten it; and this might be repeated several times with different names, all of which he would implicitly accept.[3]

Such tests were powerful and convincing to Wallace that something beyond the confines of what traditional science could explain was at work, and over time he became more and more commited to finding out just what that something was. On another occasion, with a phrenological skull at his disposal, he put "my patient in the trance, and standing close to him, with the bust on my table behind him, I touched successively several of the organs, the position of which it was easy to determine. After a few seconds he would change his attitude and the expression of his face in correspondence with the organ excited. In most cases the effect was unmistakable, and superior to that which the most finished actor could give to a character exhibiting the same passion or emotion."[4] This was no act, however, nor was it a fantasy role-playing game between mesmerist and subject. Noting that "painless surgical operations during the mesmeric trance" were being routinely practiced, it appeared to Wallace that some mysterious force was being transduced between individuals, and experimental science was the best method to determine the nature of that force.[5] He recalled another experiment in which he endeavored to eliminate suggestion as an intervening variable:

> One day I intended to touch a particular organ, and the effect on the patient was quite different from what I expected, and looking at the bust while my finger was still on the boy's head, I found that I was not touching the part I supposed, but an adjacent part, and that the effect exactly corresponded to the organ touched and not to the organ I *thought* I had touched, completely disproving the theory of suggestion. I then tried several experiments by looking away from the boy's head while I put my finger on it at random, when I always found that the effect produced corresponded to that indicated by the bust. I thus established, to my own satisfaction, the fact that a real effect was produced on the actions and speech of a mesmeric patient by the operator touching various parts of the head; that the effect corresponded with the natural expression of the emotion due to the phrenological organ situated at that part—as combativeness, acquisitiveness, fear, veneration, wonder, tune, and many others; and that it was in no way caused by the will or suggestion of the operator.[6]

Wallace's belief in the basic premises of phrenology never attenuated throughout his life, and in old age he still proudly exhibited a phrenological cranium reading done on himself by the same individuals (E. T. Hicks and J. Q. Rumball) who measured Herbert Spencer's head. He even invoked a phrenological argument in his autobiography to explain why he could not evoke memories of his parents and siblings, but could recall the environment of his upbringing so many decades before:

> I cannot find any clear explanation of these facts in modern psychology, whereas they all became intelligible from the phrenological point of view. The shape of my head shows that I have *form* and *individuality* but moderately developed, while *locality, ideality, colour,* and *comparison* are decidedly stronger. Deficiency in the first two caused me to take little notice of the characteristic form and features of the separate individualities which were most familiar to me, and from that very cause attracted less close attention; while the greater activity of the latter group gave interest and attractiveness to the everchanging combinations in outdoor scenery.[7]

Spiritualism was related to but had a different historical trajectory than phrenology and mesmerism. A revivification of spiritualism began in 1848 in New York when two youthful and spirited sisters, Margaret and Kate Fox, claimed to communicate with spirits of the dead through a series of rapping and popping noises. Although decades later they confessed to faking the whole affair through cracking their toe knuckles (Thomas Huxley took pride in being able to replicate the effect), the girls' flapdoodle triggered a social movement that quickly spread across the Atlantic to England and the continent. Psychology as a science was in its infancy, and this "dynamic psychology" in the form of spiritualism peaked around 1850, when the words "telepathy" and "medium" were first used in print. Mediums holding séances soon spread rapidly through centers of population, with diverse claims being proffered, such as the ability to contact the dead, read the past and predict the future, and produce such psychic phenomena as rapping noises and appearances of ghosts.[8] Victorian cartoonists lampooned believers, but, despite outspoken skepticism by some scientists, a gullible public was quickly swept up by the enthusiasm and excitement that surrounded such mystical phenomena, especially when endorsements came from a few respected members of the scientific community, some from the very highest levels.

Although we dislike the notion that truths, especially scientific truths, might be strongly influenced by who is doing the truth telling as much as by the quality of the evidence, the fact is that who you are and who you know sometimes matters as much as the consistency of your arguments or the quality of your evidence. Integrity, trust, reputation, fame, society memberships, and institution affiliations all converge to construct the validity of a claimant, and thus his or her claim.[9] As it was with phrenology, for a time a considerable segment of the more staid and conservative scientific community also took an interest in spiritualism. In 1882 the Society for Psychical Research was founded in London, with a membership epicenter at Cambridge University and a roster that included such renowned scientists as the physicists Sir William Crookes, Lord Rayleigh, and Sir Oliver Lodge, the noted eugenicist (and Darwin's cousin) Francis Galton, the mathematician Augustus De Morgan, the naturalist St. George Mivart, the physiologist Charles Richet, and the psychologists Frederic Myers and G. T. Fechner. Their goal was not to challenge or debunk these claims; it was to discover a scientific, naturalistic explanation for the phenomena they assumed had a basis in reality.[10] Not surprisingly, they often found what they were looking for, and thus claims became truth when sanctioned by such noted truth seekers.

As a member of the Society for Psychical Research, Wallace found himself at the epicenter of the spiritualism movement. Like the other members, once he was convinced of the validity of the claims, he sought further verification and a causal explanation. More important, he pursued a deeper natural cause that could be explained by science, even if it meant modifying the boundaries of science beyond what his more conservative colleagues might have accepted.

This quest culminated in 1866 when Wallace published a fifty-seven-page monograph, appropriately entitled: *The Scientific Aspects of the Supernatural: Indicating the Desirableness of an Experimental Enquiry by Men of Science into the Alleged Powers of Clairvoyants and Mediums.* In a revealing passage that shows the power of status in science and society, Wallace calls his readers' attention not to the superior evidence for the claim, but the superior supporters of it:

> A little enquiry into the literature of the subject, which is already very extensive, reveals the startling fact, that this revival of so-called supernaturalism is not confined to the ignorant or superstitious, or to the lower classes of society. On the contrary, it is rather among the middle and upper classes that the larger proportion of its adherents are to be found; and among those who have declared themselves convinced of the reality of facts such as have been always classed as miracles, are numbers of literary, scientific, and professional men, who always have borne and still continue to bear high characters, are above the imputation either of falsehood or trickery, and have never manifested indications of insanity.[11]

Putting Spiritualism to the Test: Ghosts, Spirits, and Mediums

In the 1860s, Wallace's interest in spiritualism that would reinforce his heretical hyper-selectionism that triggered the break from Darwin was actually a revitalization of a curiosity that had begun two decades earlier with his personal experimentation with phrenology and mesmerism. In July 1865, Wallace began to link spiritualism to phreno-mesmerism after he attended a séance at the home of a friend. The table moved and vibrated and rapping noises were heard. Could this be a manifestation of some force beyond those currently understood by science? To find out, in November 1866, Wallace began to experiment at home with a medium named Miss Nichol. Based on his earlier experiments from the 1840s in which he ruled out suggestion as a causal variable, Wallace claims (rather naively, it would seem) that he entered the inquiry "utterly inbiased [*sic*] by hopes or fears, because I knew that my belief could not affect the reality." The levitation of the corpulent Miss Nichol, along with the production of fresh flowers in the dead of winter, convinced Wallace that further investigation was necessary. Something exceptionally unusual was at work and now that Wallace was at the height of his scientific prowess he was going to find out what it was by doing what any naturalist and theorist would do: make copious observations and deductions.

Unlike so many others driven by religious motivations to confirm the existence of a spiritual world, Wallace was in search of a *natural* explanation for the supernatural. Since he was not a religious man in any traditional manner, and did not believe in a personal God, his exploration of spiritualism and the supernatural had a scientistic purpose integral to his unique worldview. *The Scientific Aspects of the Supernatural,* in fact, is one long argument that these phenomena are "not really miraculous in the sense of implying any alteration of the laws of nature. In that sense I would repudiate miracles as entirely as the most thorough sceptic."[12] Rigorous scientist that he was, Wallace began his analysis of miracles with the classic skeptic David Hume, noting that "Hume was of opinion that no amount of human testimony could prove a miracle" because "a miracle is generally defined to be a violation or suspension of a law of nature" and "the laws of nature are the most complete expression of the accumulated experiences of the human race."[13] If these spiritual events are not miracles, then what are they? According to Wallace, "The apparent miracle may be due to some yet undiscovered law of nature."[14] Just because we cannot understand or explain these occurrences does not mean they lack causes, or that the causes are miraculous. It is just that we have yet to discover the causes: "A century ago, a telegram from 3000 miles' distance, or a photograph taken in five seconds, would not have been believed possible, and would not have been credited on testimony, except by the ignorant and superstitious who believed in miracles."[15] Thus, Wallace concludes, "it is possible that intelligent beings may exist, capable of acting on matter, though they themselves are uncognisable directly by our senses."[16]

(Wallace, in fact, later devoted an entire paper to Hume, entitled "An Answer to the Arguments of Hume, Lecky, and others, Against Miracles," in which he argued that "thousands of intelligent men now living know from personal observation that some of the strange phenomena which have been pronounced absurd and impossible by scientific men, are nevertheless true. It is no answer to these, and no explanation of the facts, to tell them that such beliefs only occur when men are destitute of the critical spirit, and when the notion of uniform law is yet unborn; that in certain states of society illusions of this kind inevitably appear, that they are only the normal expression of certain stages of knowledge and of intellectual powers, and that they clearly prove the survival of savage modes of thought in the midst of modern civilisation."[17])

Because Wallace did not connect these phenomena to any religious doctrines or church dogmas, these intelligent beings, whatever their constitution, are not in any way connected with divine providence or, in Wallace's words, "acts of the Deity." In fact, Wallace argues in an interesting twist on the argument for God's existence from miracles, "the nature of these acts is often such, that no cultivated mind can for a moment impute them to an infinite and supreme being. Few if any reputed miracles are at all worthy of a God."[18]

Natural phenomena are to be explained by natural causes. Wallace's worldview was thoroughly scientistic. He was not the schizophrenic man of sense and nonsense, science and nonscience. If there were spiritualistic occurrences to be explained, the scientist could only do so through scientific means, "by direct observation and experiment," Wallace proclaimed in the empiricist mode of following the data wherever they may lead:

> It would appear then, if my argument has any weight, that there is nothing self-contradictory, nothing absolutely inconceivable, in the idea of intelligences uncognisable directly by our senses, and yet capable of acting more or less powerfully on matter. Let direct proof be forthcoming, and there seems no reason why the most sceptical philosopher should refuse to accept it. It would be simply a matter to be investigated and tested like any other question of science. The evidence would have to be collected and examined. The results of the enquiries of different observers would have to be compared.[19]

For the next forty years that is precisely what Wallace did—involving himself with the systematic examination of spiritualism, with such experiments as this one, described in a letter to a friend:

> Our seance came off last evening, and was a tolerable success. The medium is a very pretty little lively girl, the place where she sits a bare empty cupboard formed by a frame and doors to close up a recess by the side of a fireplace in a small basement breakfast-room. We examined it, and it is absolutely impossible to conceal a scrap of paper in it. Miss Cooke is locked in this cupboard, above the door of which is a square opening about 15 inches each way, the only thing she takes with her being a long piece of tape and a chair to sit on. After a few minutes Katie's whispering voice was heard, and a little while after we were asked to open the door and seal up the medium. We found her hands tied together with the tape passed three times around each wrist and tightly knotted, the hands tied close together, the tape then passing behind and well knotted to the chair-back. We sealed all the knots with a private seal of my friend's, and again locked the door. A portable gas-lamp was on a table the whole evening, shaded by a screen so as to cast a shadow on the square opening above the door of the cupboard till permission was given to illuminate it. Every object and person in the room were always distinctly visible. A face then appeared at the opening, but dark and indistinct. After a time another face quite distinct with a white turban-like headdress—this was a handsome face with a considerable general likeness to that of the medium, but paler, larger, fuller, and older—decidedly a different face, although like. We were then ordered to release the medium. I opened the door, and found her bent forward with her head in her lap, and apparently in a deep sleep or trance—from which a touch and a few words awoke her. We then examined the tape and knots—all was as we left it and every seal perfect.[20]

Believing Is Seeing

Wallace's active involvement with the spiritualist movement postdated his theory of natural selection (1858), but predated his 1870 paper "The Limits of Natural Selection as applied to Man." (And, as noted, his commitment to phrenology and mesmerism predates all of his scientific discoveries and theories.) This sequence is important in understanding how a naturalist (in the methodological sense as well as the biological) comes to believe in the supernatural. Wallace approached the study of spiritualism with his usual analytical enthusiasm. His first séance was 1865. By 1866 he had already published the monograph *The Scientific Aspects of the Supernatural,* and in 1875 he codified his thoughts on the subject in an entire book on *Miracles and Modern Spiritualism.*

Reactions among scientists to Wallace's initial public support for spiritualism in the 1866 pamphlet were mixed. Robert Chambers, author of *Vestiges of the Natural History of Creation,* received it with great "gratification" and wrote back to Wallace: "I have for many years known that these phenomena are real," and "My idea is that the term 'supernatural' is a gross mistake. We have only to enlarge our conceptions of the natural, and all will be right."[21] Inspired by Wallace, in fact, Chambers revised a later edition of the *Vestiges* to include spiritual phenomena. By contrast, Charles Darwin remained a skeptic. His cousin and brother-in-law, Hensleigh Wedgwood (Darwin married into the famous Wedgwood pottery family), recounted to Darwin his experiences of witnessing tables mysteriously rising off the floor (known as "table tipping") and an accordion that apparently played by itself. Having lost two children to disease in their youth, Darwin was not amused by those who preyed on the grieving, calling them "wicked and scandalous." After finally attending a séance with his twenty-nine-year-old son George and Hensleigh Wedgwood, Darwin recalled: "We had grand fun, one afternoon, for George hired a medium, who made the chairs, a flute, a bell, and candlestick, and fiery points jump about in my brother's dining-room, in a manner that astounded every one, and took away all their breaths. It was in the dark, but George and Hensleigh Wedgwood held the medium's hands and feet on both sides all the time. I found it so hot and tiring that I went away before all these astounding miracles, or jugglery, took place."[22] Despite his assessment that "how the man could possibly do what was done passes my understanding," a few days later he wrote in a letter: "I am pleased to think that I declared to all my family, the day before yesterday that the more I thought of all that I had heard happened at Queen Anne St., the more convinced I was it was all imposture."[23] Darwin, a scientific skeptic to the end, concluded: "The Lord have mercy on us all, if we have to believe in such rubbish."[24]

Darwin's tireless defender, Thomas Henry Huxley, who confessed that he "could not get up any interest in the subject," was finally persuaded by George Darwin to attend another Hensleigh Wedgwood–sponsored séance, this one "smaller and more carefully organized." Huxley went under cover as "Mr. Henry," noting that despite the fact that a guitar played itself and bottles moved about a table on their own, he remained skeptical. In fact, Huxley turned the séance experience into a lesson in skeptical thinking for the younger Darwin, who subsequently wrote Huxley after he told his father about the experience: "My father was delighted at my report and so am I beyond measure. It has given me a lesson with respect to the worthlessness of evidence which I shall always remember—and besides will make me very diffident in trusting myself."[25] Huxley well knew that scientists are not trained in the art of detecting conscious fraud. "In these investigations," he explained, "the qualities of the detective are far more useful than those of the philosopher. . . . A man may be an excellent naturalist or chemist; and yet make a very poor detective." After years of observing the spiritualist movement with wry cynicism, Huxley made this amusing observation on spiritual manifestations: "Better live a crossing-sweeper than die and be made to talk twaddle by a 'medium' hired at a guinea a seance."[26]

Such remarks stung, but Wallace struck back in a biting article for *The Year-book of Spiritualism for 1871,* entitled "On the Attitude of Men of Science Towards the Investigators of Spiritualism." In it he complained bitterly that "the men of science are at least consistent in treating the phenomena of Spiritualism with contempt and derision. They have always done so with new and important discoveries; and, in every case in which the evidence has been even a tenth part of that now accumulated in favor of the phenomena of Spiritualism, they have always been in the wrong." Noting that "the time-honored names of Galileo, Harvey, and Jenner, are associated with the record of blind opposition to new and important truths," and enlisting the names of prominent scientists on his side of the spiritualism ledger, Wallace opined that "the day will assuredly come when this will be quoted as the most striking instance on record of blind prejudice and unreasoning credulity."[27]

Characteristically, Wallace had a rational explanation for his colleagues' skepticism. In a paper communicated to the 1893 Psychical Congress in Chicago, entitled "Notes on the Growth of Opinion as to Obscure Psychical Phenomena during the Last Fifty Years," Wallace explained that his "first great lesson in inquiry into these obscure fields of knowledge, [was] never to accept the disbelief of great men, or their accusations of imposture or of imbecility, as of any weight when opposed to the repeated observation of facts by other men admittedly sane and honest."[28] Even when eminent scientists endeavor to impose rigorous controls over such phenomena, Wallace argued in a very revealing statement given in an 1898 interview that the attitude of the experimentalist may determine the outcome of the experiment: "Usually those who at the very beginning demand tests are the wrong kind of people to get any satisfactory result. Those who experiment in the proper spirit don't fail. Professor William Crookes, F.R.S., experimented in his laboratory for years with the greatest success. Professor Oliver Lodge, Professor W. F. Barrett, of Dublin, and others have been more or less successful."[29]

Wallace was not alone in his fascination with spiritualism and the supernatural, and he accumulated what he believed to be much empirical evidence in support of these claims. One of the stranger incidents in his long study of the subject came through his sister, Frances Sims (née Wallace), which I discovered in the archives at Oxford University when I came across a copy of *The Scientific Aspects of the Supernatural.* On the frontispiece, in the hand of Frances, is written the following (reprinted in Figure 7-1):

> This book was written by my Brother *Alfred* and with 24 others was laying on my table. They had been there 4 days and I had not had time to give them away. One morning I had been sitting at my Table writing and left the room

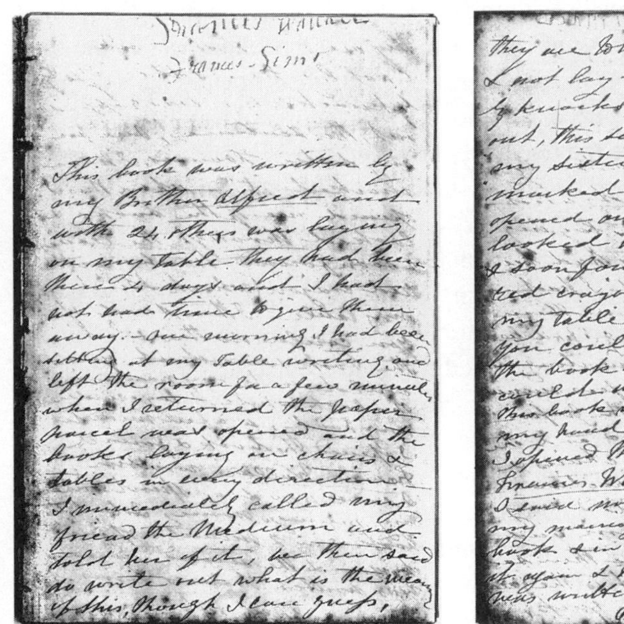

Frances Sims

This book was written by
my Brother Alfred and
with 24 others was laying
on my table they had been
there 4 days and I had
not had time to give them
away. One morning I had been
sitting at my Table writing and
left the room for a few minutes,
when I returned the paper
parcel was opened and the
books laying on chairs &
tables in every direction.
I immediately called my
friend the Medium and
told her of it, we then said
do write out what is the mean
of this, though I can guess,

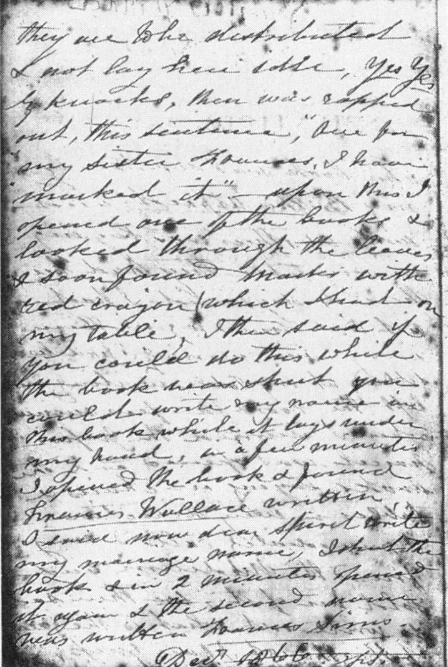

they are to be distributed
& not lay here idle, yes yes
by knocks, then was rapped
out, this sentence, "One for
my sister Frances, I have
marked it" — upon this I
opened one of the books &
looked through the leaves,
& soon found marks with
red crayon (which I had on
my table, I then said if
you could do this while
the book was shut you
could write my name in
this book while it lays under
my hand, in a few minutes
I opened the book & found
Frances Wallace written
I said now dear Spirit write
my marriage name, I shut the
book & in 2 minutes opened
it again & the second name
was written Frances Sims.
Decr 1866

Figure 7-1 On the frontispiece of a copy of Alfred Wallace's *Scientific Aspects of the Supernatural,* owned by his sister Frances, appears this inscription in her hand describing an apparently supernatural event that convinced her that her brother was right about the reality of spiritualism. (Courtesy of Hope Entomological Collections of Oxford University Museum)

for a few minutes when I returned the paper parcel was opened and the books laying on chairs & tables in every direction. I immediately called my friend the medium and told her of it, she then said to write out what is the meaning of this, though I can guess, they are to be distributed & not lay here idle. *Yes Yes* by knocks, then was rapped out, this sentence, "One for my Sister Frances. I have marked it"—upon this I opened one of the books & looked through the leaves & soon found marks with red crayon (which I had on my table). I then said if you could do this while the book was shut you could write my name in this book while it lays under my hand, in a few minutes I opened the book & found *Frances Wallace* written. I said now dear Spirit write my marriage name, I shut the book & in 2 minutes opened it again & the second name was written *Frances Sims.*

Dec. 1866. FS

What are we to make of this bizarre occurrence? One explanation is that Frances (or someone else) was attempting to perpetrate a hoax and wrote her name in the book, perhaps to enhance the publicity for the publication or add to the credibility of her brother's reputation. There is a little similarity between the names at the top of the page and those in the text, though not enough to conclude that this explanation is correct. It is possible that her "medium" friend concocted a trick to reinforce the belief of Frances and/or Alfred, though the text of the passage is unclear where the medium was at the time. Perhaps the medium was in the other room and when Frances "left the room for a few minutes" the deed was done. How then could the medium have written the names while the book was under Frances's hand? This, of course, we will never know, but there are standard techniques used by magicians, then and now, to do something very similar (called "slate writing"), so it seems reasonable to assume that Wallace and his sister were duped.[30]

Wallace's American lecture tour in 1886 proved to be yet another testing ground for both the veracity and naturalism of spiritual phenomena. His unpublished journal from the trip is filled with entries that, in a most nonchalant way, mix botanical collecting, zoological exploring, public lecturing, and spiritual séances all in the same day. The entry for Saturday, December 18, 1886, for example, is reproduced in Figure 7-2. Wallace has drawn the room, indicating where he ("AW") sat, the cabinet where the medium was encased, and the sliding doors "privately marked with pencil & found untouched after." This is vintage Wallace—the man of science conducting what he considers to be a rigid experiment, complete with controls for fraud and witnesses for corroboration:

To library & Museum—called on Williams & McIntyre—Evening to Seance at Mrs. Ross. Remarkable Exd room carefully & rooms below . . . [diagram] Below the cabinet is the heating furnace & on the ceiling air pipes hot & cold

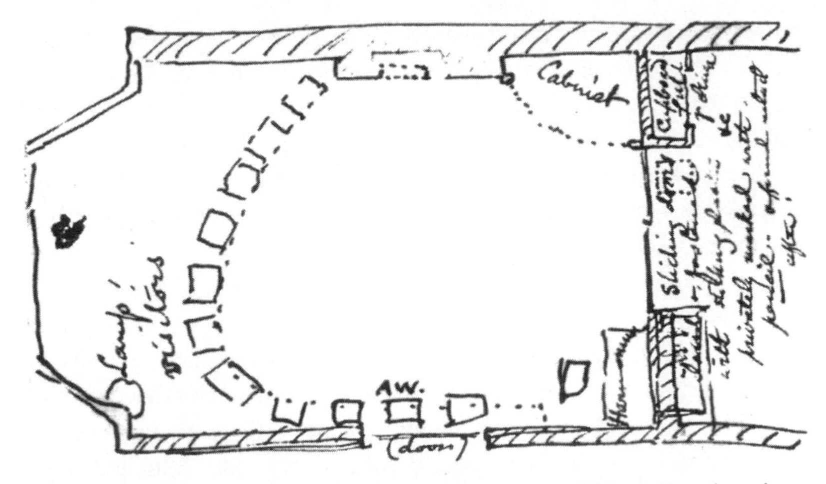

Figure 7-2 Diagram in Wallace's hand from his unpublished American journal depicting the séance he participated in along with the journal description of the room layout and additional participants. Wallace has indicated where he ("AW") sat, the cabinet where the medium was encased, and the sliding doors "privately marked with pencil & found untouched after." (Courtesy of the Linnean Society of London)

clothed with cobwebs. Room carpeted up to walls, entire . . . walls solid. Cabinet a cloth curtain with cloth top 2 ft. below ceiling. Door to next room secured but gas-lights burning in it afforded perfect security. Ten visitors—Mr. & Mrs. Ross . . . Most striking phenomena.

1. A female figure in white came out with Mrs. Ross in black, and at the same time a male figure—to mid. of room.
2. Three female figures appeared together all in white of different heights—came 2 or 3 feet in front of cabinet.
3. A male figure came out recognised by a gentleman as his son.
4. A tall indian in white moccasins came out danced and spoke, shook hands with me & others—a large strong rough hand.
5. A female figure with a baby—to entrance of cabinet. Went up and felt baby's face, nose, and hair, & kissed it—a genuine soft skinned living baby as ever I felt. Other gentlemen & ladies agreed.[31]

Spiritualism on Trial

Data never just speak for themselves. They must always be filtered through the clouded lenses of human perception, and as Wallace's commitment to the reality of spiritualism grew, so too did his willingness to accept even obvious fakes. A century before James "The Amazing" Randi exposed psychics as fakes on national television, and a half-century before Harry Houdini disguised himself to sneak into séances to uncover the tricks of the flimflam artists who pretended to talk to the dead, the nineteenth-century British zoologist and physician Edwin Ray Lankester donned the mantle of scientific skepticism when he busted the spiritualist medium Henry Slade and took him to court for defrauding a naive and unsuspecting public. Spiritualism was no longer just an academic debate. It would now come down to the testimony of expert witnesses on both sides, Wallace for the defense and, covertly, Darwin for the prosecution.

In April 1876 the American medium Henry Slade was brought to England by the founder of the Theosophical Society, Madame Blavatsky, and he soon had a sizable following of believers desperate to contact their lost loved ones through the medium's mysterious "slate writing," in which the spirits of the dearly departed would apparently communicate with those in the room by chalking their thoughts on slates. Since Lankester was once a student of Thomas Huxley he well knew his professor's skepticism of all matters spiritual, so when he heard about Slade he was inspired to confront him and, ideally, catch him in the act of conjuring. With his colleague Dr. Horatio Donkin, Lankester attended one of Slade's séances, waiting patiently and watching intensely for an opportunity to strike. That moment came when Lankester and Donkin

Figure 7-3 The British zoologist Edwin Ray Lankester testifies on behalf of the prosecution in the celebrated 1876 trial of Henry Slade, the celebrated American medium who claimed the dead communicated with the living by chalking their thoughts on slates he manipulated in a darkened room as part of a séance. Lankester holds the slates in his hand while onlookers stretch to see the seemingly miraculous writing on them. (Courtesy of Richard Milner)

noticed Slade holding the slate over the edge of the table, with his fingers on top of the slate and his thumbs underneath. It appeared to Lankester and Donkin that Slade was moving his thumbs (a thumb-tip writing device was and is a standard piece of equipment for magicians) and making noises to cover the sound of the chalk on the slate. Lankester leaned over and grabbed the slate out of Slade's fingers, exposing to the group present that an answer was given to a question yet to be asked! "You have already written on it," Lankester proclaimed. "I have watched you. You are a scoundrel and an imposter." Donkin called Slade "a damned liar," and the two scientists promptly left the room to warn awaiting customers and report Slade to the police. Slade, along with his partner Geoffrey Simmonds, were charged under the British Vagrancy Act, passed to protect people against the fraud of "unlawfully using subtle craft, means and devices to deceive and impose upon certain of Her Majesty's subjects."[32]

The stage was now set for a courtroom showdown, a direct confrontation that, because of the adversarial nature of the law, would also be decisive (at least in this particular case). A courtroom on Bow Street in London was daily "most inconveniently crowded" with dozens of reporters and curious onlookers, including, according to *The Times* of London, an unusually large number of women. In fact, as *The Times* noted on the fifth day of the trial (October 28), "one London paper sent no less than four special reporters on each occasion, a circumstance which is quite unprecedented as far as this Court is concerned." Slade's attorney, a Mr. Munton, argued that his client had been set up. Lankester and Donkin testified that they had "no private interest or feeling in the matter" and that they were merely "prosecuting in the public interest." Slade's attorney countered by noting Lankester's hostile response to a paper recently read at the British Association for the Advancement of Science (BAAS) over which his star witness, Alfred Russel Wallace, presided.

The week before Lankester confronted Slade in the séance, Wallace, at that time the acting president of the Biological Section of the BAAS, sponsored a paper under the Anthropology Section written by a physicist student of Michael Faraday named William Barrett, now Professor of Physics at the Royal College of Science in Dublin, entitled "On Abnormal Conditions of Mind" and involving "mesmerism, induced somnambulism, and telepathy."[33] *The Times* reported that Barrett and a friend "mesmerized a girl and they found that no sensation was experienced unless accompanied by pressure over the eyebrows of the subject. When the pressure of the subject was removed the girl fell back in the chair utterly unconscious and had lost all control over voluntary muscles. On reapplying the pressure she answered readily, but her acts and expression were capable of wonderful diversity, by merely altering the place on the head where the pressure was applied." With phrenology in mind, Barrett wondered "whether a careful and systematic study of them might throw some light on the localization of the functions." Another subject, a young girl "who had never been out of a remote Irish village," was allegedly able to read Barrett's thoughts accurately and gave a detailed description of a street scene and "told him the English time on a clock in London." Barrett then opined that "at one time it was customary for scientific men to deny the truth of the papers concerning mesmerism, and to turn the whole phenomena to ridicule, but now this prejudice had disappeared" under the weight of such empirical evidence.[34]

The paper was read at the BAAS meeting in Glasgow on September 12, 1876, four days after which Lankester lambasted both Slade and Wallace in *The Times:*

> I trust that you will find space for a brief account of an interview with "Dr." Slade, from which I have just returned. In consequence of the more than questionable action of Mr. Alfred Wallace, the discussions of the British Association have been degraded by the introduction of the subject of spiritualism, and the public has learnt—perhaps it is time they should—that "men of science" are not exempt as a body from the astounding credulity which prevails in this country and in America. It is, therefore, incumbent upon those who consider such credulity deplorable to do all in their power to arrest its development.[35]

(The qualified "Dr." title by Lankester was in reference to Slade's own unqualified use of it, which the prosecuting attorney George Lewis explained in court "had no further meaning than 'professor' had when it was assumed by a conjurer."[36])

Donkin fired a salvo in support of Lankester in a letter to *The Times* of the same day, recounting what he witnessed:

> A spirit message was soon written, the slate being held in opposition to the under surface of the table, the thumb alone of the medium's right hand being on the table. During the alleged writing a scratching was plainly heard, and at the same time a slight to-and-fro movement of the arm with some contraction of flexor tendons on the wrist was visible. The writing was imperfect and distorted, requiring the interpretation of an expert, and appeared on the surface of the slate, which faced downwards. The result was in accordance with the theory of the agency of a minute piece of slate-pencil probably held under the nail of the middle finger.[37]

Wallace was not alone in his sponsorship of Barrett's BAAS paper. The president of the Anthropological Institute, Lane Fox, gave Barrett his "entire support," noting "the existence of unexplained psychic phenomena that are occurring daily in private families" and that anthropology "will only do itself honour by grappling with the errors of our time" because "if gentlemen professing to be anthropologists are afflicted with a superstitious terror of the subject and are content to limit their investigations exclusively to old mounds, old scratches, or the relative position of people's toes," then science will suffer. Recalling the seventeenth-century popular belief in witchcraft that led to the drowning of women, and how such beliefs now reside only "among the lower orders," Fox explained that "it is usual to record the circumstance as a survival of ancient superstition, and a whole district has been condemned as an abyss of ignorance through the existence of one such case; but among the upper classes of society the allied belief in spiritual manifestations through the agency of media is now as widely received as witchcraft was in the seventeenth century, and is continuing to spread rapidly." Would spiritualism go the way of witchcraft? Only science could answer the question, wrote Fox: "One of the main functions of the science of anthropology consists in interpreting the past by the present, the unknown by the known. It is rarely that any popular belief is so entirely devoid of truth as to be destitute of some few grains of fact upon which the belief is founded, and the work of anthropology consists in sifting these facts from the large volume of credulity and some imposture with which they are associated." Fox then concluded: "Our study is man, and we must take him as we find him, with all his credulity and imposture, and, I may add, his unwarrantable assumption of knowledge respecting nature."[38]

Three days later, on September 19, Wallace jumped into debate, now being played out publicly in *The Times,* arguing this time from authority instead of evidence: "As to Professor Lankester's opinion as to what branches of inquiry are to be tabooed as 'degrading,' we have, on the other side, the practical evidence of such men as Lord Rayleigh, Mr. Crookes, Dr. Carpenter, and Colonel Lane Fox—none of them inferior in scientific eminence to Professor Lankester, yet all taking part in the discussion, and all maintaining that discussion and inquiry were necessary; while the close attention of a late President of the Association and of a crowded audience showed the great interest the subject excited." On the evidentiary question, Wallace questioned Lankester's observations of Slade's actions: "His account of what happened during his visit to Dr. Slade is so completely unlike what happened during my own visit, as well as the recorded experiences of Sergeant Cox, Mr. Carter Blake, and many others, that I can only look upon it as a striking example of Dr. Carpenter's theory of preconceived ideas. Professor Lankester went with the firm conviction that all he was going to see would be imposture, and he believes he saw imposture accordingly." Therefore, Wallace concluded in Slade's defense, it is "quite impossible for me to accept the explanation of Professor Lankester and Dr. Donkin as applicable to any portion of the phenomena witnessed by me."[39]

Following Wallace's letter in *The Times* that day are two more in defense of Slade, from participants at the séance in which Lankester claims that he exposed the medium. The first, penned by A. Joy of London, argues that Lankester's and Donkin's account of the séance "differs so widely from my experience that I trust you will, in common justice to Slade, allow me to state some of the points of difference," which is then done in detail. A second letter, from George Joad of Wimbledon-park, enthusiastically explains that "notwithstanding Professor Lankester's exposure (?) of Dr. Slade I still believe in him, and I just beg a few lines of your space to give my reason. I had three sittings with Dr. Slade, and at none had detected anything like imposture. After reading Professor Lankester's letter I resolved to go again; I had just returned, and will state as briefly as possible what occurred." Joan, like Joy, recounts his experiences in which Slade appeared to them to produce genuine slate writing by means other than sleight of hand.[40]

The next day, on the 20th of September, J. Park Harrison of the Royal Institution, Donkin, and Edward Cox, president of the Psychological Society of Great Britain, all weighed in on the debate in *The Times*. Harrison explained that most members of the BAAS Anthropology Section committee were "opposed to the admission of any discussion on Spiritualism" at the Glasgow meeting, and thus the presentation of Barrett's paper there should not be considered an endorsement. Donkin responded to Wallace and the others who had witnessed a Slade séance, noting that these are nothing more than "simple narratives of what they have seen, or categorical denials of what others have endeavoured to prove, and amount to nothing more than the rather *naive* statement that they have been to see a conjuring trick, and actually cannot find out 'how it's done.' " By contrast, Donkin continued in an empiricist mode meant to contrast the weaker anecdotal reasoning of believers: "The great characteristic of Professor Lankester's explanation is that it is a verified prediction, as shown in his letter to you. Any reasoning mind allows that when, in a search after scientific truth, a hypothesis leads to a true prediction, very little more evidence is required for its unqualified reception." Cox remained neutral after his visit with Slade, noting "that I could detect no imposture, nor find any explanation, mechanical or otherwise, either of the writing, the rapping, the floating chairs, or the hands," but admitted that "knowing how a clever conjurer can deceive the eye of a stranger, I should be reluctant to form an opinion until I had seen the exhibition twice or trice, so as to be enabled to keep the eye steadily upon the exhibitor, and not upon the phenomena,—watching what he is doing instead of observing what is done,—by which process alone can sleight of hand be discovered."[41]

Finally, on September 22, the man whose paper triggered the brouhaha in the first place, William Barrett, published his defense of spiritualism and Slade, reasoning that "though it is obvious that if Slade be guilty of fraud in one case he is open to suspicion in all, yet I do not think Professor Lankester's exposure by any means covers all that myself and several scientific friends have witnessed of Slade's performances." Although Barrett recalled seeing what Lankester saw in Slade's movements, "instead of forcibly interrupting Slade and discovering writing when none was supposed to be present," Barrett decided to conduct an additional test of Slade's ability:

> Taking a slate clean on both sides, I placed it on the table so that it rested above, although its surface could not touch, a fragment of slade pencil. In this position I held the slate firmly down with my elbow; one of Slade's hands was then grasped by mine, and the tips of the fingers of his other hand barely touched the slate. While closely watching both of Slade's hands, which did not move perceptibly, I certainly was much astonished to hear scratching going on apparently on the under side of the slate, and when the slate was lifted up I found the side facing the table covered with writing. A similar result was obtained on other days; further, an eminent scientific friend obtained writing on a clean slate when it was held entirely in his own hand, both of Slade's being on the table.

From these tests Barrett concluded that "I am inclined to believe other mental phenomena—such, for example, as the possibility of the action of one mind upon another, across space, without the intervention of the senses—demand a prior investigation. That cases of such mental action at a distance do really exist, I, in common with others, have some reason to believe."[42]

Slade's attorney, then, demonstrated to the court that his client not only had ample support from respected members of the scientific community, but that Lankester had publicly voiced his dissatisfaction with the fundamental premise that it was acceptable for science to investigate such phenomena. Since such prominent scientists as Fox and Wallace had endorsed spiritualism, Munton argued, and Slade was one of the most highly respected spiritualists in the world, how could his client be prosecuted under a law written to protect the public against such vagrant flimflam artists as gypsies and panhandlers? In response, prosecuting attorney Lewis called to the stand John Neville Maskeleyne, one of the greatest magicians and sleight-of-hand artists of the age. Despite the judge's warning that duplicating an allegedly psychic feat by means of magic does not prove that Slade used the same techniques, Maskeleyne explained that "slate writing is a very old trick" and followed his explanation with a dramatic courtroom demonstration of wiping a blank slate with a wet sponge, after which he induced to appear on the slate the message "THE SPIRITS ARE HERE!" Slade's partner, Simmonds, asked to examine Maskeleyne's slate, to which the magician responded sardonically: "Oh, you know all about it."[43] Lewis then pressed Slade to explain the delay in the spirit writing on the slate during the séance, as he held it over the edge of the table making distracting noises. The medium answered: "The spirits were a long time in coming." The judge then jumped into the exchange:

JUDGE: *Well, they had a long way to come! (laughter) Now were these messages, however they were produced, represented to come from the defendant's wife?*

DEFENSE ATTORNEY: *I am not called upon to prove that this writing was done by any supernatural agency.*

JUDGE: *Excuse me, but I think you are (applause).*

DEFENSE ATTORNEY: *If the defendant believed that the writing was that of his deceased wife, surely that is enough without my being called on to prove it.*

JUDGE: *This is a kind of new religion, and many people, no doubt, are sincere believers in it, but we must keep the issue before us. The question is: Did these people fraudulently represent as an act of spiritual agency certain things which were done by themselves?*[44]

Lewis then challenged Slade to have the spirit of his wife, Allison (known as "Allie" and allegedly the spirit behind most of Slade's performances), write on the inside of two slates padlocked together. Slade was not about to be trapped by such a definitive test, so he explained that Allie had grown weary of similar challenges in America and that she "had vowed never to write on a locked slate."[45]

Just before closing arguments were made, Wallace was introduced as the key witness for the defense. Wallace's testimony was supposedly that of a neutral observer, although he was clearly in the Slade camp, recalling that after attending three séances by Slade he had seen nothing "indicative of imposture," although he could not say definitively whether the slate writing was done "by spirits or some other force." Presumably this assessment applies to his final testimony about Slade's table tipping, in which Wallace reported "while their hands were together the table rose as high as they could lift. That could not have been done with the hands and feet."[46] (On the contrary, table tipping is done by wedging the lip of a shoe sole under a table leg and gently lifting one's leg while the hands on the table top hold it steady. Slate writing can also be done through sleight of hand. Two slates may be manipulated in such a manner that the viewer sees only three blank sides, the fourth containing the message written before the séance begins. At the appropriate moment of suspense and drama the medium "reveals" the writing on the fourth side to the viewer as if it had just been applied by the spirits from the great beyond. Another method involves conducting the writing during the séance—particularly useful if information on

Figure 7-4 Slate writing as a conduit to the other side became a favorite among spirit mediums in the nineteenth century. These slates, discovered by historian of science Richard Milner at the Cambridge University Library (donated by the Society for Psychical Research), are from one of Slade's séances. Lankester and the prosecution accused Slade of defrauding the public by using a thumb writer. The top slate contains Greek text from the book of Genesis describing the creation of humans and animals. The bottom slate reads: "Dear Sir—our friend could not do more—he will come again—let this be proof for this time. James Taylor." (Presumably "Taylor" was the spirit doing the writing—"spirits" were often given names.) (Courtesy of Richard Milner/Psychical Research Archive/ Cambridge University Library)

the subjects cannot be gleaned ahead of time—by means of a hidden writing device, such as a thumb-tip writer.)

A letter accompanying the spirit slates from Hensleigh Wedgwood, Charles Darwin's cousin and brother-in-law who was a believer in spiritualism, describes the séance he attended during which such slate writing was done (see Figure 7-4). Eglinton was another popular medium.

> The writing within was obtained at a sitting with Dr. Slade in the Autumn of 1876. We took two of his slates, apparently new, having the grey look of unused slates. I breathed upon them, rubbed them with my handkerchief, and putting the rubbed faces together, we tied them up fast with a piece of cord, with a fragment of pencil between them. Thus tied up the slates were laid flat on the table without having been put underneath or removed for a moment from under my eyes. I placed both my hands upon them & Slade one of his. Presently we heard the writing begin. I bent down my ear to listen to it & we both remarked that it did not sound like writing, but like a succession of short strokes. My first impression was that they could not make the pencil work. But it went on too long for that.
>
> At last the sound entirely changed giving me the impression of rapid writing in a running hand. When I came to open the slates I found that on one side was written the 26th verse of the 1st chap. of Genesis in Greek of the Septuagint versus & on the other a short message in English. The Greek letters being each written separately were what had given the broken sound of the first part of the writing.
>
> As the writing can be rubbed off with the slightest touch it plainly could not have existed in an invisible state upon the slate when well rubbed with my pocket handkerchief, to be subsequently brought out by the heat of my hand, as some have absurdly supposed. There is the same confusion between n & X that I have been in most of the other slate-writing either by Slade or Eglinton. (Courtesy of Richard Milner.)

Munton took Wallace's cue, summing up his case in defense of Slade by noting that Lankester and Donkin had never been able to prove that the medium had made the writings himself, concluding only that some "mysterious agency" or "unknown force" was at work. Who could say what it was? After all, he noted in calling forth Galileo (and echoing Wallace's 1871 analogy), "the pioneers of every new movement which clashed with the prejudices of the day have been subject to persecution" and that "what is laughed at today might be very differently regarded tomorrow."[47]

It was a good argument, but the judge ruled against Slade, primarily on the testimony of Lankester and Donkin: "The whole case turns upon the evidence of the two last-named persons, which, in a few words, is to the effect that they saw Slade's hands move as if he was writing, and that on snatching the slate from him immediately afterwards, and before it was placed in the position in which the spirits were to write, and without any sound as if writing, they found words written upon it. If this be true, it involves the inference that Slade wrote the words himself, and that therefore he could not think the spirit of his wife had written them." Expanding on the meaning of the Vagrancy Act, Judge Flowers explained that slate writing—at least as it was practiced by Slade at the séance attended by Lankester and Donkin—is a form of palmistry, also an offense under the law: "Palmistry is defined in 'Richardson's Dictionary,' thus, 'divination by inspection of the hands, from the roguish tricks of the pretenders to this art; to palm is to trick or play a trick, to impose, to pass, or practise a trick, imposition, or delusion.' More restricted, to palm is to hold and keep in the palm, to touch with the palm, and to handle. The trick imputed to Slade consists in falsely pretending to procure from spirits messages written by such spirits on a slate held under the table by Slade for the purpose, such message having previously been written by himself. Such a trick seems to me to be 'a subtle craft, means, or device' of the same kind as fortune-telling. In such instance the imposter pretends to practise a magical, or at least an occult art." Therefore, Flowers concluded, "Upon the whole, I think that an offence against the Vagrancy Act has been proved, and considering the great mischiefs likely to result from such practices—mischiefs which those who remember the case of Home, also a professional medium, cannot consider unsubstantial—I feel I cannot mitigate the punishment the law imposes, and, therefore I sentence the defendant to three calendar months' imprisonment in the House of Correction, with Hard labour."[48] The verdict, however, was appealed and overturned on a technicality in the Vagrancy Act involving the definition of "palmistry," and Slade was sent packing for home.

Darwin's Secret War on Spiritualism

In addition to Wallace's participation in the Slade trial, one of the most interesting discoveries about it (made by historian of science Richard Milner) is Darwin's own discreet but important involvement. In the Charles D. Warner Library of the World's Best Literature, Milner discovered Lankester's 1896 introductory memoir to Darwin's writings, in which he gives this account of a letter Darwin wrote him: "When I prosecuted Slade the spiritualistic imposter, and obtained his conviction at Bow Street as a common rogue, Darwin was much interested . . . he considered [it] to be a public benefit and that he should like to be allowed to contribute ten pounds to the cost of the prosecution [equivalent of a month's wages for a workingman]. He was ever ready in this way to help by timely gifts of money what he thought to be a good cause."[49] As Milner notes, "The fact of this monetary gesture, buried in an obscure reminiscence by Lankester, places Darwin and Wallace squarely on opposite sides of one of the most sensational trials of the nineteenth century."[50] In fact, Milner persuasively shows that the image of Darwin as a conciliatory recluse avoiding confrontation at all costs is something of a myth. "Darwin had his finger on the pulse of everything that was happening in the scientific, political, and economic spheres of London, though it was up to Huxley and others discreetly to carry out any actions deemed necessary on 'controverted questions' (the title of one of Huxley's books). The 'Saint of Science' (as Darwin was nicknamed by one early biographer) was constantly protected by a tight familial circle; in their zeal to bequeath history an unblemished image they routinely blurred evidences of his passions, antagonisms, and private crusades." Milner gives numerous examples of such machinations, "particularly about his 'secret war' with Spiritualism," in which Darwin "personally attempted to wreck the careers" of two mediums besides Slade, Frank Herne and Charles Williams.[51]

This Darwinian "revisionism" (as Milner calls it) is important because it shows the social (and sometimes underhanded) side of science, and that both Darwin and Wallace used their powerful influence in the service of their preferred beliefs and ideologies, however well or poorly they were supported by science. A case in point is the comparative psychologist George Romanes, a student of the renowned eugenicist Francis Galton and a longtime friend and colleague of both Darwin and Wallace. Romanes found himself temporarily interested in spiritualism, even attending a few séances held by Charles Williams, the man whose career Darwin apparently tried to undermine. He wrote Darwin of the experience, requesting that he keep his involvement a secret. Darwin wrote back, promising that the secret would be "never mentioned to a human being," but that he could not be too supportive as "I fear I am a wretched bigot on the subject."[52] Romanes also confided in Wallace, who, of course, was far more enthusiastic and supportive. Fifteen years later, however, when Wallace turned to him for support, Romanes either had a change of heart on spiritualism, was embarrassed at his youthful association with it, or both, and publicly castigated Wallace for his beliefs. In a letter dated July 18, 1892, Wallace complained that Romanes holds his spiritualistic beliefs in secret. "He thinks no one knows it. He is ashamed to confess it to his fellow-naturalists; but he is not ashamed to make use of the ignorant prejudice against belief in such phenomena, in a scientific discussion with one who has the courage of his opinions, which he has not."[53]

On courage of opinion Wallace was without peer, ashamed of nothing and prejudiced only against those he perceived to be dogmatically closed-minded to what he believed to be unambiguous factual proof of a remarkable phenomenon. Perhaps his admiration and respect for Darwin would have been lessened had he known about the behind-the-scenes intrigues that went on in his cherished world of spiritualism.

Beaten by the Facts

It was Wallace's confidence that these spiritual experiences represented a real connection to a nonmaterial world that reinforced his belief that natural selection was inadequate as a causal agent to explain the origin of the human mind. In an 1874 *Fortnightly Review* article on "A Defence of Modern Spiritualism," Wallace argued that man is not just a physical being, but "a duality, consisting of an organized spiritual form."[54] In a letter to E. B. Poulton, February 22, 1889, he wrote (with a hint of doubt appropriate for his scientism):

> I (think I) *know* that non-human intelligences exist—that there are *minds* disconnected from a physical brain,—that there *is,* therefore, a *spiritual world.* This is not, for me, a *belief* merely, but *knowledge* founded on the long-continued observation of facts,—& such *knowledge* must modify my views as to the origin & nature of human faculty.[55]

In an article on "Spiritualism" in *Chambers' Encyclopaedia,* Wallace defined spiritualism as "a science based solely on facts: it is neither speculative nor fanciful. On facts and facts alone, open to the whole world through an extensive and probably unlimited system of mediumship, it builds up a substantial psychology on the ground of strictest logical induction."[56] Wallace's brand of spiritualism was strictly confined to his scientific worldview, as evidenced earlier and in this autobiographical passage from *Miracles and Modern Spiritualism:*

> From the age of fourteen I lived with an elder brother, of advanced liberal and philosophical opinions, and I soon lost (and have never since regained) all capacity of being affected in my judgments, either by clerical influence or religious prejudice. Up to the time when I first became acquainted with the facts of spiritualism, I was a confirmed philosophical sceptic, rejoicing in the works of Voltaire, Strauss, and Carl Vogt, and an ardent admirer (as I am still) of Herbert Spencer. I was so thorough and confirmed a materialist that I could not at that time find a place in my mind for the conception of spiritual existence, or for any other agencies in the universe than matter and force. Facts, however, are stubborn things. My curiosity was at first excited by some slight but inexplicable phenomena occurring in a friend's family, and my desire to knowledge and love of truth forced me to continue the inquiry. The facts became more and more assured, more and more varied, more and more removed from anything that modern science taught, or modern philosophy speculated on. The facts beat me.[57]

By contrast, and to show that he was not uncritically accepting of all such spiritual and supernatural claims, Wallace thoroughly rejected reincarnation, and stated so publicly in an article in *The London* published in 1904, entitled "Have We Lived on Earth Before? Shall We Live on Earth Again?" Wallace based his analysis (grounded in science, of course) on the Theosophical doctrine that "all, or almost all, souls in due course became re-incarnated for purposes of development in the grades of spirit existence. If this theory be true, it undoubtedly follows that, speaking broadly, we have all lived on earth before, and shall live on earth again, at all events till man is far more advanced morally and intellectually than he is now. But is it true?" No, he argues, turning to the laws of evolution and heredity to show that the claim of reincarnation "can appeal to no direct evidence in its support" and that there is, in fact, "a considerable body of evidence which renders it in the highest degree improbable." Considering Wallace's favorable evaluation of spiritualism and its assumption of an afterlife, one could be forgiven for thinking that he would have woven reincarnation into his spiritual worldview. But, in fact, he lashed out against it with uncharacteristic causticity: "The whole conception of re-incarnation appears to me as a grotesque nightmare, such as could only have originated in ages of mystery and superstition." As he always did, Wallace's spiritual foundation was grounded in science. "Fortunately, the light of science shows it to be wholly unfounded."[58]

"Something That Surpasses Them All"

It should be clear by now that Wallace's abandonment of natural selection did not arise from his experiences with spiritualism and séances in 1865. As we have already seen, Wallace's experiments with phrenology and mesmerism, in time definitively linked to spiritualism, began two decades before that, and a letter at the British Museum, dated April 9, 1864, from Dr. W. B. Carpenter, shows that his interest and belief in spiritualism predates the aforementioned time scale of his conversion to the phenomenon: "I quite agree with you that the influence of one or [illegible] upon another through a force capable of acting at a distance, producing the phenomena of community of sensation, thought-reading, etc., is quite conceivable; and I have several times thought that I had satisfactory evidence of such an action."[59] This entire volume of letters at the museum, in fact, consists of correspondence to and from Wallace on spiritualistic matters. Reading through hundreds of them it becomes clear that for Wallace spiritualism was more than just a problem that could not be explained by natural selection. There was something sublime about the whole subject, as he indicates in a vitriolic letter to the editor of the *Pall Mall Gazette,* May 1868, in response to a letter by Mr. Lewes, who attacked those who believe in spiritualism as "reckless" and "unwarranted" (Wallace's description). Wallace fired back: "I am sorry that the author of the "History of Philosophy" should have written at all on a subject of which he knows so little as he does of spiritualism . . . I find in the philosophy of spiritualism something that surpasses them all,—something that helps to bridge over a chasm whose borders they can not overpass—something that throws a clearer light on human history and on human nature."[60]

These passages reveal the importance of spiritualism in the development of Wallace's thoughts on human evolution, though the reasons for the shift were complex and mitigated by *both* his hyper-selectionism and spiritualism. Further, this belief was part of the larger worldview of a heretic scientist whose temperment and personality drove him toward the temptation to transcend the materialistic world of a blind watchmaker in such a way that his supernatural muses were both products of and helped produce this worldview. Wallace's belief in the supernatural was caused by a number of important agents interacting over time from his youth to his final days. These include: a working-class background, self-education (and thus lack of pressure to conform to the status quo), experiences in the Mechanics' Institutes with fringe and heretical ideas, youthful tinkering with mesmerism and phrenological readings, discovery of the then-radical theory of natural selection and its subsequent payoff of scientific fame and confidence, taking his own theory to the extreme of hyper-selectionism that forced him to find a purpose for everything in nature, personal experiences with séances and spirit mediums that convinced him of the reality of a spiritual world, the need to incorporate these experiences into his scientific worldview, and the final leap from the natural to the supernatural when his science failed to explain by the laws of nature what he knew to be true. His subsequent studies and observations, then, only served to confirm the validity of the knowledge claims, which led to more studies and observations, and so on in an autocatalytic feedback loop: the variables influencing the development of ideas not only interact with each other (with their potency changing over time), they become locked into an information system feedback loop with the thinker, where the ideas feed back into the culture and change the variables themselves, which in turn affect the ideas, that alter the variables, and so on. Throughout his long life and up to the very end, Wallace maintained this borderlands position between the natural and the supernatural, all the while believing he was doing good science in both worlds.

8

Heretical Thoughts

In his sweeping two-volume vista of *A History of European Thought in the Nineteenth Century,* John Theodore Merz opens with a classic statement of the intellectual historian of his generation on the power of individual Thought (his emphasis) in history:

> Behind the panorama of external events and changes which history unfolds before our view there lies the hidden world of desires and motives, of passions and energies, which produced or accompanied them; behind the busy scenes of Life lie the inner regions of Thought. That which has made facts and events capable of being chronicled and reviewed, that which underlies and connects them, that which must be reproduced by the historian who unfolds them to us, is the hidden element of Thought. Thought, and thought alone, be it as a principle of action or as the medium of after-contemplation, is capable of arranging and connecting, of combining what is isolated, of moving that which is stagnant, of propelling that which is stationary. Take away thought, and monotony becomes the order.[1]

Merz's heroic depiction of the power of ideas is appealing to those who cut their teeth on John Herman Randall's *The Making of the Modern Mind,* which historian of science Richard Olson identifies as the pinnacle of positivist historiography, or the view of Western history seen "predominantly in terms of a progressive unfolding of truth and freedom grounded in the constant advance of scientific knowledge."[2] In the second half of the twentieth century, however, intellectual history and the history of science underwent a startling metamorphosis toward the anti-positivist view held by extreme relativists and those who endorsed the sociological "strong programme" that denied a privileged position for any knowledge claim. Their interest in understanding the influence of culture on thought, as important as that has been in broadening our understanding of the history of ideaas, has nevertheless obscured (and in some cases denied) the impact of thought on culture. As we will see in this case study of the interaction of thought and culture through a single individual, both are integral to understanding how science works.

The Law of Higgledy-Piggledy

Wallace's name will forever be linked to that of Charles Darwin, and probably remain subservient to it since Darwin, more than Wallace, created a research program in evolutionary biology that continues unabated to this day. Yet even as we elevate the importance of Wallace in this story, particularly his role in the discovery and description of natural selection, we would do well to remember that Darwin and Wallace differed dramatically in their theories of evolution. As we have already seen, even a narrow version of Darwin's theory applied only to nonhuman animals differed from Wallace's, and, when it came to human evolution, there was no such species as a pure "Darwinian."[3] Such differences within a single paradigm revealed the complexity of the theory of evolution and the different of levels of its applications. Throughout the 1860s and 1870s it became clear to both men that there were irreconcilable differences of opinion on natural and sexual selection that led to monumental confrontations with ramifications for how science was understood as a process to operate.[4]

At stake in the great evolution debates of the nineteenth century was nothing less than the methods of science itself. In the early part of the century, when Darwin and Wallace were coming of age, English scientists became intensely interested what it is they were doing when they were doing science. Out of these debates a number of questions arose: Are scientific laws formed as concepts in the mind or discovered in nature (concepts versus percepts)? Is there a difference between the logic of discovery and the logic of justification (how a discovery happened versus how it is presented)? What is the relationship of mathematical axioms and observational experiences (truth by thinking versus truth by seeing)? How can one distinguish occult qualities from theoretical entities (gravity versus attractive objects)? And most important, what is the difference between deduction and induction (from general to specific versus from specific to general)? A number of influential works were published that fueled the debate, starting notably with John Herschel's *Preliminary Discourse on the Study of Natural Philosophy* in 1830. A more thorough treatise was presented by William Whewell in 1837 in his *History of the Inductive Sciences,* fleshed out three years later in *The Philosophy of Inductive Sciences, Founded Upon their History.* These treatises were subsequently countered by John Stuart Mill in 1843 in his *System of Logic, Ratiocinative and Inductive, Being a Connected View of the Principles of Evidence, and the Methods of Scientific Investigation,* with Whewell countering with *Of Induction, with Especial Reference to Mr. J. Stuart Mill's System of Logic* in 1849.[5] At the heart of all this verbiage was *induction*—what it was and how it was used in science. Although definitions varied, it was roughly understood to mean arguing from the specific to the general, from observations to conclusions, from data to theory. Additionally, for Herschel and Mill, induction was reasoning from the known to the unknown, and the discovery of general laws within specific facts, verified empirically. For Whewell, induction was the superimposing of concepts on facts by the mind, even if they are not empirically verifiable.

Kepler's laws of planetary motion were a classic case study for these early philosophers of science. For Herschel and Mill, Kepler discovered these laws through careful observation and induction. For Whewell, the laws were self-evident truths that could have been known a priori. By the 1860s, as the theory of evolution was gaining momentum and converts, Herschel and Mill carried the day, not so much because they were right and Whewell was wrong, but because empiricism was becoming integral to the understanding of how good science is done. Ironically—considering how much specific data both Darwin and Wallace compiled before going public with their theories—the *Origin of Species* was vituperously attacked by Herschel and Whewell for being too conjectural. Herschel called the theory the "law of higgledy-piggledy."[6] The case of Whewell is especially ironic because of his identification of a process in science he called *consilience of inductions.* To prove a theory, Whewell argued, one must have more than one induction, or a single generalization drawn from specific facts. One must have multiple inductions that converge on one another, independently but in conjunction. Whewell said that if these inductions "jump together" it strengthens the plausibility of a theory: "Accordingly the cases in which inductions from classes of facts altogether different have thus jumped together, belong only to the best established theories which the history of science contains. And, as I shall have occasion to refer to this particular feature in their evidence, I will take the liberty of describing it by a particular phrase; and will term it the Consilience of Inductions."[7]

The irony is that the theory of evolution is arguably the most consilient theory ever generated, and Whewell rejected it, going so far as to block the book from being shelved at the library at Trinity College, Cambridge. Consilience, in fact, is a technique employed by all historical scientists. Cosmologists use evidence from astronomy, astrophysics, planetary geology, and physics to reconstruct the history of the universe. Geologists piece together the history of the Earth through a convergence of evidence from geology and numerous related Earth sciences. Archaeologists reformulate the history of civilization using artwork, written sources, and other site-specific artifacts and data from temporal corresponding sites. Evolution is confirmed by the fact that so many different lines of evidence converge to a single conclusion. Independent sets of data from geology, paleontology, botany, zoology, biogeography, comparative anatomy, physiology, and many other sciences each point to the conclusion that life has evolved. This is one of the most powerful convergences of evidence ever compiled. As Darwin noted in his autobiography, "Some of my critics have said, 'Oh, he is a good observer, but has no power of reasoning.' I do not think that this can be true, for the *Origin of Species* is one long argument from the beginning to the end, and it has convinced not a few able men."[8]

But not all able men. Adam Sedgwick complained that Darwin had "departed from the true inductive track" because his theory was "not based on a series of acknowledged facts pointing to a *general conclusion,*—not a proposition evolved out of facts, logically, and of course including them." Richard Owen complained of the book's lack of "inductive foundations." Mill was generally supportive, calling the *Origin of Species* "another unimpeachable example of a legitimate hypothesis. What he terms 'natural selection' is not only a *vera causa,* but one proved to be capable of producing effects of the same kind with those which the hypothesis ascribes to it." But the theory of evolution was not inductive, Mill wrote. "It is unreasonable to accuse Mr. Darwin (as has been done) of violating the rules of Induction. The rules of Induction are concerned with the conditions of proof. Mr. Darwin has never pretended that his doctrine was proved. He was not bound by the rules of Induction, but by those of Hypothesis."[9]

In response to his critics Darwin pleaded that he "worked on true Baconian principles and without any theory collected facts on a wholesale scale," but it was special pleading at best. Just before its publication, in fact, he confided to the American botanist Asa Gray: "What you hint at generally is very, very true: that my work will be grievously hypothetical, and large parts by no means worthy of being called induction, my commonest error being probably induction from too few facts." He wrote to Lyell, "I have heard, by a roundabout channel, that Herschel says my book 'is the law of higgledy-piggledy.' What this exactly means I do not know, but it is evidently very contemptuous. If true this is a great blow and discouragement."[10] Darwin knew perfectly well what Herschel meant. And he knew how best to conduct historical science. What was not clear to him or anyone else at the time was what it is that scientists in general were doing.

Let Theory Guide Your Observations

Once the *Origin of Species* was published in 1859, Darwin's confidence grew as converts confessed their faith to the new doctrine. At the BAAS annual meeting in which the *Origin* was savaged for being too theoretical and that he should have just let the facts speak for themselves, Darwin's response that "all observation must be for or against some view if it is to be of any service!" revealed a philosophy of science that included a balance between data and theory.[11] Of course, there were a lot of metaphysical, deductive ideas floating around the cultural landscape of which Darwin was quite critical, so he cautioned about finding the right balance between theory and data. "I would suggest to you the advantage, at present, of being very sparing in introducing theory in your papers; *let theory guide your observations,* but till your reputation is well established, be sparing of publishing theory. It makes persons doubt your observations."[12]

What was important to Darwin was not induction, but verification by subsequent observation. "I had therefore only to verify and extend my views by a careful examination of coral reefs," he once explained in reference to his correct deduction of the evolution of coral reefs in which, in an act of brilliant historical science, Darwin reasoned that the different types of coral reefs did not represent different entities with different causes, but the same entity at different stages of development. And he did this before ever seeing one! "No other work of mine was begun in so deductive a spirit as this; for the whole theory was thought out on the west coast of S. America before I had seen a true coral reef."[13] But, more important, when Darwin did see a coral reef it confirmed his theory. In other words, theory came first, then the data. It was not pure induction, but it was good science, at least to some people, including Thomas Huxley, who must have been at his wit's end when he penned this harangue against the philosophers who pontificated on science, but had never practiced it themselves: "There cannot be a doubt that the method of inquiry which Mr. Darwin has adopted is not only rigorously in accord with the canons of scientific logic, but that it is the only adequate method. Critics exclusively trained in classics or in mathematics, who have never determined a scientific fact in their lives by induction from experiment or observation, prate learnedly about Mr. Darwin's method, which is not inductive enough, not Baconian enough, forsooth for them."[14]

Alfred Russel Wallace is a case in point of someone who devoted years to determining scientific facts by induction from observation, but who, in fact, converted to evolution long before he ever set foot in the tropics. "The human mind cannot go on for ever accumulating facts which remain unconnected and without any mutual bearing and bound together by no law,"[15] he wrote.

Of course, he did not have the specific theory of natural selection before the specific data of species diversity and geographical isolation, but he did have the general theory of evolution.

Is the theory of evolution inductive or deductive? Let's follow Darwin's advice and let theory guide our observations. We begin by recognizing that this is a complex theory with multiple components. Evolutionary biologist and historian of science Ernst Mayr has identified at least five:[16]

1. *Evolution:* Change through time.
2. *Descent with modification:* The mode of evolution by branching common descent.
3. *Gradualism:* Change is slow, steady, stately. *Natura non facit saltus.* Given enough time evolution can account for the origin of new species.
4. *Multiplication of speciation:* Evolution produces not just new species, but an increasing number of new species.
5. *Natural selection:* The mechanism of evolutionary change can be subdivided into five steps:[17]
   A. Populations tend to increase indefinitely in a geometric ratio.
   B. In a natural environment, however, population numbers stabilize at a certain level.
   C. There must be a "struggle for existence" since not all organisms produced can survive.
   D. There is variation in every species.
   E. In the struggle for existence, those variations that are better adapted to the environment leave behind more offspring than the less well adapted individuals, also known as differential reproductive success.

This process of natural selection, when carried out over countless generations, gradually leads varieties of species to develop into new species. Within the natural selection paradigm, points A, B, and D are observations, C and E are inferences. C follows from A and B, and E follows from all three observations. Is this the inductive process? As outlined in this simplistic format, yes, it is the process of induction at work since the generalizations follow from the observations. But science rarely (if ever) works in such a contrived manner and it is usually only after enough time has elapsed and the foundational principles of a theory are established, that the founder, or historians of science, can take the long view and present it as such a linear system of thought. In reality, the scientific process is a very messy one, with guesses, hypotheses, and theories in constant interaction with data, facts, and observations, in what is best described as the *hypothetico-deductive* method. Historian of science Frank Sulloway debunked the myth of the Galápagos Islands as Darwin's epiphany,[18] and we have already seen what a long and convoluted road it was for Wallace to arrive at the theory's driving force. Science is an exquisite amalgam of data and theory, nature and mind, and rarely does the chronological sequence of actual discovery match the logical sequence of later description.

The impact of the theory of evolution on the general culture is so pervasive it can be summed up in a single observation: we live in the age of Darwin. As one of the half dozen most culturally jarring theories in the history of science, the Darwinian revolution changed both the science and the culture in ways that reveal clearly where Darwin and Wallace differed:

1. The replacement of essentialism and the view of species as Platonic types with ever-changing entities with no fixed essence.
2. The replacement of a static creationist model with a fluid evolutionary model.
3. The replacement of intelligent design by a supernatural force with natural design by natural selection.
4. The replacement of the view of God as necessary with the view of God as optional.
5. The replacement of anthropocentrism with the view of humans as just another species.
6. The replacement of teleology and the view of the cosmos as having direction and purpose with contingency and the view of the world as a conjuncture of events without purpose.[19]

Not everyone who accepted evolution agreed with all six of these tenets, and more could be added to the list (and some deleted) and still fit the evolutionary paradigm. From the moment the *Origin of Species* landed in the hands of booksellers and readers on November 24, 1859 (all 1,250 copies were already subscribed to by the retail trade, so the book immediately went into a second printing), all six of these tenets were challenged by critics, and even supporters did not necessarily subscribe to all of them. Of all people to dissent, however, as the co-discoverer of the theory's main mechanism, Alfred Russel Wallace stood out above most others and caused Darwin considerable consternation. And of their many disputations, none was more portentous than that over human evolution, particularly the evolution of the mind. Although Wallace accepted the first four tenets, he clung to a unique form of anthropocentrism and teleology (points 5 and 6), and reshaped them to fit his own unique blend of science and philosophy. To understand further how and why Wallace diverged from Darwin and most other evolutionists of his time, we shall examine a number of related thoughts and theories he held as part of his larger worldview (following the outline of the Historical Matrix Model—*Hyper-selectionism, Mono-polygenism, Egalitarianism,* and *Environmental determinism*). For Wallace, the specific theory of evolution by natural selection was just one component in his grander philosophy of scientism, and became part of his attempt to integrate his disparate interests into a unified whole.

More Darwinian Than Darwin: Hyper-Selectionism

The bane of biography is trying to capture the essence of an individual while also acknowledging the variation and diversity of thought and action through a lifetime. Although Wallace became so committed to the adaptationist program that his view of natural selection could be described as a form of *hyper-selectionism,*[20] it is rarely noted that Wallace began his studies as a non-adaptationist. In his 1856 paper "On the Habits of the Orang-utan in Borneo," for example, Wallace asserts that "many animals are provided with organs and appendages which serve no material or physical purpose. The extraordinary excrescences of many insects, the fantastic and many-coloured plumes which adorn certain birds, the excessively developed horns in some of the antelopes, the colours and infinitely modified form of many flower-petals, are all cases, for an explanation of which we must look to some general principle far more recondite that a simple relation to the necessities of the individual. Naturalists are too apt to *imagine,* when they cannot *discover,* a use for everything in nature; they are not even content to let 'beauty' be sufficient use, but hunt after some purpose to which even *that* can be applied by the animal itself."[21]

Within two years, however, Wallace devised an answer that did not need to be imagined. With the discovery of natural selection in 1858 Wallace became a strict selectionist and adaptationist, his confidence in the mechanism of natural selection growing more robust over time. In 1867, in a *Westminster Review* paper on "Mimicry and other Protective Resemblances among Animals" he spelled out a "necessary deduction from the theory of Natural Selection, namely—that none of the definite facts of organic nature, no special organ, no characteristic form or marking, no peculiarities of instinct or of habit, no relations between species or between groups of species, can exist but which must now be, or once have been, useful to the individuals or races which possess them."[22]

So, before 1858 Wallace was a non-adaptationist. By 1867 he was an uncompromising adaptationist, a hyper-selectionist. Why?[23] One possible answer (by contextual inference) is that Wallace became a proselytizer of his own theory. Plausible scenarios assert themselves: Once he had a mechanism by which everything might be explained, through post hoc reasoning nearly everything *was* explained. It is simple to take any current structure or function of an organism, then reflect back in time to imagine how that mechanism might have given an organism a certain survival or reproductive advantage.

It works equally well with natural or sexual selection, though for Wallace, natural selection was far more important than sexual selection, and his examples focus on the former, as we see in his historical retrospection. In his important 1864 paper "The Origin of Human Races and the Antiquity of Man Deduced from the Theory of 'Natural Selection,' " Wallace deduced the following: "By a powerful effort of the imagination it is just possible to perceive him [man] at that early epoch existing as a single homogeneous race without the faculty of speech, and probably inhabiting some tropical region." He continued his inferential reconstruction of past events based on present principles in the just-so mode of storytelling: "As he ranged farther from his original home, and became exposed to greater extremes of climate, to greater changes of food, and had to contend with new enemies, organic and inorganic, useful variations in his constitution would be selected and rendered permanent, and would, on the principle of 'correlation of growth,' be accompanied by corresponding external physical changes."[24]

For Wallace, if a structure appears purposeless, or we are unable to understand how natural selection could have accounted for its existence, we simply lack the requisite knowledge to explain it. Here Wallace employs a form of teleological natural selection in which everything has a purpose. He said as much in his ultimate account (and tribute to Darwin, who never took it this far) of the mechanism and process of evolutionary change, entitled simply *Darwinism:* "The assertion of 'inutility' in the case of any organ . . . is not, and can never be, the statement of a fact, but merely an expression of our ignorance of its purpose or origin."[25] It was a natural step, then, for Wallace to apply his theory to man, which he did sooner and more directly than Darwin.

In June 1865 Wallace published an article on the direct application of natural selection to humans, but within three years he had determined that natural selection could not account for *everything,* as he previously thought. In the article "The Limits of Natural Selection Applied to Man," for example, Wallace could not fathom how "Man's capacity to form ideal conceptions of space and time, of eternity and infinity" could have been produced by natural selection. Wallace insisted that Darwin would have to convince him how rudimentary or latent mental attributes "in the lowest races can have been developed by survival of the fittest,—can have been of *use* to the individual or the race."[26] Furthermore, and in line with his comment from the orangutan paper of 1856 where he critiqued naturalists for searching for purpose in "beauty," Wallace himself looks for but cannot find a purpose, in that "his delicate and yet expressive features, the marvellous beauty and symmetry of his whole external form" were not advantageous to the organism, and could even be disadvantageous.

Clearly, Wallace's shift must be considered in the light of his rigid hyper-selectionism, especially keeping in mind the several other nodes of failure (for Wallace) of natural selection, such as the hand, speech, and the shape of the external bodily form. A savage, Wallace reasoned, had no need for such a perfect hand or sophisticated organs of speech, and since advanced civilization depends mightily on these two structures (and natural selection does not select for structures in advance—no "pre-adaptation"), they too had to be designed by a "Higher Intelligence." Furthermore, hairless skin, erect posture, and the general beauty of the human body were not only useless to savages, but could even be harmful. The unequivocal conclusion was that natural selection could not have been the shaping mechanism for these unique human features.

By the time he published *Darwinism* in 1889, Wallace was aware of how much he had diverged from Darwin on this issue, yet still maintained that his version was the most scientifically tenable "Darwinian doctrine," as he called it:

> Although I maintain, and even enforce, my differences from some of Darwin's views, my whole work tends forcibly to illustrate the overwhelming importance of Natural Selection over all other agencies in the production of new species. I thus take up Darwin's earlier position, from which he somewhat receded in the later editions of his works, on account of criticisms and objections which I have endeavoured to show are unsound. Even in rejecting that phase of sexual selection depending on female choice, I insist on the greater efficacy of natural selection. This is pre-eminently the Darwinian doctrine, and I therefore claim for my book the position of being the advocate of pure Darwinism.[27]

Here Wallace alludes to another deep theoretical issue dividing him and Darwin—the relative role of sexual selection in shaping organisms in general, and humans in particular. Whereas Darwin allocated a large portion of *The Descent of Man* to the importance of sexual selection in the development of species, Wallace attenuated the role of sexual selection because he viewed natural selection more as a struggle for existence against the environment, not for winning more copulations, and that sexual selection can and does lead to the evolution of harmful features and thus would have been eliminated. Thus, his rejection of sexual selection as a potent force in evolution reinforced his overemphasis on natural selection. If one accepts Wallace's initial premise of rigid hyper-selectionism, and allows for the speculative just-so storytelling of how organisms came to be, it becomes clear how he reached the conclusions he did. Today such a line of reasoning would be a non sequitur, but when read in the light of Wallace's logical and consistent arguments, one gleans a purer understanding of the man's thoughts.

(Today this is called the problem of incipient stages. For example, a fully developed wing has obvious survival advantages, but what good would the incipient or intermediate stages be? A 50 percent or even 80 percent developed wing would be aerodynamically unsound and useless for flight, and therefore eliminated by natural selection. Likewise, a fully developed brain clearly has survival advantages in a modern technological society, but, Wallace argued, what good are the incipient stages of a large brain in a natural, preindustrial environment? The solution in the case of the wing is that the incipient stages were not poorly developed wings but well-developed something-elses. The wing probably evolved as a thermoregulator for certain ectothermic—cold-blooded—organisms. As species increased in size—E. D. Cope's "rule of phyletic size increase" states that body size tends to increase within evolutionary lineages—the thermoregulating structures, with no pre-adaptational foresight, became aerodynamically sound flight structures. Likewise, for reasons not fully understood, as the human lineage increased in size from *Australopithicus* to us, the brain also increased, well beyond the brain/body ratio exhibited by other primates. Tool use, bipedalism, hunting, and infant care that free the hands for more "intelligent" usages are a few explanations considered today, but the problem has by no means been solved.)

Around this time Wallace began a long and thoughtful correspondence with his friend and colleague from Oxford University Museum, the professor of zoology E. B. Poulton. In these letters we see that Wallace's belief in the power of natural selection was not fleeting, but one he held throughout his life. Most of these letters focus on matters botanical and zoological, with occasional exchanges on evolutionary theory and even spiritualism. For example, on February 22, 1889, Wallace requested that Poulton review the proofs of his book, *Darwinism,* including his controversial views on human evolution, knowing full well that this "would only horrify you still more. I am quite aware my views as to Man, will be,—as they have been—criticised."[28] Poulton makes reference to this letter in the obituary notice he wrote on Wallace's death in 1913, noting Wallace's firm conviction in "non-human intelligences . . . that are *minds* disconnected from a physical brain;—that there *is,* therefore, a *spiritual world.*" Wallace's hyper-selectionism had profound philosophical consequences.

These letters reveal a developing friendship between Wallace and Poulton, with Wallace as the grand old man of science having outlived most of his earlier contemporaries, and Poulton as the younger protégé who looked up to Wallace. The letters are a fascinating glimpse into some of the most contentious debates of the age, including Mendelism, mutationism, acquired characteristics, and the relative roles of natural and sexual selection.

In 1900 Hugo De Vries, Karl Franz Joseph Correns, and Erich von Tschermak independently and nearly simultaneously rediscovered the work of Gregor Mendel on genetics, virtually lost or ignored for over three decades. The new science of Mendelian genetics was born, and the following year De Vries published *The Mutation Theory,* in which he introduced the idea that changes in species occur in leaps, which he called mutations, directly challenging Darwin's and Wallace's conviction that *natura non facit saltus.*

Many thought that Mendelian genetics and De Vries's mutationism posed a serious challenge to Darwinian gradualism (which was driven primarily by natural selection with an occasional nod toward Lamarckian use inheritance when needed). Wallace, the hyper-selectionist, would have none of this backsliding (thus supporting his own claim of being more Darwinian than Darwin), nor would he accept the early evidence coming in from research on Mendelian genetics or mutationism. In numerous letters to Poulton he stated so in no uncertain terms. In addition to rejecting Darwin's theory of sexual selection in humans in a letter he penned on February 22, 1889, Wallace wrote to Poulton on February 1, 1893, challenging Darwin's belief in Lamarckianism, and in a most curious manner. It seems that Darwin was concerned that his children would inherit his acquired disease from the voyage of the *Beagle.* Wallace notes "that Darwin's constant nervous stomach irritation *was* caused by his 5 years sea sickness. It was thoroughly established before, and in the early years of his marriage, and, on his *own* theory his children *ought* all to have inherited it. Have they?—and if not it is a fine case!"[29]

Wallace's disdain for any derailment from the main track of natural selection was boundless. In letter after letter to Poulton, Wallace's hyper-selectionism rings through unmistakably clear:

> September 8, 1894: Neither he [Bateson] nor Galton appear to have any adequate conception of what Natural Selection is, or how impossible it is to escape from it.
>
> August 5, 1904: What a miserable abortion of a theory is "Mutation," which the Americans now seem to be taking up in place of Lamarckism "superseded." Anything rather than Darwinism!
>
> July 27, 1907: I am glad to hear you have a new book on "Evolution" nearly ready and that in it you will do something to expose the fallacies of the "Mutationists" and "Mendelians," who pose before the world as having got *all* wisdom, before which we poor Darwinians must hide our diminished heads! "Mutation," as a theory is absolutely nothing new—only the assertion that new sp. originate *always* in sports—for which the evidence adduced is the most meagre and inconclusive of any ever set forth with such pretentious claims! "Mendelism" *is* something new, & within its very limited range, *important,* as leading to conceptions as to the causes & laws of *heredity*—but only misleading when adduced as the *true origin of species* in Nature—as to which it seems to me to have no part whatever.

> May 12, 1910: My view is, that Nature works on so vast a scale, with countless *millions* of varying individuals, and that the *changes,* requiring fresh *adaptations,* are so *slow,* that natural selection is able to effect the adaptation in the enormous majority of cases if not in all.[30]

Wallace openly discussed his belief in the absolute power of natural selection every chance he could. To Joseph Hooker, on November 10, 1905, Wallace told him he was "*extremely pleased,* and even greatly *surprised,* in reading your letters to Bates, to find that, at that early period (1862) you were *already* strongly convinced of three facts which are absolutely essential to a comprehension of the method of organic evolution, but which many writers, *even now,* almost wholly ignore." In reviewing these "three facts," Wallace offers a succinct summary of his theory and emphasis on natural selection:

> They are (1) the *universality* and large *amount* of normal variability—(2) the *extreme* regions of natural selection,—and (3) that there is *no* adequate evidence *for,* and very much *against,* the inheritance of acquired characters. It was only some years later when I began to write on the subject & had to think out the exact mode of action of natural selection, that I myself arrived at (1) and (2) and had ever since dwelt upon them in season & out of season, as many will think—as being absolutely essential to a comprehension of organic evolution. I have *never* seen the *sufficiency* of normal variability for the modification of species more strongly or better put than in your letters to Bates. Darwin himself never realised it, and consequently played into the hands of the "discontinuous variation" and "mutation" men, by so continually saying "*if* they vary"—"*without variation* nat. select. can do nothing"—etc. etc.[31]

To others Wallace vented his frustration with the willingness of scientists to reject natural selection in favor of these new theories. Wallace thanked A. Smith Woodward, on April 21, 1907, for a paper on "Relations of Paleontology to Biology," but found it unappealing, "since you adopt a view of Evolution that seems to me, not only altogether *unnecessary* for a clear comprehension of the facts, but also one that is altogether *erroneous*—the so-called 'Mutation" theory of DeVries." Wallace then presents his reasoning: "I consider DeVries' theory so completely wrong and so wholly opposed to any sound reasoning on the facts of variation and of the struggle for existence, that I am amazed to see how many of the younger Biologists have adopted it, and have supported it by ludicrously exaggerated claims and utterly inconclusive reasoning. The theory utterly fails to account for the marvellous & intricate *adaptations* in organised beings, which *normal variation, rapid increase* & the *severity* of the struggle for life, inevitably bring about."[32]

Wallace's reasoning in rejecting DeVries (through Woodward's interpretation) is not unreasonable considering the enormous span of deep geological time and the differential use of language to describe the speed of change. "Much of the error on this subject is due to the use of terms that are unjustified by knowledge," Wallace explained. "In both your papers you continually speak of 'sudden' changes of structure or type. But how can you possibly have any such acceptable knowledge of *lapse of time* in remote geological epochs as to justify the use of such a term? It implies the very thing you admit to be non-existent—a continuous *geological record!* Your 'sudden,' may be a lapse of 100,000, or many millions of years, during which most physical and biological changes may have occurred necessitating the changes of organisation that you describe."[33] (The problem of describing geological time periods with language fixed in human time scales is a nontrivial one and the source of some of today's controversy over the nature of evolutionary change—slow and steady, or stability punctuated by bursts of rapid change? Wallace has articulated a linguistic as well as scientific problem that has not been resolved. When geologists speak of "rapid speciation," they may mean tens or hundreds of thousands of years.)

As late in his life as age eighty-nine, Wallace continued to harp on the universal power of natural selection and the apparent inability of his critics to understand his position in the matter. On February 8, 1911, Wallace wrote Sir W. T. Thiselton-Dyer, thanking him for "your views on my new book" (*The World of Life*), and summarizing what he sensed the reviewers missed:

> Hardly one of my critics (I think absolutely *not one*) has noticed the distinction I have tried and intended to draw between *Evolution,* on the one hand, and the fundamental *powers* and *properties* of *Life—growth, assimilation, reproduction, heredity,* etc., on the other. In "Evolution" I recognise the action of "natural selection" as *universal* and capable of explaining all the facts of the continuous development of species from species, "from Amoeba to Man." But this, as Darwin, Weismann, Kerner, Lloyd-Morgan, and even Huxley—have seen, has nothing whatever to do with the basic *mysteries* of *life—growth,* etc. etc.[34]

As natural selection *cannot* explain these "mysteries of life," Wallace concludes, consistent from the time he first developed his theory of the necessity of a "Higher Intelligence" in 1869 to the end of his life, "It is *here* that I state guidance, & organising power are essential."[35]

Even Wallace's anti-vaccination campaign, waged for several decades in his later life, has been documented and described by one historian as an outgrowth of Wallace's "assumption that nothing exists in nature that is not useful."[36] Wallace, like Spencer, believed in the natural state of man's existence in society, and that government intervention generally had the effect of upsetting the balance of *harmonia naturae* since Wallace "believed in an innate harmony within nature, in a perfect and absolute balance between its governing rules and its products" that vaccination would destroy.[37] Wallace did extensive research to demonstrate that not only did vaccination not prevent disease, it actually increased the number of deaths as a consequence of the haphazard dosages and methods by which inoculations were given. He argued that the state's statistics "proving" the effectiveness of vaccination were seriously flawed. Where deaths due to smallpox, for example, apparently decreased after the introduction of vaccination into a community, Wallace showed that the death rate was either already on the decline before vaccination, or eventually increased despite vaccination, neither trend included in the state's published data. In a booklet he wrote in 1904 for the Anti-Vaccination League with the maximally descriptive title *Summary of Proofs That Vaccination Does not Prevent Small-pox but Really Increases It,* Wallace concluded that "the figures go increasing and decreasing so suddenly and so irregularly, that by taking only a few years at one period, and a few at another, you can show an increase or a decrease according to what you wish to prove."[38] Further, not only did Wallace illustrate through statistical analysis that populations did not benefit from vaccination, he argued that individuals took greater risk by being vaccinated than from catching the disease naturally.

Convinced that vaccination disturbed the proper order of biological systems, Wallace ratcheted up his attack on the Vaccination Act through a monograph he wrote in 1898 entitled *Vaccination a Delusion; Its Penal Enforcement a Crime: Proved by the Official Evidence in the Reports of the Royal Commission.* Vaccination was not only an attack on the natural order, he argued earlier before a Royal Commission on vaccination, it was an assault on individual liberty by the state: "From the moment when, through the great influence of the medical profession, a medical dogma was enforced by penal law, it became a question of personal liberty. When almost every week I read of men fined or imprisoned for refusing to subject their children to a surgical operation which they (and I) believed to be, not only useless, but injurious and dangerous, I felt impelled to aid, if ever so little, in obtaining a repeal of a cruel and tyrannical law."[39] He even wrote the former Prime Minister Gladstone: "I take the liberty of sending you my pamphlet on the vaccination question, in the hope that you will be able to examine it during your comparative leisure at Bournemouth, and that, if you find it as conclusive as I believe it to be, you will give the great weight of your name to a public statement to that effect."[40] In time, *Vaccination a Delusion* found its way into Parliament, winning over enough members to see the approval of a bill recognizing as legitimate the anti-vaccination arguments, and making it legal for fathers to object to forced vaccination of their children.

Finally, in two letters to Francis Galton, Wallace's hyper-selectionism comes out both overtly and covertly. On March 6, 1895, Wallace penned a postcard to Galton, apparently in response to a letter regarding his opinion of Galton's theory, in which Wallace explained: "That any species be formed without the aid of natural selection, or that any *'specific characters'* in the true sense of the term, should be non-adaptive, seems to me quite impossible & even (to me) unthinkable."[41] Wallace's belief in the importance of individual variation within a species as a major component of natural selection led him to apply Galton's new science of statistical analysis using frequency curves to describe the variation, or lack thereof, within a species. Late in 1893, Galton apparently invited Wallace to be a member of a committee to analyze statistical research on populations, a request Wallace found interesting; but on December 1, 1893, he confessed to Galton: "I cannot follow the formulas & tables in their mathematical form. The only thing I can clearly comprehend are the *diagrams & curves* showing variability in various ways." His general mathematical knowledge may have been limited, but his analysis of specific mathematical applications to his own field of evolutionary biology most certainly was not, as indicated in the following lengthy passage, reproduced in Figure 8-1 to show his own diagrams of what a frequency curve would look like when the importance of individual variation is considered in the context of natural selection:

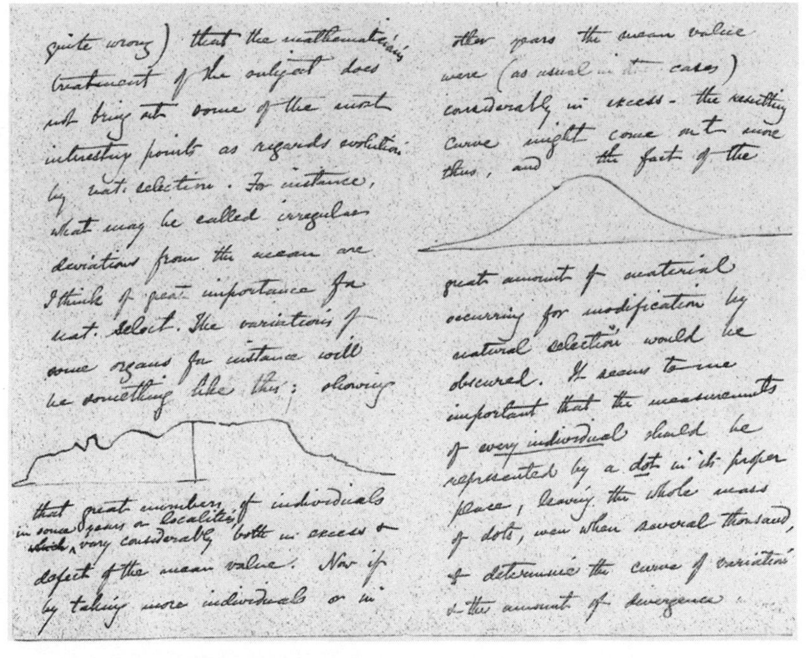

quite wrong) that the mathematical treatment of the subject does not bring out some of the most interesting points as regards evolution by nat. selection. For instance, what may be called irregular deviations from the mean are I think of great importance for nat. select. The variations of some organs for instance will be something like this: showing

that great numbers of individuals in some years or localities vary considerably both in excess & defect of the mean value. Now if by taking more individuals or in

other years the mean value were (as usual in these cases) considerably in excess – the resulting curve might come out more thus, and the fact of the

great amount of material occurring for modification by natural selection would be obscured. It seems to me important that the measurements of every individual should be represented by a dot in its proper place, leaving the whole mass of dots, even when several thousand, to determine the curve of variation & the amount of divergence . . .

Figure 8-1 Wallace's frequency curves applied to individual and species variability, in a letter to Francis Galton, December 1, 1893. (Courtesy of University College London)

> It seems to me (though I may be quite wrong) that the mathematicians' treatment of the subject does not bring out some of the most interesting points as regards evolution by nat. selection. For instance, what may be called irregular deviations from the mean are I think of great importance for nat. Select. The variations of some organs for instance will be something like this; showing [diagram] that great numbers of individuals in some years or localities, vary considerably both in excess & defect of the mean value. Now if by taking more individuals or in other years the mean value were (as usual in these cases) considerably in excess—the resulting curve might come out more thus, and the fact of the [diagram] great amount of material occurring for modification by natural selection would be obscured. It seems to me important that the measurements of *every individual* should be represented by a *dot* in its proper place, leaving the whole mass of dots, even when several thousand, to determine the curve of variation & the amount of divergence in each direction. This would be understood by the most unmathematical intellect![42]

One Species or Many?: Mono-Polygenism

In the middle of the nineteenth century one of the debates among anthropologists and naturalists was whether the human races were local varieties of one species, or separate species entirely. Advocates of these two positions were known as *monogenists* or *polygenists,* respectively. Part of the problem in resolving the debate stems from the subjective human element of those who see similarities and thus tend to classify similar organisms into one species (lumpers), and those who see differences and thus classify organisms into more than one species (splitters). The *monogenist/polygenist* debate falls under this rubric, asking a most fundamental question: is humankind one species or many?

Early-nineteenth-century pre-Darwinian *monogenists* generally believed that all present human races were the product of a slow deterioration from the perfect creation in the Garden of Eden at the beginning of time. This degeneration, however, was not believed to have been equal among races. In the racial ranking of most nineteenth-century intellectuals, some races fell further from the brood creation than others—blacks most, whites least, with Egyptians and American Indians (and others of varying shades of nonwhite) in between. The *polygenists,* on the other hand, did not need a single womb from which all races were born. Rather, there were multiple "Adams," each race descending from its particular progenitor and represented by currently separate biological species. Most British anthropologists were monogenists influenced by James Prichard. But the polygenists found their champions in such respected scientists as Samuel Morton and Paul Broca. Further, polygenist research established academic credibility for racial policies in both the nineteenth and twentieth centuries, as historian George Stocking notes in his panoramic history of Victorian anthropology: "On the basis of skeletal and cranial evidence, polygenists insisted that blacks were physically distinct and mentally inferior; on the basis of the racial representations on 'ancient Egyptian monuments' they argued that races had remained unchanged throughout the major portion of human history; on the basis of the mortality of whites in tropical areas they hypothesized that different races were aboriginal products of different 'centers of creation' and could never fully 'acclimate' elsewhere; on the basis of anecdotal evidence they asserted that the hybrid offspring of blacks and Europeans were only partially interfertile."[43]

Skull size and brain capacity were important criteria (among others) for racial separation, and Morton's *Crania Americana* (1839), as well as Broca's *On the Phenomenon of Hybridity in the Genus Homo* (1864), lent further support to the polygenist position. The evidence was compelling and their arguments presented with force, with the number of human species varying with the divers polygenists.[44]

It was on the issue of brain size and intelligence that Wallace began to depart from the polygenists, based on a very un-Victorian notion of romantic primitivism that arose out of his anthropological experiences in South America and Malaya.[45] We see in a letter to Darwin on August 30, 1868, for example, Wallace admonishing G. H. Lewes, who "seems to me to be making a great mistake in the *Fortnightly,* advocating *many distinct* origins for different groups, and even, if I understand him, distinct origins for some allied groups, just as the anthropologists do who make the red man descend from the orang, the black man from the chimpanzee."[46]

Wallace's interest in the monogenist/polygenist debate began soon after his return from the Malay Archipelago in 1862. By 1864 he presented his paper "The Origin of Human Races and the Antiquity of Man" to the Anthropological Society of London, in whose journal it was published the same year. Wallace first set out the monogenist/polygenist argument by taking his hyperselectionism to its logical extreme. He begins with a brief outline of the debate: "The one party positively maintaining that man is a species and is essentially one—that all differences are but local and temporary variations, produced by the different physical and moral conditions by which he is surrounded; the other party maintaining with equal confidence that man is a genus of many species, each of which is practically unchangeable, and has ever been as distinct, or even more distinct, than we now behold them." Wallace recognized that evidence alone would not settle the issue, as "each [monogenist and polygenist] will persist in looking only at the portion of truth on his own side of the question, and at the error which is mingled with his opponent's doctrine." Therefore, it was Wallace's goal "to show how the two opposing views can be combined, so as to eliminate the error and retain the truth in each."[47]

The conciliatory Wallace argued that natural selection operated on the physical body of man long before a mind with consciousness existed. The races, represented by a "protoman," were physically fully developed before civilization began, and therefore man is one species. Once the brain reached a certain level, however, natural selection would no longer operate on the body because man could now manipulate his environment. The creation of mind had attenuated the effectiveness of natural selection (and therefore the process of evolution): "In the rudest tribes the sick are assisted, at least with food; less robust health and vigour than the average does not entail death. The action of natural selection is therefore checked; the weaker, the dwarfish, those of less active limbs, or less piercing eyesight, do not suffer the extreme penalty which falls upon animals so defective."[48]

With this alteration of natural law, Wallace argued, came a shift from individual to group selection. While individuals would be protected by the group from the ravages of nature, groups themselves might continue evolving, especially those with high intelligence, foresight, sympathy, a sense of right, and self-restraint: "Tribes in which such mental and moral qualities were predominant would therefore have an advantage in the struggle for existence over other tribes in which they were less developed—would live and maintain their numbers, while the others would decrease and finally succumb."[49] Wallace argued that the harsher, more challenging climate of northern Europe had produced "a hardier, a more provident, and a more social race" than those from more southern climates. Indeed, he pointed out, European imperialism, particularly British, was causing whole races to disappear "from the inevitable effects of an unequal mental and physical struggle."[50] Wallace then answers the question of man's unity or separateness with a mono-polygenism synthesis:

> Man may have been—indeed I believe must have been—once a homogeneous race; but it was at a period of which we have as yet discovered no remains—at a period so remote in his history that he had not yet acquired that wonderfully developed brain, the organ of the mind . . . at a period when he had the form but hardly the nature of man, when he neither possessed human speech, nor those sympathetic and moral feelings which in a greater or less degree everywhere now distinguish the race. If, therefore, we are of opinion that he was not really man till these higher faculties were fully developed, we may fairly assert that there were many originally distinct races of men; while, if we think that a being closely resembling us in form and structure, but with mental faculties scarcely raised above the brute, must still be considered to have been human, we are fully entitled to maintain the common origin of all mankind."[51]

The answer, then, is all in how the question is asked and the terms defined.

Ever the grand synthesizer, Wallace finishes his mono-polygenist blending with a flare of teleological purposefulness and his egalitarian hope for the future of humanity shaped via environmentally determined selection: "If my conclusions are just, it must inevitably follow that the higher—the more intellectual and moral—must displace the lower and more degraded races; and the power of 'natural selection,' still acting on his mental organisation, must ever lead to the more perfect adaptation of man's higher faculties to the conditions of surrounding nature, and to the exigencies of the social state. While his external form will probably ever remain unchanged, except in the development of that perfect beauty which results from a healthy and well organised body, refined and ennobled by the highest intellectual faculties and sympathetic emotions, his mental constitution may continue to advance and improve, till the world is again inhabited by a single nearly homogeneous race, no individual of which will be inferior to the noblest specimens of existing humanity."[52]

Savaging Civilization: Egalitarianism

When Herbert Spencer read Wallace's 1864 paper "The Origin of the Races of Man," he immediately wrote Wallace and told him: "Its leading idea is, I think, undoubtedly true, and of much importance towards an interpretation of the facts. . . . I think it is quite clear, as you point out, that the small amounts of physical differences that have arisen between the various human races are due to the way in which mental modifications have served in place of physical ones."[53] Integrating his unique blend of mono-polygenism with his egalitarian preferences, Wallace believed that all human groups were biologically (and therefore innately) equal, since, from his anthropological fieldwork he had concluded that physical evolution had ceased and the basic hardware was equivalent throughout all races. From his studies of phrenology, it was clear that the structures of the brain were no different between Europeans and so-called "primitives." As he wrote in 1864: "In the brain of the lowest savages, and, as far as we know, of the prehistoric races, we have an organ . . . little inferior in size and complexity to that of the highest type."[54]

Throughout his travels Wallace was struck by both the abilities and moralities of the indigenous peoples he encountered. Writing from Borneo in 1855, Wallace observed: "The more I see of uncivilised people, the better I think of human nature on the whole, and the essential differences between so-called civilised and savage man seem to disappear."[55] And in the 1865 article on "How to Civilize Savages," Wallace sarcastically observed that "the poor savage must be sorely puzzled to understand why this new faith, which is to do him so much good, should have had so little effect on his teacher's own countrymen. The white men in our colonies are too frequently the true savages, and require to be taught and Christianised quite as much as the natives."[56] Wallace offered further empirical evidence of this claim in his observation that natives could be trained in advanced cultural tendencies such as music: "Under European training native military bands have been formed in many parts of the world, which have been able to perform creditably the best modern music."[57]

Wallace was a racial egalitarian in an age of Victorian imperialism that included the polygenist racial dominance of a Eurocentric worldview. In fact, Wallace's trip to America awakened him to the evils of American racism (not to mention what he perceived as the evils of capitalism—thus his reference to white slavery in the passage that follows), which was still prevalent two decades after the Civil War: "We gave them slavery both white and black, a curse from the effects of which they still suffer and out of which a wholly satisfactory escape seems as remote as ever." Yet, through scientifically sound social legislation and policy (via "the teachings of Herbert Spencer"), "It is to America that the world looks to lead the way towards a just and peaceful modification of the social organism, based upon a recognition of the principle of Equality of Opportunity, and by means of the Organization of the Labour of all for the Equal Good of all."[58]

Estimable Characters: Environmental Determinism

Though Wallace would never have referred to himself as a biological or environmental determinist as such, it is interesting to denote the development of his thoughts along this spectrum. It would seem that in the first half of his life Wallace showed moderate biological deterministic leanings, while in the second half, particularly after his anthropological studies and cross-cultural experiences in the Malay Archipelago, he shifted to a more environmental deterministic position. In his autobiography, for instance, Wallace explains that before "middle age" he had developed an elitist attitude toward uneducated and "commonplace" people who took his "reserve and coldness as rudeness." One friend told him that he was unable "to tolerate fools gladly." But his experiences living with native peoples, compared with later observations of his own English culture, "modified" his "views of life" such that "later on, as I came to see the baneful influence of our wrong system of education and of society, I began to realize that people who could talk of nothing but the trivial amusements of an empty mind were the victims of these social errors and were often in themselves quite estimable characters."[59] This passage also serves as another example of why travel is a good proxy for measuring the personality trait of openness to experience. The exposure to other people and environments that comes with travel makes it more difficult to be closed and intolerant.

In addition, Wallace's involvement in spiritualism further reinforced an egalitarian, environmental determinist position: "Later on, when the teachings of spiritualism combined with those of phrenology led me to the conclusion that there were no absolutely bad men or women, that is, none who, by a rational and sympathetic training, and a social system which gave to all absolute equality of opportunity, might not become useful, contented and happy members of society, I became much more tolerant."[60]

Because of the bidirectional influence of Wallace's thoughts, we may understand Wallace's view of the role of natural selection as simultaneously supporting and well supported by his egalitarianism and environmental determinism. In his 1864 paper, for example, Wallace asserted that natural selection ends where civilization begins: in the earliest stages of tool use and the domestication of plants. "From the moment when the first skin was used as a covering, when the first rude spear was formed to assist in the chase, the first seed sown or shoot planted, a grand revolution was effected in nature, a revolution which in all the previous ages of the earth's history had had no parallel, for a being had arisen who was no longer necessarily subject to change with the changing universe—a being who was in some degree superior to nature . . . as he knew how to control and regulate her action, and could keep himself in harmony with her, not by a change in body, but by an advance of mind. . . . Man has not only escaped 'natural selection' himself, but he actually is able to take away some of that power from nature."[61]

Genius and Eccentricity

We began this chapter with a bald statement on the power of thought in history by Theodore Merz, and have shown throughout how Wallace's thoughts on hyper-selectionism, mono-polygenism, egalitarianism, and environmental determinism all influenced his perception of the world around him, both of nature and of society. In his discussion of "The Scientific Spirit in England," Merz claims that "surely the advance of the highest kind of thought will always depend upon the unfettered development of the individual mind, regardless of established habits, of existing forms of expression, or of adopted systems," and that "England, the country of greatest individual freedom, has been the land most favourable to the growth of genius as well as eccentricity, and has thus produced a disproportionate number of new ideas and departures."[62]

There is no denying the relative extent of nineteenth-century British individualism, or the genius and eccentricity of Alfred Russel Wallace, but as we will next see, the established habits, existing forms of expression, and adopted systems of Victorian England most certainly *did* fetter, interact with, and in many ways determine the thoughts of Wallace and his contemporaries. Science, since it is conducted by scientists, cannot help being firmly embedded in a culture that shapes its methods and findings, which in turn feed back into the culture in a self-driving feedback cycle.

9

Heretical Culture

A knotty problem in the psychology of science is understanding why some scientists break out of the paradigmatic mold to launch or lead a new revolution and other scientists do not. Further, since most new ideas in science, as in other human endeavors, are unproductive or simply wrong, how does a scientist know when to challenge the status quo and when to follow it? This is what Thomas Kuhn called the "essential tension" in which "only investigations firmly rooted in the contemporary scientific tradition are likely to break that tradition and give rise to a new one."[1] In other words, one must understand the rules of the game of science in order to violate them.

The tension arises in the conflict and uncertainty of knowing when tradition should give way to change. In many scientific revolutions the successful scientist seemingly holds competing attitudes of traditionalism and iconoclasm at the same time.[2] The tension can become especially high when the commitment to the traditional view is challenged by evidence to the contrary. Iconoclastic scientists are willing to abandon one set of commitments for another of their own creation.[3] But change in science is not entirely driven by commitments, since these commitments are themselves made to particular claims that are usually based on some empirical data. The advantage of the hypothetico-deductive method over pure induction (which is chimerical in any case) is grounded in the fact that there are no absolute Truths to be discovered and that all observations are, in fact, a product of discovery and *description.* Therefore, we must be constantly searching for and testing new hypotheses. This means that the scientist must be vigilant in seeking error in his own research as well as that of others, and that the scientist's work is never done. Thus, scientists who change their mind are simply reflecting the culture of science itself.

Still, some hypotheses in science are borne out by more evidence than other hypotheses, and some theories are superior to others, so one cannot jump from commitments to positions haphazardly. Why some scientists change their minds more than others is a question for psychologists and social scientists to answer, and this biographical study of one scientist is an attempt to understand the difference between a scientist and a heretic scientist. But before we delve further (beyond the Prologue discussion) into the mind and personality of Alfred Russel Wallace, we need to round out his heretical thoughts with his heretical culture to see what social forces and historical trends most strongly shaped the development of his ideas and, especially, how this led to the integration of his scientistic worldview. These forces (following the outline of the Historical Matrix Model) include *Teleological Purposefulness, Scientific Communal Support, Anthropological Experiences,* and *Working-Class Associations.*

Ernst Mayr has written extensively about the process of scientific creativity and receptivity in the history of evolutionary thought, noting that Darwin himself thought hard on the problem, admitting that he was constantly "speculating" about everything he observed, asking questions and tossing out answers to see what would take. "Another characteristic of successful scientists is flexibility," Mayr notes, "a willingness to abandon a theory or assumption when the evidence indicates that it is not valid." He also suggests that "all great scientists . . . have a considerable breadth of interest. They are able to make use of concepts, facts, and ideas of adjacent fields in the elaboration of theories in their own fields. They make good use of analogies and favor comparative studies."[4] This observation is an apt description of Wallace's style as a scientist and thinker.

The *Raison d'Être* of the World: Teleology and Belief in the Perfectibility of Nature

From the earliest Greek philosophers such as Plato, Aristotle, and the Stoics, through medieval theologians such as Thomas Aquinas, into the early modern natural theologians such as William Paley, and into the nineteenth century when Darwin and Wallace were formulating their theories, there existed a pervasive belief that nature is purposeful and designed. The tradition even has biblical roots in the book of Ecclesiastes: "To every thing there is a season, and a time to every purpose under the heaven: A time to be born, and a time to die." Aristotle was the first to recognize in living organisms a teleology, or "end-directed" purpose, that was not present in inorganic matter. That something he called *eidos,* roughly meaning end, goal, or ultimate cause, one of his four different levels of causality. Since he believed that purposeful-looking structures and behaviors could only have come about through either chance or design, it seemed inconceivable to him that the good fit of organisms to their environment could be attributed to chance. "There is purpose, then," he concluded, "in what is, and in what happens in Nature."[5]

Thomas Aquinas offered proof of the almighty's existence through a teleological argument of design: "We see that things which lack knowledge, such as natural bodies, act for an end, and this is evident from their acting always . . . in the same way, so as to obtain the best result. Hence it is plain that they achieve their end not by chance, but by design." This design element constitutes the fifth of five proofs of God: "Now whatever lacks knowledge cannot move towards an end, unless it be directed by some being endowed with knowledge and intelligence, as the arrow is directed by the archer. Therefore some intelligent being exists by whom all natural things are ordered to their end; and this being we call God."[6]

The classic statement of what has since become known as the watchmaker argument—if there is a watch there must be a watchmaker, if there is a world there must be a worldmaker—was made by the Archdeacon of Carlisle, William Paley, in his 1802 work *Natural Theology.* Paley begins with a simple and obvious example with which the reader cannot help agreeing: "In crossing a heath, suppose I pitched my foot against a *stone,* and were asked how the stone came to be there; I might possibly answer, that, for any thing I knew to the contrary, it had lain there for ever: nor would it perhaps be very easy to show the absurdity of this answer." He then contrasts this with a more complex example that logically would seem to require a different answer altogether, one that involves design and purpose: "But suppose I had found a *watch* upon the ground, and it should be inquired how the watch happened to be in that place; I should hardly think of the answer which I had before given,—that for any thing I knew, the watch might have always been there." He then asks rhetorically: "Yet why should not this answer serve for the watch as well as for the stone? Why is it not as admissible in the second case, as in the first? For this reason, and for no other, viz. that, when we come to inspect the watch, we perceive (what we could not discover in the stone) that its several parts are framed and put together for a purpose, e.g. that they are so formed and adjusted as to produce motion, and that motion so regulated as to point out the hour of the day."

Cleverly, Paley now shifts to a counterfactual argument in which he suggests that the structure could not have been otherwise without losing its purpose altogether. "[I]f the different parts had been differently shaped from what they are, of a different size from what they are, or placed after any other manner, or in any other order, than that in which they are placed, either no motion at all would have been carried on in the machine, or none which would have answered the use that is now served by it." This is a probability argument for complex designs, where the more complex the object the less likely it is to have come about through unaided natural forces. The conclusion, to Paley and most of his contemporaries anyway, was obvious and compelling: "[T]he inference we think is inevitable, that the watch must have had a maker; that there must have existed, at some time, and at some place or other, an artificer or artificers who formed it for the purpose which we find it actually to answer; who comprehended its construction, and designed its use."[7] Paley then devotes the rest of his treatise on examples of complex and apparently designed objects from nature, including and especially the eye, which henceforth became the canonical exemplar for creationists of a magnificantly complex, obviously purposeful, and beautifully designed structure that has the handiwork of the deity throughout.

Not everyone accepted the design argument, however. A half-century before Paley put the watchmaker argument on the intellectual map Voltaire lampooned it in his fictional *Candide,* in which Dr. Pangloss, a professor of "metaphysico-theology-cosmolonigology," through reason, logic, and analogy "proved" that this is the best of all possible worlds: " 'Tis demonstrated that

Figure 9-1 William Paley, architect of the watchmaker argument for the intelligent design of the world. (From *Natural Theology,* 1802, frontispiece)

things cannot be otherwise; for, since everything is made for an end, everything is necessarily for the best end. Observe that noses were made to wear spectacles; and so we have spectacles. Legs were visibly instituted to be breeched, and we have breeches."[8] The absurdity of this argument was intended, since Voltaire firmly rejected the Panglossian paradigm that all is best in the best of all possible worlds. At every level of life, culture, and history Voltaire saw that it was anything but the best of all possible worlds and, in fact, to work for change to make it so was an underlying current through the Enlightenment.

Not bothering to couch his critique in fiction, David Hume began by admitting that the watchmaker argument is compelling in that "the author of Nature is somewhat similar to the mind of man; though possessed of much larger faculties."[9] Hume, however, was an atheist, and as such he rejected the design argument by contending that human-crafted objects are not necessarily analogous to the universe, because we can see artifacts being constructed and therefore gain visual and visceral contact with the artificer. We have no such experience with the deity. Furthermore, Hume argued, what about problem of evil, as well as the fact that many of the objects in nature are not so intelligently designed? "This world, for aught he knows, is very faulty and imperfect, compared to a superior standard." Hume then suggests, sarcastically, that perhaps this creation "was only the first rude essay of some infant Deity, who afterwards abandoned it, ashamed of his lame performance."[10] Besides, Hume continued, the idea of an orderly world with everything in its rightful place only seems that way because of our experience of it as such. We have perceived nature as it is, so for us, this is how the world *must* be designed.

Leashed to the teleological argument from design was the naturalist's hierarchical "great chain of being," or a ladder of progress from stones to angels, which itself has a history dating back to Aristotle, richly inculcated into the medieval cosmological worldview, reaching a peak in the nineteenth century.[11] Within this hierarchical system the mind had earned humans the ranking of a superior rung high up on the ladder; before Wallace's century, "savages" were assumed to share the same mental capacities as Europeans, but during the first half of the nineteenth century the hereditarian view dominated and the various human races were classified as separate rungs representing different levels of mental capacity.[12] In the polygenists' theory of human evolution, black savages and white savages were psychologically equal until evolution began to select for superior brains, with white savages being selected for cultural progress and black savages selected against. This evolutionary process, they argued, created a gap between primates and the most advanced civilized humans; "savages" became the missing link in the evolutionary chain.[13]

Even after Darwin and Wallace introduced natural selection as the driving force behind evolutionary change, with gradualism embodied in Darwin's favorite latin dictum *natura non facit saltus,* the teleology of the great chain remained relatively intact. In fact, it subtly supported the imperialism of an ever-expanding Victorian culture that, in the course of a century of "saltwater diplomacy," built an empire on which the sun would never set. Slow and steady wins the race. The daily grind pays off in the long run. The theory of evolution provided an intellectual justification for those on the top rungs to justify their position as naturally superior to those below, who obviously ended up there through the dictates of nature, not politics.

Wallace's unique blend of evolutionary hyper-selectionism, teleological purposefulness, and supernatural spiritualism led him to conclude that through "different degrees of spiritual influx" supernatural forces had "come into action" at least three times in the history of life: (1) in the initial origin of organic life; (2) in the creation of sensation and consciousness in higher animals; and (3) in the shaping of certain human faculties, such as morality and cultural intelligence. "Neither natural selection or the more general theory of evolution can give any account whatever of the origin of sensational or conscious life. . . . But the moral and higher intellectual nature of man is as unique a phenomenon as was conscious life on its first appearance in the world, and the one is almost as difficult to conceive as originating by any law of evolution as the other."[14]

In *Darwinism,* Wallace began conservatively, explaining how he and Darwin had laid the groundwork in a closely parallel fashion. But by the final chapter of the book, entitled "Darwinism Applied to Man," Wallace framed his teleological evolution to contrast those who "maintaining that we, in common with the rest of nature, are but products of the blind eternal forces of the universe, and believing also that the time must come when the sun will lose his heat and all life on the earth necessarily cease . . . who are compelled to suppose that all the slow growths of our race struggling towards a higher life, all the agony of martyrs, all the groans of victims, all the evil and misery and undeserved suffering of the ages, all the struggles for freedom, all the efforts towards justice, all the aspirations for virtue and the well being of humanity, shall absolutely vanish."[15] In response to this "baseless fabric of a vision," Wallace fired off his baldest statement of teleological purposefulness and the perfectibility of nature to date, setting the philosophical pattern for his subsequent stance on a number of fringe and heretical causes:

> As contrasted with this hopeless and soul-deadening belief, we, who accept the existence of a spiritual world, can look upon the universe as a grand consistent whole adapted in all its parts to the development of spiritual beings capable of

> indefinite life and perfectibility. To us, the whole purpose, the only *raison d'etre* of the world—with all its complexities of physical structure, with its grand geological progress, the slow evolution of the vegetable and animal kingdoms, and the ultimate appearance of man—was the development of the human spirit in association with the human body.[16]

This, then, was Wallace's *raison d'être*: a belief in a purposeful cosmos that under the direction of a higher intelligence inexorably led to the appearance of humans who were capable of perfectibility and would, in time, achieve immortality of the spirit. It was a consilient worldview that tied together his many and diverse interests and commitments, ideologies and philosophies, and was ultimately grounded in a unique form of Wallacean scientism. Its origin dates back to the night lectures he attended as a young man at the Mechanics' Institutes, but whose ultimate congealing was the result of countless experiences and ideas he encountered throughout his varied and adventurous life voyage. Despite Wallace's obvious intelligence and creativity that led him to see the anomalies of the accepted scientific paradigm as evidence for a new theoretical model of nature with regards to the origin of species, Wallace was unable to extricate himself from the general nineteenth-century progressivism of most intellectuals, or to look beyond the positivistic vision of humanity culturally advancing toward a more elevated intellectual and moral level.[17] Twenty years earlier, in the paper that caused Darwin so much grief over man, Wallace was "forced to conclude that it is due to the inherent progressive power of those glorious qualities which raise us so immeasurably above our fellow animals, and at the same time afford us the surest proof that there are other and higher existences than ourselves, from whom these qualities may have been derived, and towards whom we may be ever tending."[18] Two decades later, in a *Fortnightly Review* article "Evolution and Character," Wallace minced no words: "My view . . . was, and is, that there is a difference in kind, intellectually and morally, between man and other animals."[19]

Wallace's teleology also led him to argue for the uniqueness of humans not only on earth, but in the entire cosmos as well, a subject on which he characteristically wrote an entire book. In *Man's Place in the Universe* Wallace notes the extreme improbability that every contingent step of evolutionary change from basic bacteria to big brains could possibly have been repeated somewhere else: "The ultimate development of man has, therefore roughly speaking, depended on something like a million distinct modifications, each of a special type and dependent on some precedent changes in the organic and inorganic environments, or in both. The chances against such an enormously long series of definite modifications having occurred twice over . . . are almost infinite."[20]

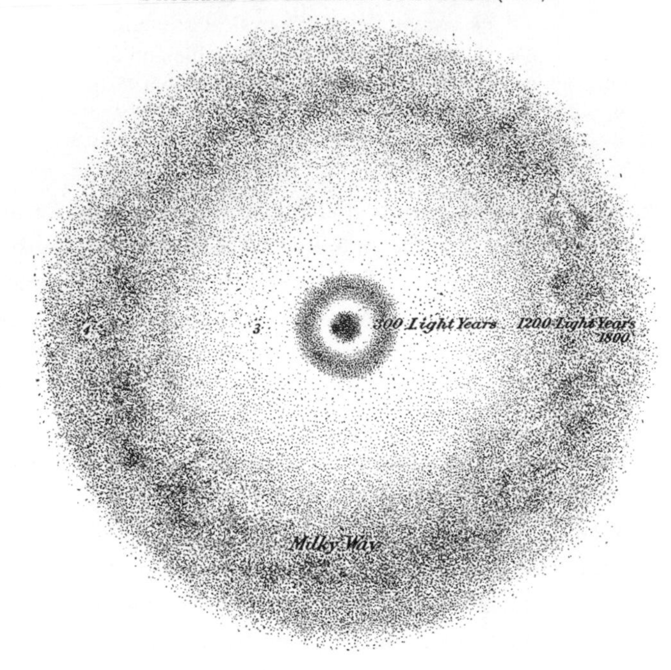

1. Central part of Solar Cluster.
2. Sun's Orbit (Black spot).
3. Outer limit of Solar Cluster.
4. Milky Way.

Figure 9-2 The universe in 1903 was small enough at only 3,600 light-years across (nearly seven orders of magnitude smaller than we think it is today) that Wallace could reasonably argue that the earth is unique in the cosmos as the only planet inhabited by intelligent life. (From *Man's Place in the Universe,* 1903, 300)

Continuing his synthesis, in 1910 Wallace penned the maximally teleological *The World of Life; A Manifestation of Creative Power, Directive Mind and Ultimate Purpose.* "This purpose, which alone throws light on many of the mysteries of its modes of evolution, I hold to be the Development of Man, the one crowning product of the whole cosmic process of life-development."[21] Thus, not only are humans unique in the cosmos, they are also ultimately designed by a higher intelligence, for how else could such a complex being come about? It was an argument that would be echoed throughout the twentieth century, from such religious traditions as the creationists' arguments for the origin and development of the universe and man, to such scientistic discourses as embodied in the anthropic cosmological principle and modern design arguments for directed evolution.[22] In this teleological sense, Wallace was a pioneer and antecedent to these modern intellectual trends and movements.

On the Side of the Angels: Support from the Scientific Community

The rejection of a purely materialist natural selection operating on the human mind was by no means unique or confined to Wallace. In addition to a priggish Victorian society uncomfortable with the idea of man as beast, the scientific community generally failed to rally behind Darwin in his application of natural selection to the human mind. The theologically minded American naturalists Asa Gray and George Frederick Wright, for example, who hailed from the school of "Christian Darwinists," saw evolution as a shaping force guided by the hand of God. Behind all natural law was divine providence.[23]

St. George Jackson Mivart, a highly respected naturalist and contemporary of Wallace and Darwin, supported the validity of natural selection as a creative force in the shaping of the human body, but doubted it could account for man's unique psychological nature. In numerous publications, including *Man and Apes* (1873), *Contemporary Evolution* (1876), *Nature and Thought* (1882), and *The Origin of Human Reason* (1889), Mivart firmly placed himself "on the side of the angels" when it came to the human soul and intellect, which he felt could only be accounted for by supernatural infusion.[24]

The naturalist George Henslow agreed with Wallace and said that man's intellect and moral fortitude "cannot have been evolved solely by Natural laws." In such works as *Genesis and Geology* (1871) and *The Theory of Evolution of Living Things* (1873), Henslow concluded that "some special interference of the Deity" was necessary to give us our moral and religious and intellectual nature.[25]

At the Darwin Correspondence Project at Cambridge there is archived a snippet of a letter, a postscript actually, sent to and saved by Darwin, that supports Wallace's teleology. The excerpt is undated and unsigned:

> I quite agree with you that Wallace's sketch of natural selection is admirable. I was therefore not opposed to his idea, that the Supreme Intelligence might possibly direct variation in a way analogous to that in which even the limited powers of man might guide it in selection, as in the case of the breeder and horticulturist. In other words, as I feel that progressive development or evolution cannot be entirely explained by natural selection, I rather hail Wallace's suggestion that there may be a Supreme Will and Power which may not abdicate its functions of interference, but may guide the forces and laws of Nature.[26]

The letter was from Lyell, dictated to his wife on May 5, 1869, in response to Darwin's letter of the previous day in which he winced at Wallace's heresy: "What a good sketch of natural selection! but I was dreadfully disappointed about Man, it seems to me incredibly strange . . . and had I not known to the contrary, would have sworn it had been inserted by some other hand."[27] Lyell sided with Wallace on this debate. In fact, Lyell predates Wallace, at least in print, on this front. In 1863, six years before Wallace announced his heresy, Lyell published *The Geological Evidences of the Antiquity of Man, with Remarks on Theories of the Origin of Species by Variation.*[28] Here Lyell argued that he could not fathom how anything but a God could have created the human mind. Lyell was not only an important ally because of his status among both the general public and the scientific community, he and Wallace were best of friends. Wallace looked back "upon my friendship with Sir Charles Lyell with unalloyed satisfaction as one of the most instructive and enjoyable episodes in my life-experience," and said that "among the eminent men of science with whom I became more or less intimate during the period of my residence in London, I give the first place to Sir Charles Lyell," ahead even of Darwin.[29]

In fact, for many years Lyell was unable to let go of the traditional picture of man created in God's image, and Darwin lamented this, as he explained to his colleague and confidant Joseph Hooker: "The Lyells are coming here on Sunday evening to stay till Wednesday. I dread it, but I must say how much disappointed I am that he has not spoken out on species, still less on man. And the best of the joke is that he thinks he has acted with the courage of a martyr of old."[30] Darwin believed Lyell to be inconsistent in his reasoning, particularly with regard to the evolution of mind, and this time he let it be known directly. "You . . . leave the public in a fog," he wrote Lyell.[31] Lyell did not help relations with Darwin when he mischaracterized the theory, as in *The Antiquity of Man* where Lyell referred to natural selection as a modification of Lamarckian evolution—an analysis that Darwin found deplorable.[32]

Thus, in the early years in the development of his theory of the co-evolution of body, mind, and spirit, Wallace had at least as much support as Darwin did for his purely materialistic research program, and his allies were among the intellectual elite of the time.

Intelligent and Noble Races: Anthropological Experiences

In May 1855, when he was slogging his way through the heat and humidty of Sarawak in the Malay Archipelago, trying to understand the relationship between geographic isolation and species diversity, Wallace found a few moments to note an observation on the indigenous people: "The more I see of uncivilized people, the better I think of human nature on the whole, and the essential differences between civilized and savage man seem to disappear."[33]

Wallace does not often figure prominently in histories of anthropology, but he did, in fact, write extensively on the subject in prominent scientific journals, including a dozen papers in the *Journal of the Anthropological Society of London.* Of his 747 published papers a full 12 percent, or 90 papers, can be classified in the anthropological sciences (which include archaeology, ancient history, primatology, and linguistics). And of this remarkable outpouring of words the paper cited more than any other was Wallace's 1864 "The Origin of Human Races and the Antiquity of Man Deduced from the Theory of 'Natural Selection.' " Many of his books include long sections and even whole chapters that can be considered serious enthnographic studies. He even served a term as president of the Department of Anthropology, Section D, Biology, of the BAAS, to which he delivered the annual address in 1866, encouraging his colleagues not "to neglect any facts relating to man, however trivial, unmeaning, or distasteful some of them may appear to us" and that "we must treat all these problems as purely questions of science, to be decided solely by facts and by legitimate deductions from facts."[34] No mention whatsoever is made of spiritualism, but it must be what he had in mind as this lecture was given just twelve days after the first installment of his article on "The Scientific Aspect of the Supernatural" appeared in *The English Leader.* Whatever his motives, however, Wallace's anthropological experiences deeply influenced both his science and his social attitudes.

Wallace began observing and reflecting on human origins as early as 1848 when he disembarked the *Mischief* and began his trek into the Amazon. Rousseau's pure and unsullied "noble savage" was there for Wallace to compare to the "nominally" civilized Indians as well as to his own English culture. "The . . . most unexpected sensation of surprise and delight was my first meeting and living with man in a state of nature—the absolute uncontaminated savage," he later recalled. "In every detail they were original and self-sustaining as are the wild animals of the forests, absolutely independent of civilization, and who could and did live their own lives in their own way, as they had done for countless generations before America was discovered." Like Darwin before him, Wallace was incredulous—shocked really—at the novelty of first contact. "I could not have believed that there would be so much difference in the aspect of the same people in their state and when living under European supervision."[35]

Yet, by the time he reached the Malay Archipelago, he was not so sure who was the more advanced: "If these people are not savages, where shall we find any? Yet they have all a decided love for the fine arts, and spend their leisure time in executing works whose good taste and elegance would often be admired in our schools of design!" Wallace even does an anthropological reversal, allowing himself to become the observed, rather than the observer, as he explaind in his Malaya journal entry for April 6, 1857: "I found the tables turned upon me & was become even as the Zulus or Aztecs which I had been one of the gazers at in London. I was to the Arru Islanders a new & strange variety of man, & had the pleasure of affording them in my own person an instructive lesson in comparative Ethnology."[36]

We get an additional glimpse of Wallace's egalitarianism influencing his ethnographic observations in his copy of Darwin's *Descent of Man,* which contains a marginal commentary on British imperialism in South Africa against the Boers and hints at suggesting who the savages and barbarians really are. Darwin wrote:

> The main conclusion arrived at in this work, namely that man is descended from some lowly organised form, will, I regret to think be highly distasteful to many. But there can hardly be a doubt that we are descended from barbarians. The astonishment which I felt on first seeing a party of Feugians on a wild and broken shore will never be forgotten by me, for the reflection at once rushed into my mind—such were our ancestors. These men were absolutely naked and bedaubed with paint, their long hair was tangled, their mouths frothed with excitement, and their expression was wild, startled, and distrustful. They possessed hardly any arts, and like wild animals lived on what they could catch; they had no government, and were merciless to every one not of their own small tribe.[37]

Wallace underlined the words noted in the final sentence and in the margin next to this line he penned: "We are? Boers!"

Wallace is not so naive as to think the native peoples he encountered in the Amazon or the Malay Archipelago were approaching the civility of his fellow countrymen. It was their *potential* for cultural development, given the time and direction, that he saw. He recognized immediately the superstitious nature of their aboriginal lives, as he noted in this amusing journal entry for April 20, 1857:

> I have no doubt that to the next generation or even before I myself will be transformed into a magician or a demigod, a worker of miracles & a being of supernatural knowledge. They already believe that all the animals I preserve will come to life again, & to their children it will be related that they actually did so. An unusual spell of fine weather commencing just at my arrival has made them believe I can control the seasons. . . . My very writing materials & books are to them weird things, & were I to choose to mystify them by a few simple experiments, with lens magnet, etc. hundreds of miracles would in a few years cluster about me & the next European visitors would hardly believe that a poor English naturalist who had resided a few months among them, could have been the original of the supernatural being to whom so many marvels were attributed.[38]

There is little doubt that Wallace went to both the Amazon and the Malay Archipelago, at least in large part, to understand the origins of humanity and society. The subtitle of his book, *A Narrative of Travels on the Amazon and Rio Negro,* affirms this: "With An Account of the Native Tribes, and Observations on the Climate, Geology, and Natural History of the Amazon Valley." In Chapter XVII, "On the Aborigines of the Amazon," Wallace prefaces his observations by noting that "these truly uncivilised Indians are seen by few travellers, and can only be found by going far beyond the dwellings of white men, and out of the ordinary track of trade." Like Darwin in Tierra del Fuego, Wallace experienced the blunt force of humans in the wild, but his description has the ring of the noble savage to it. After comparing them to the "intelligent and noble races of north America" he waxed poetic about their physicality: "Their figures are generally superb; and I have never felt so much pleasure in gazing at the finest statue, as at these living illustrations of the beauty of the human form. The development of the chest is such as I believe never exists in the best-formed European, exhibiting a splendid series of convex undulation, without a hollow in any part of it."[39] And to the Amazonian Indians he compares the natives of Aru, in Malaya, in similar flowery prose meant more to prescribe than describe:

> Here, as among the Dyaks of Borneo & the Indians of the Upper Amazon, I am delighted with the beauty of the human form, a beauty of which stay at home civilised people can never have any conception. What are the finest Grecian statues to the living moving breathing forms which every where surround me. The unrestrained grace of the naked savage as he moves about his daily occupations or lounges at his ease must be seen to be understood. A young savage bending his bow is the perfection of physical beauty. Few persons feel more acutely than myself any offence against modesty among civilised folk, but here no such ideas have a moment's place; the free development of every limb seems wholly admirable, & made to be admired. Tight fitting garments of every kind are disgusting, they hide or distort all the beauty of the human form, while they produce feelings of indelicacy. There is no medium between the nakedness of the savage & the flowing costume of the East. Either exhibit man as a noble & majestic being, all others as a ridiculous animal.[40]

But Wallace, always the scientist, is cautious about drawing conclusions too quickly from limited observations. "In my communications and inquiries among the Indians on various matters, I have always found the greatest caution necessary, to prevent one's arriving at wrong conclusions." Prejudice comes natural. Postjudice requires extra effort, as he discovered on further inquiry. "They are always apt to affirm that which they see you wish to believe, and, when they do not at all comprehend your question, will unhesitatingly answer, 'Yes.' I have often in this manner obtained, as I thought, information, which persons better acquainted with the facts have assured me was quite erroneous."[41]

Wallace's observations and musings on humans in the Amazon, even at this early stage in his development as a scientist (not yet thirty years old), shows his awareness of and sensitivity to important anthropological issues, such as the debate about whether similarity of cultural customs is caused by diffusion or independent parallel development: "One of the singular facts connected with these Indians of the Amazon valley, is the resemblance which exists between some of their customs, and those of nations most remote from them. The gravatana, or blow-pipe, reappears in the sumpitan of Borneo . . . while many small baskets and bamboo-boxes, from Borneo and New Guinea, are so similar in their form and construction to those of the Amazon, that they would be supposed to belong to adjoining tribes." Wallace's explanation, touching on the question of cultural origins, hints at the historian's problem in teasing out the relative influence of contingency and necessity in the past. "It will be necessary to obtain much more information on this subject, before we can venture to decide whether such similarities show any remote connection between these nations, or are mere accidental coincidences, produced by the same wants, acting upon people subject to the same conditions of climate and in an equally low state of civilisation; and it offers additional matter for the wide-spreading speculations of the ethnographer."[42]

More important to Wallace than the physical makeup or cultural development of these Indians, however, was their malleability. He was struck by how quickly native peoples can be changed by civilization: "In the neighbourhood of civilisation the Indian loses many of his peculiar customs,—changes his mode of life, his house, his costume, and his language,—becomes imbued with the prejudices of civilisation, and adopts the forms and ceremonies of the Roman Catholic religion."[43] Although Wallace, unlike so many of his contemporaries, did not think this was such a good influence, it demonstrated to him the environmental plasticity of humans—and thus perfectibility—of man. If people are all born equal, then social inequalities must be caused by environmental injustices, for which many nineteenth-century intellectuals believe a socialized state would be best suited to correct—a position Wallace was well disposed to adopt.

No Individual Inferior: Socialism and Working-Class Associations

Wallace's anthropological experiences with, and ethnographic research on, the Indians of Amazonia and Malaysia were reinforced by beliefs whose origin can be traced to his upbringing and later working-class associations, and especially the influence of prominent socialist thinkers. Although the young Alfred started life in a middle-class home, his father's financial failings more often than not left the family in dire economic straits. When Alfred was six, the family was forced to move five times due to difficulty in paying the rent. When he was thirteen, his father declared bankruptcy, and the following year he was sent to live with his older brother John, who promptly introduced him to the Society for the Diffusion of Useful Knowledge, where he was first exposed to socialist writings.

First among equals in this intellectual cohort was Robert Owen, whom Wallace considered "the real founder of modern socialism," and of whom he wrote: "I have always looked upon Owen as my first teacher in the philosophy of human nature and my first guide through the labyrinth of social science."[44] Nearly three-quarters of a century later he wrote to a friend: "I am just now reading Robert Owens' Autobiography. What a marvellous man he was! A most clear-seeing socialist and educator ages before his time, as well as one of the most wonderful *organisers* the world has seen."[45] Two days later he elevated his praise: "I go even further and consider Owen one of the *first* as well as one of the *greatest* men of the 19th century, an almost ideally perfect character but too far in advance of his time."[46] Obviously in a reflective mood, after reading Prince Peter Kropotkin's autobiography, *Memoirs of a Revolutionist,* Wallace confirmed his own socialistic thinking when he made this comparison: "His early life—its childhood I mean—allowing for immense differences of rank, wealth and country—was, in *essentials* (education, play, etc.) not unlike my own and affords another indication of how wonderfully alike is human nature under all external changes."[47]

If Owen planted the socialist seed in Wallace, the noted evolutionist and polymathic synthesist Herbert Spencer nurtured it into full development. In Wallace's influential 1864 paper "The Origin of Human Races and the Antiquity of Man," after telling the reader that "the general idea and argument of this paper I believe to be new," he went on to give credit—which he did throughout his professional writings, sometimes to a fault—to the idea's source: "It was, however, the perusal of Mr. Herbert Spencer's works, and especially 'Social Statics,' that suggested it to me, and at the same time furnished me with some of the applications."[48] That same year Wallace wrote to Darwin and told him Spencer is "as far ahead of John Stuart Mill as J.S.M. is of the rest of the world, and, I may add, as Darwin is of Agassiz."[49] Wallace even wrote Spencer directly to thank him because "the illustrative chapters of your 'Social Statics' produced a permanent effect on my ideas and beliefs as to all political and social matters."[50] Appreciative of the recognition, and acknowledging Wallace as an intellectual ally, Spencer hired Wallace to read the proofs of *Principles of Sociology.* Decades later, in reflecting on the century in its final year, in a paper entitled simply "Evolution" that included a synopsis of the field's most influential thinkers, Wallace called Spencer's *First Principles* "the greatest intellectual achievement of the nineteenth century" in

Figure 9-3 Alfred Russel Wallace in 1878 at age fifty-five, shortly after publishing *The Geographical Distribution of Animals*, which became the foundation of the science of biogeography, and the same year he wrote "Epping Forest, and How to Deal with it," which cost him the job opportunity as superintendent of Epping Forest because of his radical philosophy of conservationism in which he told investors that they would not be allowed to develop the land for commercial use. (From *My Life,* 1905, v. II, 98)

that it synthesized "all human knowledge of the universe into one great system of evolution everywhere conforming to the same general principles."[51] Finally, in what was surely a rare act of sycophancy on Wallace's part, he named his first son Herbert Spencer Wallace, on which Darwin remarked sardonically: "I heartily congratulate you on the birth of 'Herbert Spencer,' and may he deserve his name, but I hope he will copy his father's style and not his namesake's."[52]

While both Spencer and Wallace shared a similar vision of a classless society through which the greatest happiness for the greatest number would,

Figure 9-4 Herbert Spencer, a powerful influence on the development of Wallace's social theories. Wallace even named his firstborn son after him. (Courtesy of the Henry E. Huntington Library and Art Gallery, San Marino, CA)

in time, be reached, they differed in the mechanism that would produce that society. Spencer saw social adaptation through a Lamarckian inherited effects of habit, whereas Wallace, of course, based his theory strictly on natural selection. Spencer and Wallace also held similar goals for the development of a system of ethics based on science instead of religion. In *The Principles of Ethics* Spencer claimed that his "ultimate purpose, lying behind all proximate purposes has been that of finding for the principles of right and wrong in conduct at large, a scientific basis."[53]

The parallels between Spencer and Wallace are remarkable and telling. Both lived long lives and were contemporaries throughout most of the nineteenth century (Spencer, 1820–1903; Wallace, 1823–1913). In addition to common beliefs in a society and moral system structured on sound scientific principles, Spencer also cut his teeth on phrenology through Johann Spurzheim. Like Wallace, he had his own phrenological reading done by Hicks and Rumball, and even attempted to construct a more accurate scientific instrument for such readings, described in his autobiography, itself published at nearly the same time as Wallace's.[54] Both Spencer and Wallace admired the anarchist/biologist Prince Peter Kropotkin, and, ultimately, believed evolution to be teleological in nature (whether by natural selection or use-inheritance), where, for Spencer, "evolution can end only in the establishment of the greatest perfection and the most complete happiness."[55] Based on these shared theoretical premises, Spencer also showed much interest in Wallace's later attempts at social activism, such as the Land Nationalization Society. But Spencer apparently objected to the extent of state intervention that Wallace proposed. "As you may suppose," he wrote to Wallace, "I fully sympathize in the general aims of your proposed Land Nationalisation Society; but for sundry reasons I hesitate to commit myself, at the present stage of the question, to a programme so definite as that which you send me. The question is surrounded with such difficulties that I fear anything like a specific scheme for resumption by the State will tend, by the objections made, to prevent recognition of a general truth which might otherwise be admitted."[56]

Finally, Spencer and Wallace had similar thoughts on the role of government in bringing about their utopian state. Spencer, for example, observed the failings of governmental intervention in the old Elizabethan Poor Laws, which imposed a tax on the parish to give aid to such needy people as widows, orphans, the sick, and deprived. The problem, as Spencer saw it, was that a welfare system creates a welfare class—government subsidies rewarded these poor women to have more children, and frequently illegitimate ones at that. This, Spencer thought, disturbed the natural, self-correcting balance that the evolution of society was supposed to create. In an article he wrote for the appropriately named journal *Nonconformist,* "The Proper Sphere of Government," Spencer asked rhetorically: "In short, do they want a government because they see that the Almighty has been so negligent in his arrangements of social laws that everything will go wrong unless they are continually interfering?" He gave his answer without qualification. "No; they know, or they ought to know, that the laws of society are of such a nature that minor evils will rectify themselves; that there is in society, as in every other part of creation, that beautiful self-adjusting principle which will keep everything in equilibrium; and, moreover, that as the interference of man in external nature destroys that equilibrium, and produces greater evils than those to be remedied, so the attempt to regulate all the actions of a people by legislation will entail little else but misery and confusion."[57]

Wallace's socialism was a confusing blend of government intercession and laissez-faire inactivity. At times he argued for more government intrusion, usually when he did not think that changes would come about naturally as a result of biological or cultural evolution. By contrast, in an analysis similar to those of his ideological mentor, Wallace argued that bureaucratic mediation disturbed the natural flow of evolution that should someday drive society to an ultimate government-free state:

> While his external form will probably ever remain unchanged, except in the development of that perfect beauty which results from a healthy and well organized body, refined and ennobled by the highest intellectual faculties and sympathetic emotions, his mental constitution may continue to advance and improve till the world is again inhabited by a single homogeneous race, no individual of which will be inferior to the noblest specimens of existing humanity. Each one will then work out his own happiness in relation to that of his fellows; perfect freedom of action will be maintained, since the well balanced moral faculties will never permit any one to transgress on the equal freedom of others; restrictive laws will not be wanted, for each man will be guided by the best of laws; a thorough appreciation of the rights, and a perfect sympathy with the feelings of all about him; compulsory-government will have died away as unnecessary (for every man will know how to govern himself), and will be replaced by voluntary associations for all beneficial public purposes.[58]

It was a naively utopian vision that fit well into Wallace's scientism and his teleological view of evolution, and further reinforced his predilection for social egalitarianism; the theme continues in his writings that the mind can evolve only so far through biological evolution. If the human mind evolves along with the body, and natural selection was inequitable in its treatment of organisms (which Wallace knew it is—nature is amoral), then cultural inequalities would be biologically determined and stiflingly unalterable. Wallace's own cultural background and social experiences, and his later rise to fame and acceptance among society's elite, provided additional empirical counterevidence that there was more to the story than simple materialistic evolution.

Wallace disagreed with those who suggested that man is still evolving physically, because these polygenists saw such evolutionary change as asymmetrical, favoring the "advanced" European race over the "savages" of primitive lands. In his 1864 article on human races, Wallace argued that "it still continues to be asserted or suggested that because we have been developed physically from some lower form, so in the future we shall be further developed into a being as different from our present form as we are different from the orang or the gorilla. My paper shows *why* this will not be; *why* the form and structure of our body is permanent, and that it is really the highest type now possible on the earth."[59] Wallace, however, was not satisfied that *civilization* had advanced to its highest state. "We most of us believe that we, the higher races, have progressed and are progressing. If so, there must be some state of perfection, some ultimate goal, which we may never reach, but to which all true progress must bring us nearer." For Wallace, we may live in the best world possible (as his teleology insisted must be so), but we do not live in the best of all possible worlds: "What is this ideally perfect social state towards which mankind ever has been, and still is tending? Our best thinkers maintain that it is a state of individual freedom and self-government, rendered possible by the equal development and just balance of the intellectual, moral, and physical parts of our nature."[60]

Furthermore, Wallace believed that "savages," on at least one level, were more advanced than civilized peoples: "Now it is very remarkable that among people in a very low state of civilisation, we find some approach to such a perfect state. . . . There are none of those wide distinctions, of education and ignorance, of wealth and poverty, master and servant, which are the product of our civilisation." Here his socialism (in its more traditional sense) rings loud and true: "There is none of that widespread division of labour which, while it increases wealth, produces also conflicting interests; there is none of that severe competition and struggle for existence, or for wealth, which the dense population of civilised countries inevitably creates."[61]

Therefore, and in seeming direct contradiction to his previous statement, in order for society to continue advancing to its perfect state, selection must continue operating, but on an organized, social level, not the random selectionism of nature. In 1890 Wallace published an article on "Human Selection" in the *Fortnightly Review,* and in 1892 one on "Human Progress, Past and Future" in the Boston *Arena.* In both papers he supported "the gradual improvement of the race" (culturally) and opposed "the various artificial processes of selection advocated by several English and American writers," refering to Francis Galton's new science of "eugenics," or the selective breeding for "good traits." While advocating a socialistic form of egalitarianism, Wallace "showed that the only method of advance for us . . . is in some form of natural selection . . . that can act alike on physical, mental, and moral qualities [and] will come into play under a social system which gives equal opportunities of culture, training, leisure, and happiness to every individual."[62] The confusion over Wallace's unique brand of socialism caused him to attempt a clarification in a letter to William Tallack, April 20, 1899, in a discussion on crime and the reform of criminals (the only goal of punishment, he thinks), in which he concludes (knowing, as usual, that his reader will not agree with him): "Of course you will think these ideas dreadfully wild, impractible and socialistic. They are so, no doubt. But then I *am* a socialist."[63]

Wallace, along with Spencer in *Social Statics,* argued that societies progress toward perfection along a hierarchy, with the goal of a state in which everyone might fulfill their purpose without harming others. This ideal state would be a classless society in which people would work for each other, and in the process would yield a utilitarian goal of the greatest good for the greatest number. It must follow, he argued, that "the more intellectual and moral— must displace the lower and more degraded races; and the power of 'natural selection' [cultural, not organic], still acting on his mental organisation, must ever lead to the more perfect adaptation of man's higher faculties to the conditions of surrounding nature, and to the exigencies of the social state . . . till the world is again inhabited by a single nearly homogeneous race, no individual of which will be inferior to the noblest specimens of existing humanity."[64]

Freeland

Wallace's theorizing on socialist reforms for society was no mere academic speculation. He actively sought political and economic reform through legislation, yet was careful to pick and choose his battles. In 1870, for example, he wrote a piece in *Nature* on "Government Aid to Science," surely a source of funding he would fight to increase. Not so, because "though I love nature much I love justice more, and would not wish that any man should be compelled to contribute towards the support of an institution of no interest to the great mass of my countrymen, however interesting to myself." So, although he supported a national education program open to all citizens, government aid to any single program not accessible to the general populace he opposed. "The schools, the museums, the galleries, the gardens, must all alike be popular (that is, adapted for and capable of being fully used and enjoyed by the people at large), and must be developed by means of public money to such an extent only as is needful for the highest attainable popular instruction benefit. All beyond this should be left to private munificence, to societies, or to the classes benefited, to supply." And, characteristically, Wallace formulated his thoughts into a general maxim applicable to situations beyond this specific question: "The broad principle I go upon is this,—that the State has no moral right to apply funds raised by the taxation of all its members to any purpose which is not directly available for the benefit of all."[65]

The mirror image of this principle—that the State has a moral duty to regulate its members to any purpose directly harmful to the benefit of all—was reflected in Wallace's public stance on the coal question under discussion in England in the early 1870s: to what extent should the government regulate the free market in coal? In a lengthy letter to the editor of *The Daily News,* "Free-trade Principles and the Coal Question," Wallace first reasoned by way of analogy that essential commodities for all, which were controlled by only a few, could quickly and easily be restricted, such as a country whose water supply was in the hands of a few owners who then allocated it unjustly, "thus rendering the remainder of the country almost uninhabitable." Likewise coal, which had become an essential fuel commodity "of comfort or misery, even of life or death, to millions of the people whose happiness it is our first duty to secure." Wallace's sense of social justice even extended to the unborn, noting how environmental destruction through coal mining robbed not only this generation of fertile land, but deprived "future generations of any of the advantages we have derived from them," and therefore it is "clearly our duty to check the further exhaustion of our coal supplies by at once putting export duties on coal and iron in every form, very small at first, so as not to produce too sudden a check on the employment of labour, but gradually increasing, till, by stimulating an increased production in other countries, they may no longer be required."[66]

Of all precious commodities, however, none took center stage for Wallace more than land. Although he wrote numerous pamphlets, magazine articles, and letters on a variety of social issues and reform, he dedicated one book, *Land Nationalization,* to the topic of what needed to be done to achieve social stability and economic equality. He was the founding president of the Land Nationalisation Society, whose basic philosophy was that no one really owns land (land is permanent, human life is tenuous), yet landowners have an unequal share of wealth and power, therefore the government should nationalize all privately held land and redistribute its *use* to the public. In the Preface to *The Case for Land Nationalisation* by Joseph Hyder, Wallace wrote that because "no individual *can* absolutely *own* land. . . . The very least that can be done is for Parliament to recognise that existing land-holders and their living heirs have no more than a *life interest* in the land they are permitted to hold, and that they shall in no case be compensated for more than the lowest net value of that life interest." Once this is done, Wallace argued, "only will this great injustice and spoliation of the people be gradually and beneficially redressed, with full regard to the fundamental rights of all to the use and enjoyment of their native land."[67]

Wallace even applied this principle to the Church of England, calling for religious reform by declaring "that existing Church Property of every kind is National Property, and that no portion of it must under any circumstances be alienated, either for the compensation of supposed or real vested interests, or to the uses of any sectarian body; and further, that the parish churches and other ecclesiastical buildings must on no account be given up, but be permanently retained, with the Church property, for analogous purposes to those for which they were primarily established—the moral and social advancement of the whole community." Once again Wallace's sense of fairness led him to reason that the Church of England had a certain religious duty that was being compromised by sectarian interests and religious prejudices that were a result of its enormous power derived through its "venerable antiquity; to its intimate association with our great Universities; to its establishment by law and its position in the Legislature; and to its possession of the cathedrals and parish churches, which from time immemorial have been the visible embodiments of the religion of the country." It was the latter that tweaked Wallace's ire—the ownership of property. "The clergy of the Church of England owe their chief influence for good in their respective parishes to their connnection with these permanent and often venerable buildings."[68] A modest revolutionary in this case, Wallace called for reform of the Church through various measures linked to their property, not dissolution of the Church.

In 1900 Wallace sought to purchase, along with a number of investors, a "Joint Residential Estate" outside of London, "in a healthy district, with picturesque surroundings, which can be permanently preserved," as a type of planned community "to secure many social, residential and material advantages, in a rural retreat, that could not be obtained by single-handed or individual effort."[69] The project never got off the ground, but it shows Wallace's active involvement in attempting to live by his principles, as does a fantastic plan for a utopian community called "Freeland" that was to be established in Africa. In his presidential address at the annual meeting of the Land Nationalisation Society, June 23, 1892, Wallace discussed and promoted a book entitled *Freeland; a Social Anticipation,* by Dr. Theodor Hertzka, a writer on political economy in Vienna. The novel's setting is central Africa, where a new colony is established "based on the free use by all of the nation's land, and also of its accumulated capital; together with the very ingenious arrangement by which the wealth creating opportunities of all are equalised."[70] Not long after, Wallace was nominated vice-president of the Executive Committee of the Freeland Movement (Hertzka was president), whose goal was to colonize the "unoccupied highlands surrounding Mount Kenya, in the interior of Equatorial Africa." In a letter to the editor accompanying a two-page printed announcement of the Freeland Colony, Wallace and Hertzka did a similar mailing to all newspapers in England with the hopes of raising money to establish the "Freeland Colony." In essence, these utopian entrepreneurs proposed "to establish a community on the basis of perfect economic freedom and justice, a community which shall preserve the independence of its members, and shall secure to every worker the full and undiminished enjoyment of that which he produces. By placing the means of production at the disposal of the workers, we shall enable them, without exception, to work in the most advantageous manner."[71]

The letter requested contributions to be sent to any of the listed banks, addressed to "The Freeland Colony," at the address for the "British Freeland Association." In the statement itself, Wallace discusses the general philosophy, as outlined in the novel *Freeland,* and notes that "the idea of this scheme has already excited so much enthusiasm that within a little more than two years of the publication of the first edition of *Freeland* (early 1890) Associations

PRIVATE and CONFIDENTIAL.

PROPOSAL as to a JOINT RESIDENTIAL ESTATE.

1. In order to secure the advantages of a Country Home, in a healthy district, with picturesque surroundings, which can be permanently preserved, and within one to two hours of London, it is proposed that a few persons shall purchase jointly, an estate of from 100 to 300 acres or more.

2. During the summer of 1900 Dr. A.R. Wallace and Mr. A.C. Swinton examined a large number of estates, and this year, Dr. A.R. Wallace and Mr. A. Roland Shaw have inspected several more, but without finding one in all respects suitable, though some were very attractive.

3. Believing that further research will soon lead to success, it would then be necessary to act promptly. The above-named gentlemen must therefore ascertain how many of their friends are prepared to join in the purchase, of course conditionally on their approval of the estate chosen.

4. They believe that a suitable estate may be found at a cost of from £10,000 to £12,000 and if that amount can be obtained from ten or even from twenty persons, they will proceed energetically with the search. The three gentlemen named are prepared to advance £3,000 towards the purchase (or their proportionate share if less is needed), and will be glad to receive, promptly, the names of those who will join to complete the amount. Sums of £500 would however, be accepted, with a proportionate interest in the estate.

5. The estate being purchased, Dr. A.R. Wallace, as being the originator of the plan and having spent much time in the search, will take, as his portion, the dwelling house and adjacent grounds, and will then act as the Agent and Surveyor for the group of purchasers till the whole estate is occupied. Mr. A. Roland Shaw will undertake the business correspondence and negotiations with sellers and purchasers, and other business details, and for these services will be entitled to the choice of a site.

6. It is proposed to mark out as many residential beneficial sites, as there are persons who contribute towards the purchase, these sites to be so selected and arranged as to be approximately equal in value to the sums subscribed. According to size of the estate and other circumstances, these sites would probably range in acreage from two up to ten acres.

7. Part of the estate, including most of the woods and wilder portions, would be reserved as a natural park for the enjoyment of all the residents alike.

8. Surplus land would be disposed of for the joint benefit of the beneficial owners - that is, those who joined to purchase the estate - in any way they consider advisable.

9. It is however, understood that nothing more than their advances for the purchase of the estate is to be divided among the purchasers. Any further revenue or profit that may accrue shall be expended upon the estate for the benefit of all its residents.

10. By some such plan of co-operation as above outlined, it is considered possible to secure many social, residential and material advantages, in a rural retreat, that could not be obtained by single-handed or individual effort. Such co-operative effort may also result in enhancing the value of the land purchased by reason of the improvements and privileges provided for, and in speedily returning the purchase money to the several beneficial owners.

11. Persons receiving this statement are requested to write soon to A. Roland Shaw Esq., 12 Suffolk Street, Pall Mall, London, S.W., stating their wish to become one of the purchasers, and to what extent they are prepared to subscribe.

So soon as the undersigned ascertain what support they receive, they will be in a position to select one or more estates for the inspection of the subscribers, and, if generally approved, secure it as speedily as possible.

Alfred R. Wallace,
(SIGNED) -- A. Roland Shaw,
A.C. Swinton.

May 15th, 1901.

Figure 9-5 Wallace's proposal for a joint residential estate to be initially purchased with A. Roland Shaw and A. C. Swinton, May 15, 1901. The plan was to solicit numerous other investors to fund the development of a cooperative land project to test Wallace's socialistic theories of land ownership and management. The project never came to fruition. (Courtesy of the Zoological Society of London)

have been formed in over a score of Continental cities for the purpose of founding an International Colony on these lines." Elsewhere Wallace discussed his attempts with the British Royal Navy and Army to help secure the land in Africa, but beyond these few archival artifacts there is nothing else published on the movement, or even how far along in negotiations or finances they progressed toward the founding of such a colony. There is no record of a British Freeland Association at any of the general British archives in London, no mention is made of it in Marchant's collection of Wallace's letters (except for Wallace's having read *Freeland* in 1892), and it is conspicuous by its absence in Wallace's autobiographical work *My Life,* written just a few years after the scheme unraveled. There are two amusing anecdotes associated with the project, the first being in a letter of November [?] 1892, from one J. Brailsford Bright, M.A., who suggests to Wallace that England institute the Freeland Colony because she is so good at colonizing foreign lands (and thus English should be the official language of the new society); and two, in a letter written on January 22, 1894, Hertzka tells Wallace that while "Freeland" should have no permanent political structure, they will need a temporary autocratic government to get things started, and he, Hertzka, volunteered to be head of state.[72]

The whole incident is emblematic of Wallace's attempt to structure his life based on his principles, derived through what he considered to be sound, scientific reasoning and careful, rational discourse. Wallace's forays into the realm of spiritualistic séances, and his working-class socialist associations with Owen and Spencer, grafted onto such schemes as establishing a Freeland Colony in Africa, reveal a personality bursting with youthful energy and idealism even in the final decades of a very long life in which countless blows and setbacks would have hardened the hearts of less noble men. Wallace was more than a heretic scientist. He was a heretic personality, interested in and willing to get involved with any number of fringe elements in science and society. And he attempted to integrate them all into a unified scientific worldview. His remarkable tenacity in the face of almost total failure to achieve his goals in the social realm—particularly striking when compared to his accomplishments in the natural sciences—simply reinforces the observation that temperament drives intellect, and that personality is the motor of the mind.

10

Heretic Personality

In the annals of science one would be hard-pressed to find a more affable individual (who is also controversial) than Alfred Russel Wallace. After getting to know the man through a thorough reading of his letters and correspondence, papers, manuscripts, and books, one cannot help liking him. Recalling the generosity shown him as a young scientist by those already established (e.g., Lyell, Darwin, Hooker, Huxley), for example, Wallace later in his life returned the favor to the next generation of budding naturalists—his friend and editor of his letters, James Marchant, recalled that "Wallace loved to give time and trouble in aiding young men to start in life, especially if they were endeavoring to become naturalists. He sent them letters of advice, helped them in the choice of the right country to visit, and gave them minute practical instructions how to live healthily and to maintain themselves. He put their needs before other more fortunate scientific workers and besought assistance for them."[1] On Wallace's death in 1913, his friend Edward Poulton, writing in *Nature,* concluded: "The central secret of his personal magnetism lay in his wide and unselfish sympathy. It might be thought by those who did not know Wallace that the noble generosity which will always stand as an example before the world was something special—called forth by the illustrious man with whom he was brought in contact [Charles Darwin]. This would be a great mistake. Wallace's attitude was characteristic . . . to the end of his life."[2]

In this final assessment of Wallace's lifelong attitude is the key to a deeper understanding of the man, because personality traits and temperament tend to hold steady throughout the life of an individual and, while subject to the whims and vagaries of social conditions and circumstances, dramatically shape responses to those environmental states. Wallace's humble origins and self-made career, for example, may help to explain this generosity and kindness. He understood the difficulties of most people and could relate to their struggles. His varied and diverse education and experiences also shaped a separatist personality and created an independent thinker—good for creativity in breaking out of a paradigmatic mold (e.g., his discovery of natural selection), but making him more gullible to unusual claims (e.g., spiritualism), especially when mainstream science failed to account for such human skills as mathematical reasoning, aesthetic appreciation, and spiritualism, leading him to conclude that there must be a higher intelligence that guided and enhanced nature's selective hand.

Wallace's heretical view of the human mind was just one of many that he underwrote as part of the worldview of a heretic scientist, bent on extreme independence of thought and maverick tendencies. In addition to tackling countless scientific problems within his own field of biology, Wallace frequently went beyond the boundaries of natural history. He dabbled in astronomy, in which he considered the possibilities of life elsewhere in the universe; he studied medicine, when he investigated and then rejected vaccination on the grounds that there was no data to support it; and he was a serious amateur geologist, rejecting the theory that the continents drift around the planet because there was no driving mechanism (but he did reject the "land bridges" theory, popular at the time to explain the worldwide distribution of species). Wallace also brought the full weight of his scientific expertise to bear on many social issues and problems, including ecological conservation, overpopulation, war, poverty, unemployment, money, political representation, the House of Lords, land nationalization, labor strikes, the women's movement, individualism and collectivism, and morality, on all of which he authored numerous articles and several books. He took a public stand and exposed himself to both the rewards of attempting to solve such social conundrums as well as the ridicule of critics who did not agree with his solutions.

Wallace was willing (and quite scientifically able) to test any and all natural and supernatural claims. If he thought the evidence supported a theory or conjecture, Wallace was willing to go to any length to lend his credibility and toss his weight behind it. When the evidence did not support a claim, Wallace was as vociferous in his rejection as any of his skeptical colleagues. This tendency was never more evident than when he accepted spiritualism (which he felt had substantial evidentiary backing) and defended it with all his literary and scientific power, while simultaneously rejecting reincarnation for lacking any cogency of logic or empirical evidence. Taking Wallace as a case study in finding the essential tension between conservatism and openness in science, he usually erred on the side of the latter, preferring to risk being right rather than playing it safe and possibly missing out on a scientific revolution.

For Wallace, the decision was an intellectual one based on his perception of the boundaries of science and how its methodologies were defined. Of course, no one is purely objective in these matters, as personality can play a key role in shaping receptivity to ideas and preferences for where those boundary lines should be drawn. Thus, the life and personality of Wallace are themselves a test case for resolving the boundary problem in science—where do we draw the line between science and pseudoscience, and science and nonscience? More important, why are some people drawn to the fringe side of the boundary, willing to risk careers (and sometimes lives), gambling on revolutionary ideas, while others greatly prefer the more conservative approach of playing it safe until a consensus is reached? The answer to these questions can primarily be found through the study of the psychology of personality, as well as social psychology and sociology, since such intellectual preferences are driven by an interaction of internal traits and external states. As we explored in the Prologue on the psychology of biography, it is clear that temperament is a powerful force in human history, itself shaped by a number of variables, including genetics, birth order, sibling rivalry, parent identification, parental conflict and separation, family dynamics, peer groups, and mentors. How temperament plays itself out in history can only be seen in specific anecdotes that serve as narrative data, as it were, for or against this view of psychohistory. Here we shall explore more deeply the personal life of Wallace to flesh out the personality analysis with a synthesis of descriptions, stories, and events in his life that round out this profile of him.

The Personality of a Heretic Scientist

What was Alfred Russel Wallace like as a person? A number of stories on and interviews with Wallace that appeared in the final decades of his life help flesh out his personality and temperament. During his lecture tour of America in 1886 he was interviewed by *The Sunday Herald* of Boston, when he was there to present his "Lowell Lectures on the Darwinian Theory." He was described as "a man considerably above the medium height, is not at all a typical Englishman, has a slight stoop that shortens his height and is rising 60 years of age. He wears glasses and has a fresh countenance. His hair is white rather than gray and his beard is worn rather heavy, and is nearly white. He has a venerable look, and might be taken to be older than he is." The interviewer's portrayal of his personality is very American: "His face lights up in conversation, and there is nothing in his manner or features to distinguish him from an American. He has the bearing of an ordinary citizen rather than that of a scientist, but there is a strong individuality beneath the quiet exterior, and, after the first steps of acquaintance are entered upon, he reveals himself as a very agreeable gentleman. His presence is so good, and his enunciation is so clear for an Englishman, that he ought to be easily heard by his audience, which, at least on Monday night, will be as distinguished as any that has greeted an English lecturer before the Lowell Institute for some time."[3]

In 1893 *The Daily Chronicle* published an interview with Wallace regarding his views on women's rights and the role of women in evolution. The author began with this description of Wallace's home and the spectacular view he enjoyed as he neared his seventy-first birthday: "Three miles of lonesome road, cut through a pine forest, separates the home of Dr. Wallace, at Parkstone, from fashionable Bournemouth. The house itself, standing on a slight elevation, commands a fine view across the sea to Swanage and the Purbeck Hills." After entering the home, the interviewer "entered a cosy retreat, in the lower part of the house, ranged around with books and pictures, the chairs suggestive of comfort and the well-littered tables of much study and research." The naturalist's study, of course, "looked through the stretch of windows flanking the outer side of the room to the garden beyond, rising gradually upwards until it joined the distant wood. Then the lamp was lighted, the blinds drawn down, and the great scientist seated himself in his special armchair, drawn up close to the blazing fire, and proceeded to discourse upon the subject of natural selection, in which, as an original thinker, he stands unequalled save by Darwin."[4]

Five years later a journalist for *The Bookman* noted that "Dr. Wallace's travels and adventures in early life seem to have hardened his physique. No symptom of feebleness, physical or mental, is perceptible. With his tall substantial figure, still erect but for slight 'scholar's stoop,' his head thickly covered with smooth white hair, Dr. Wallace's appearance is at once robust and dignified." Like most descriptions of Wallace, this author "is charmed by the native simplicity and modesty of his speech and demeanour; he seems never to regard himself as one of the notable men of the century. Despite some bronchial trouble and a slight tendency to asthma, Dr. Wallace's general health is good; he says he now feels as well as he has done any time during the last twenty years. He rarely takes alcohol, and has never smoked. Although he has not forsworn meat, he believes that vegetarianism is sound in principle, and will ultimately become universal." Here we get a glimpse into the personal life of Wallace, still remarkably active at the age of seventy-five: "Dr. Wallace continues to contribute occasionally to the reviews and magazines. He usually does his writing in the forenoon. He believes in taking plenty of recreation, and has several hobbies, gardening among others. For many years he has cultivated every plant that he can get to grow in his garden. In his conservatory he has a great variety of orchids. His indoor hobby is chess. He likes music, but only very grand music." Tellingly, when asked to compare his personal habits to those of Darwin, Wallace responded: "Darwin was a continuous worker at his one great subject; I am not. I should not be happy without some work, but I vary it with gardening, walking, or novel reading. Even when in the midst of writing a book I never cease to read light literature." And what did Wallace read for recreation? "He spoke of Miss Jane Barlow, who was recently in Parkstone, as 'one of the most delightful writers of the day.' He particularly enjoyed 'Irish Idylls.' He said that H. Seton Merriman's 'Sowers' was one of the most striking, vivid pictures of Russia he had ever read." Interestingly, we see Wallace's optimistic, nearly utopian view of human nature, in his confession that "I dislike the whole pessimist school of writers. I have read two of Hall Caine's books, 'The Manxman' and 'Son of Hagar,' and they are full of misery, horror, pain, trouble. I hate it; that is not human nature."[5]

In the final two years of his life Wallace granted several interviews, one of which labeled him as "The Last of the Great Victorians." Frederick Rockell of *The Millgate Monthly* "pushed open the gate, which bore the inscription, 'Old Orchard–A. R. Wallace'—and found myself in a luxurious garden where flowers of many varieties contested in a friendly rivalry of shape, colour, and perfume. As I entered the porch, two merry children ran out of the house into the garden, and I realised that in the winter of his life the great scientist was still closely in touch with the innocence and fragrance of childhood."[6] He remained so to the end, along with his sanguine personality described by W. B. Northrop in the New York-based magazine *The Outlook* just days after his death in 1913 as "a man of great modesty. It is seldom that greatness in this world is allied to humility; but Dr. Wallace possessed self-abnegation to a rare degree." Northrop continued with this romanticized portrayal of Wallace's home life: "He occupied a small tract of land called the Old Orchard, not far from the little village of Broadstone, one of the prettiest hamlets of Kent, about five hours' ride southwest from London. His house was of the rambling English country type, and stood on a knoll commanding a view of the town of Poole and its pretty harbor. Here Dr. Wallace spent the evening of his days, devoting his spare time, when not writing books and magazine articles, to raising chickens, gardening, cross-country walking, and playing chess with neighbors who chanced to call. Up to within a year or two ago Dr. Wallace had been assisted in his work by Mrs. Wallace, who helped to prepare all his manuscripts and to read the proofs of his various books and articles. Dr. Wallace, like our Mark Twain, did all of his work with a pen, and never cultivated dictating to stenographers or using a typewriter. He made it a point to turn out each day about six thousand words—a high average for literary production."[7]

Such quaint descriptions are supportive of the description of Wallace as a highly agreeable personality. In fact, Wallace described his own "natural disposition" as "reflective and imaginative," which he attributed to "the quiet and order of my home, where I never heard a rude word or an offensive expression." This, he said, "was intensified by my extreme shyness." Personality and temperament, of course, do not a great scientist make. To explore this further, I asked the expert raters who had assessed Wallace's personality (results presented in the Prologue), "How would you describe Alfred Russel Wallace's unique intellectual style? In other words, what are his strengths and weaknesses as a scientist? In particular, I am interested in answering the question of how and why a world-class scientist like Wallace was so interested in fringe and heretical sciences and social causes."[8] The answers were revealing. Linnean Society archivist and librarian Gina Douglas noted: "I think he had a very open mind . . . a person with a very broad outlook." Darwin biographer Janet Browne quibbled: "If 'fringe' is redefined more broadly I think Wallace does not look nearly so 'heretical.' " Historian of science Michael Ghiselin made this modern comparison: "There is nothing unusual about Wallace's interest in such matters. Had he been at Berkeley in the 1960s he would have been opposed to fluoridation and 'into' acupuncture and communal life styles. Such interests made him open to novelty but there is a serious tradeoff if one is a bit gullible as was he." Linnean Society director John Marsden wondered: "Is there an answer to this? Most Nobel laureates I have been acquainted with (around seven or eight) have had severely flawed personalities. I don't think Wallace was that bad. In fact, he seems to have been a pretty decent sort. A number of people of this kind were attracted to socialism, not least Marx and a variety of intellectuals. Look at the alternatives! Hardly wonderful. As you note he had this weakness for psychics. Perhaps those who live in glass houses should not throw stones!"

Darwin biographer and social historian James Moore objected to the quantification of Wallace's personality: "To answer your question would require a long exposition of Victorian social history, including the social history of science and the sociology of knowledge. The question is a good one, both precisely and eloquently answerable in the terms of these essential disciplines. I've done as you asked, though I have to say that I think Sulloway's method is profoundly unhistorical (I told him so) and next to useless for understanding Wallace. The other thing I would say is that most of the responses are based on educated guesswork and hunch. You will end up with a composite view, not of ARW, but of what experts guess, suppose, or presume about him. In other words, it will not be a composite of a real person, as in Galton's composite photographs. Still, I look forward to your findings as to what the experts think, even though ARW will turn out to be bigger, more various, and more awkward than them all!" Perhaps, but then why did ten distinct expert raters all so consistently rate Wallace's personality? Can anyone doubt that such personality traits exist? And if they do, are they not in some manner measurable? And if they are, should we not apply the best methods of science to make that measurement? And if we should, why wouldn't the world's leading historians on a subject be qualified to so make that measurement?

Wallace archivist and biographer Charles Smith's comments were especially insightful in integrating Wallace's personality with his intellectual style, philosophy of science, and social causes:

> In general, I would say that Wallace had an extremely keen mind: both in the sense that he was able to firmly come to grips with the essentials of just about any subject he took an interest in, and in the sense that he was then able to use a very powerful capacity for logical reasoning to apply that knowledge fruitfully in whatever direction it might take him. Further, he used a highly developed ethical and moral sense as a means of logical assessment: basically, the notion that any kind of social strategy that created elemental wrongs at the level of the average citizen could not be viable in the long run, and could be discounted.

Smith begged to differ with my portrayal of some of Wallace's interests as "fringe," adding this assessment of the boundary problem and Wallace's role in determining what constitutes legitimate science:

> As far as his "fringe" interests go, I would suggest that they were then (and in many cases still are) considered "fringe" only because others had not yet caught up with the thinking involved (or in addition, in the case of spiritualism, still cannot devise adequate tests of related matters, or refuse to entertain the notion that the basic idea may be correct, though anthropomorphized). I would describe Wallace as an *absolutely* "fearless" thinker, but not a foolhardy one. To me, his strongest weakness as a thinker was his tendency to too absolutely trust some kinds of physical data as being finally diagnostic: thus, his errant conclusions in some aspects of biogeography, astronomy and glaciology. I realize he is also criticized for being gullible in his dealings with mediums; however, it seems to me that *some* of his experiences (especially those which took place in his own quarters) are difficult to easily discount. I am a reasonably skeptical person; still, it seems to me (as someone who has spent a good deal of time over the years considering the evolutionary process and related systems concepts) that Wallace's model of evolution, incorporating social and spiritualist components, is more on target than anyone else's.[9]

Historian of science Richard Milner considered Wallace's personality to be the source of both his strengths and weaknesses with this rather different assessment of Wallace's forays into the unknown: "I don't see how anyone who has seriously studied the man's character can fail to be deeply impressed by the contradictions and paradoxes of his personality. I think he was an extraordinarily good man with an uncommonly trusting attitude. Because he was so honest and straightforward, he thought that everyone was like that. He could not conceive that men who spoke like gentlemen interested in exploring the frontiers of knowledge could be calculating deceivers, who lied to his face for monetary gain. (I'm talking here about the various Spiritualist imposters and conmen, of course.)" As for the contrast of Wallace's personality with Darwin's, Milner made this insightful observation that supports modern theories of happiness as a temperament independent of social status or wealth: "I think the contrast between Wallace and Darwin at the end of their lives is most interesting. Darwin became depressed and melancholy, could no longer find any joy in scenery, art, or music, and peered gloomily into the coming darkness, where thoughts and personality would cease forever. Wallace lived much longer and happily, despite his penury, and cheerfully looked forward to his adventures in the Spirit World. Another glorious expedition into the Unknown." Finally, Milner, like Moore, rejected any attempt to quantify Wallace's life and work: "Yet the fundamental paradoxes of his nature are puzzling. I doubt that your quantitative analysis or attempts to find a simple correlation in his relationship with parents or siblings will throw much light on the matter. Wallace remains an enigma. Can we agree to disagree on how to 'solve' the complexities of this strange, delightful, brilliant, and noble man? He will not be pinned, boxed and labelled like one of his Papillary butterflies."[10]

Of course, to naturalists who came before Darwin and Wallace nature seemed as enigmatic and complex as human personality, but it was precisely because Darwin and Wallace pinned, boxed, and labeled nature that they were able to discern the pattern within the noise. We can do no better than follow the precepts of such eminent naturalists in our exploration of the natural history of personality, starting with a heretic personality, *or the unique pattern of relatively permanent traits that makes an individual open to subjects at variance with those considered authoritative.* This description well fits Wallace, who routinely maintained opinions on a variety of subjects typically at odds with the received authorities. A heretic personality is an individual, like Wallace, who differs from the majority in his openness to and support of ideas considered heretical, while also maintaining anti-authoritarian, pro-radical sympathies. These traits, being "relatively permanent," are not temporary conditions, or "states" of the environment, the altering of which changes the personality. The heretic personality, like any other personality trait, tends to act consistently over most environmental settings, throughout much of a lifetime.

Wallace became interested in heretical theories as a very young man, investigating, for example, phrenology, and considered controversial biological problems such as the mutability of species. This was not, however, a temporary flirtation with anti-authoritative ideas by a young, undisciplined mind. In midlife, after codiscovering with Darwin their innovative (and at the time moderately heretical) theory on the origin of species by means of natural selection, Wallace began experimenting with spiritualism and many other controversial beliefs. What establishes Wallace as a genuine heretic personality was that he demonstrated a unique pattern of relatively permanent traits that caused him to maintain opinions on a variety of subjects throughout his life at variance with those considered authoritative. The following two incidents, both of which occurred well into his later years, provide an exemplar of the heretic personality in action.

Challenging the Flat-Earthers

On January, 12, 1870, Alfred Wallace read the following advertisement in the journal *Scientific Opinion:*

> The undersigned is willing to deposit from £50. to £500., on reciprocal terms, and defies all the philosophers, divines, and scientific professors in the United Kingdom to prove the rotundity and revolution of the world from Scripture, from reason, or from fact. He will acknowledge that he has forfeited his deposit, if his opponent can exhibit, to the satisfaction of any intelligent referee, a convex railway, river, canal, or lake.[11]

The undersigned was John Hampden, who had become convinced by Samuel Birley Rowbotham's book, *Earth Not a Globe,* that the earth is immovable and flat, with the North Pole at the center and the sun orbiting a toasty-warm 700 miles above the plane of the planet. All proofs of the earth's sphericity, such as the earth's rounded shadow on the moon in a lunar eclipse, were written off as scientific propaganda constructed by unskeptical Copernicans. (The rounded shadow, flat-earthers argue, is because the earth is both round and flat, like a saucer, but not spherical. Likewise, modern flat-earthers believe satellite images from space are merely photographs of the upper side of the flat, rounded plane.)

Hampden took up the flat-earth cause with proselytizing enthusiasm. He obtained permission from Rowbotham to arrange for the publication of a pamphlet extracted from his book, and printed by William Carpenter, who was shortly to enter the story in another capacity. The advertisement soon followed the publications, and, unfortunately for Wallace, the challenge proved too tempting to resist. Soon after, Wallace became embroiled in an incident that, he later claimed, "cost me fifteen years of continued worry, litigation, and persecution, with the final loss of several hundred pounds." Wallace confessed that the blame was entirely his: "And it was all brought upon me by my ignorance and my own fault—ignorance of the fact so well shown by the late Professor de Morgan—that 'paradoxers,' as he termed them, can never be convinced, and my fault in consenting to get money by any kind of wager." Wallace later admitted that this was "the most regrettable incident in my life."[12]

His sense of challenge piqued (and his pockets rather empty), Wallace wrote his friend, the renowned geologist Charles Lyell, "and asked him whether he thought I might accept it. He replied, 'Certainly. It may stop these foolish people to have it plainly shown them.' " Hampden suggested using the Old Bedford Canal in Norfolk because it had a straight stretch of six miles between two bridges. Wallace agreed, suggesting that John Henry Walsh, editor of *Field* magazine, act as chief referee. Hampden agreed to Walsh as judge and witness, in addition to which each man would bring along a personal referee. Wallace was accompanied by one Dr. Coulcher, "a surgeon and amateur astronomer." Hampden brought with him none other than William Carpenter, the printer of Hampden's flat-earth literature. Walsh held the money to be given to the winner.

On the morning of March 5, 1870, Wallace set up three objects—a telescope, a disk, and a black band—along the six-mile stretch of the Old Bedford Canal, eighty miles north of London, such that "if the surface of the water is a perfectly straight line for the six miles, then the three objects . . . being all exactly the same height above the water, the disc would be seen in the telescope projected upon the black band; whereas, if the six-mile surface of the water is convexly curved, then the top disc would appear to be decidedly higher than the black band, the amount due to the known size of the earth."[13]

As the diagrams in Figure 10-1 show, Wallace's experiment clearly "proved that the curvature was very nearly of the amount calculated from the known dimensions of the earth." Not surprisingly, Hampden refused to even look through the telescope, trusting to his personal referee, William Carpenter, who claimed that he saw "the three were in a straight line, and that the earth was flat, and he rejected the view in the large telescope as proving nothing." Walsh, on the other hand, as the official referee, declared Wallace the winner, and published the results in the March 26, 1870, issue of *Field.* Wallace's victory was well deserved and he badly needed the money (throughout most of his life Wallace was short of funds and in search of work). But Hampden promptly wrote Walsh, "demanding his money back on the ground that the decision was unjust, and ought to have been given in his favour." According to Wallace, the law in England at that time was that "all wagers are null and

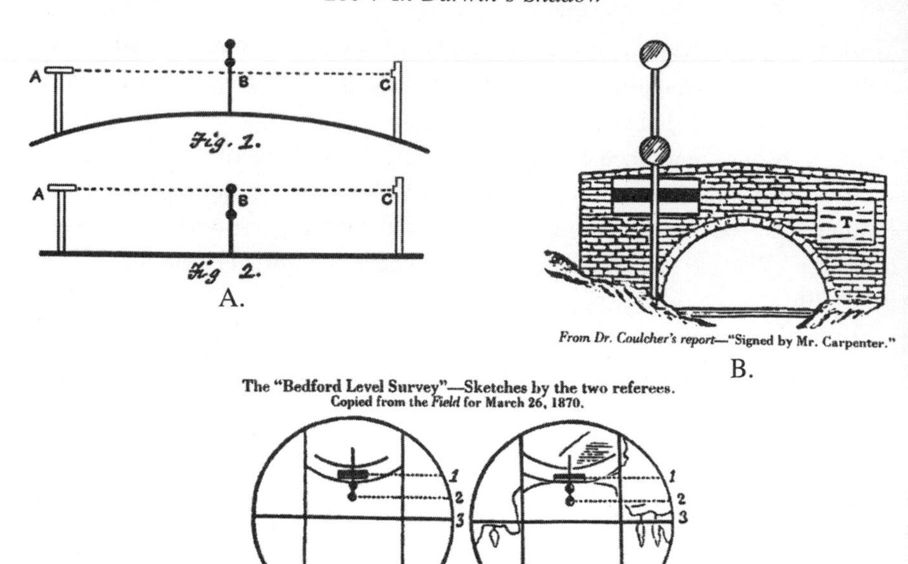

These two views, as seen by means of the *inverting* telescope, are exact representations of the sketches taken by Mr. Hampden's Referee, and attested by Dr. Coulcher as being correct in both cases: first, from Welney Bridge; and secondly, from the Old Bedford Bridge.

Figure 10-1 Wallace challenges the flat-earthers, March 5, 1870, at the Old Bedford Canal.
A. A telescope, a disk, and a black band were all placed at exactly the same height above the water along a straight six-mile stretch of the Old Bedford Canal. If the six-mile surface of the water is convexly curved, then the top disk will appear higher than the black band. (Fig.1) If the surface of the water is a perfectly straight line for the six miles, the three objects will be at exactly the same level and the disk will be seen through the telescope as superimposed upon the black band. (Fig. 2)
B. The view through the telescope of the canal bridge, showing the disks and black band. (Courtesy Royal Geographical Society.)
C. Illustration of two views through the telescope demonstrating the curvature of the Earth. These views, as seen by means of the inverting telescope, are exact representations of the sketches taken by Mr. Hampden's Referee, and attested by Dr. Coulcher as being correct in both cases: first, from Welney Bridge; and second, from the Old Bedford Bridge.

void" and "the loser can claim his money back from the stakeholder if the latter has not already paid it away to the winner. Hence, if a loser immediately claims his money from the stake-holder, the law will enforce the former's claim on the ground that it is his money."[14]

Wallace's loss of the newly won 500 pounds (a workingman's wages for one year) turned out to be the least of his problems. Hampden became a one-man nuisance in Wallace's life, initiating a series of abusive letters to the presidents and secretaries of the scientific societies of which Wallace was a member, such as the following to the president of the Royal Geographical Society on October 23 and 26, 1871:

> If you persist in retaining on your list of members a convicted thief and swindler, one A. R. Wallace, of Barking, I am obliged to infer that yr Society is chiefly made up of these unprincipled blackguards, who pay you a stipulated commission on their frauds, & secure the confidence of their dupes by their connexion with professedly respectable associations.
>
> In spite of the bluster of the whole English press, J. H. Walsh, of the Field and A. R. Wallace F.R.G.S. are still being posted as a couple of rogues and swindlers, and will continue to be so if their insolent supporters were as thick as tiles on the houses. Pray inform them that no amt or kind of exposure that can possibly suggest itself, will hear cease till every Socty is ruined to which they respectively belong.[15]

Not content to libel Wallace in public, Hampden even wrote this remarkably caustic personal letter to his wife, Annie, which Wallace kept in order to bring suit against him:

> Madam—If your infernal thief of a husband is brought home some day on a hurdle, with every bone in his head smashed to pulp, you will know the reason. Do you tell him from me he is a lying infernal thief, and as sure as his name is Wallace he never dies in his bed. You must be a miserable wretch to be obliged to live with a convicted felon. Do not think or let him think I have done with him.[16]

This was no bluff—nor was Hampden done. Wallace struck back with libel charges and lawsuits, for which Hampden was arrested and jailed on several occasions. But for the next fifteen years he tormented Wallace with letters, newspaper articles, leaflets, and the like. Wallace became concerned for more than just his reputation, as he indicated in a letter of May 17, 1871, to R. MacLachlau:

> I return Hampden's letters. I have actioned him for Libel, but he won't plead, and says he will make himself bankrupt & won't pay a penny. As the man is half mad I don't want to indict him criminally & infuriate him, & so I suppose he will continue to write endless torrents of abuse as long as he lives.[17]

And so he did. On October 24, 1871, Hampden even got a petition read into the record at the Royal Geographical Society from his supporters, claiming, among other things, that "Mr. Hampden has been most urgent and importunate for the fullest and freest investigation of facts, the other side has been equally

HAMPDEN, John 1871 Oct 23

If you persist in retaining on your list of members a convicted thief and swindler, one A.R. Wallace, of Woking, I am obliged to infer that ye Society is chiefly made up of these unprincipled blackguards, who pay you a stipulated commission on their frauds, & secure the confidence of their dupes by their connexion with professedly respectable associations. John Hampden

POST CARD

1 Savile Row W

THE ADDRESS ONLY TO BE WRITTEN ON THIS SIDE.

HALFPENNY

CHIPPENHAM

R

To The Secretary
R. Geografical Soc
~~15, Whitehall Place~~
London

S.E.

HAMPDEN, John 1871 Oct. 26

In spite of the bluster of the whole English press, J.H. Walsh, of the Field and A.R. Wallace F.R.G.S. are still being posted as a couple of rogues and swindlers, and will continue to be so. Their insolent supporters are as thick as tiles on the housetops. Pray inform them that no act or kind of exposure that can possibly suggest itself, will ever cease till every party is ruined to which they respectively belong.

persistent in resisting and opposing all such practical tests as might in any way disturb the verbal decision of Mr. Walsh."[18]

Wallace eventually recovered the 500 pounds, but "the two law suits, the four prosecutions for libel, the payments and costs of the settlement, amounted to considerably more than the 500 pounds . . . besides which I bore all the costs of the week's experiments, and between fifteen and twenty years of continued persecution—a tolerably severe punishment for what I did not at the time recognize as an ethical lapse."[19] The difference, of course, between Wallace and Hampden was the extent to which they would allow the evidence to answer a question of nature. Despite his scientism, however, Wallace's response to Hampden was one that would never have been made by his more conservative colleagues Darwin and Lyell. Nevertheless, Wallace's personality dictated his need to take on such a radical claim. His were the actions of a heretic personality. Traits usually trump states. Fascinated with all ideas on the radical edge being what it was, Wallace simply had to take up the cause regardless of the cost, which was substantial.

Leonainie: In Search of the Lost Poem of Poe

One of the more peculiar surprises to be found in the vast literature holdings of Wallace, that is further emblematic of his remarkable capacity for heresies

---

Figure 10-2 (*Opposite*) Two postcards from flat-earther John Hampden to the Royal Geographical Society in an attempt to libel Wallace, after Wallace won a bet proving that the earth is round. (Courtesy Royal Geographical Society)

Postcard from Hampden dated October 23, 1871:

If you persist in retaining on your list of members a convicted thief and swindler, one A. R. Wallace, of Barking, I am obliged to infer that yr Society is chiefly made up of these unprincipled blackguards, who pay you a stipulated commission on their frauds, & secure the confidence of their dupes by their connexion with professedly respectable associations.

John Hampden

Postcard from Hampden postmarked October 26, 1871:

In spite of the bluster of the whole English press, J. H. Walsh, of the Field and A. R. Wallace F.R.G.s. are still being posted as a couple of rogues and swindlers, and will continue to be so if their insolent supporters were as thick as tiles on the houses. Pray inform them that no amt or kind of exposure that can possibly suggest itself, will hear cease till every Scty is ruined to which they respectively belong.

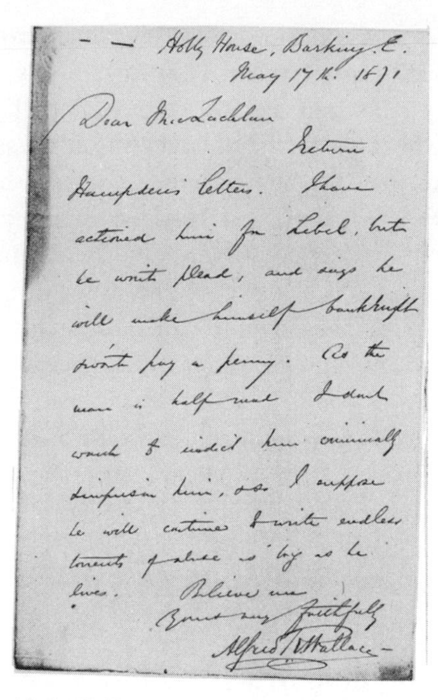

Holly House, Barking. E.
May 17th. 1871

Dear McLachlan

I return Hampden's letters. I have actioned him for Libel, but he won't plead, and says he will make himself bankrupt & won't pay a penny. As the man is half mad I don't want to indict him criminally & imprison him, & so I suppose he will continue to write endless torrents of abuse so long as he lives.

Believe me
Yours very faithfully
Alfred R. Wallace

Figure 10-3 Wallace's letter about the flat-earther John Hampden, May 17, 1871, reveals his frustration with the whole affair. (Courtesy Royal Geographical Society)

I return Hampden's letters. I have actioned him for Libel, but he won't plead, and says he will make himself bankrupt & won't pay a penny. As the man is half mad I don't want to indict him criminally & imprison him, & so I suppose he will continue to write endless torrents of abuse so long as he lives.
Believe me
Yours very faithfully
Alfred R. Wallace

of all stripes, is a 1966 publication entitled *Edgar Allan Poe: A Series of Seventeen Letters Concerning Poe's Scientific Erudition in Eureka and His Authorship of Leonainie.* The author of this tiny monograph (eighteen pages) was none other than Alfred Russel Wallace, who penned fifteen letters (and two extracts never mailed) to one Ernest Marriott, Esq., between October 29, 1903, and March 23, 1904. The incident in question—a rediscovered poem of Edgar Allan Poe supposedly written "at the Wayside Inn in lieu of cash for one night's board and lodging"—is illustrative of Wallace's vivid imagination and willingness to jump to conclusions on the scantiest of evidence.[20]

The story, as I have been able to reconstruct it, is as follows. Sometime around 1893, just seven years after a lecture tour of America (discussed in the next chapter), Wallace received a letter from his brother living in California, which included a poem entitled "Leonainie," allegedly written by Poe. Wallace, however, was "occupied with other matters" and thus "made no enquiry how he got it, but took it for granted that he had copied it from some newspaper." Ten years later, on November 3, 1903, Wallace wrote to Ernest Marriott (with no explanation offered of Marriott's role, other than he was an attorney) to inquire about confirmation of the claim: "I think you will agree with me that it is a gem with all the characteristics of Poe's genius." Wallace also made a bizarre reference in this letter about the last poems of Poe, "The Streets of Baltimore" and "Farewell to Earth," which Wallace believed were written *after* Poe's death "through another brain," and while they are "in my opinion fine and deeper & grander poems than any written by him in the earth-life . . . they are deficient in the exquisite music & rhythm of his best known work."

With typical enthusiasm for all matters heretical, Wallace threw himself into an intense study of Poe's writings, obsessed with finding out if "Leonainie" was indeed his long-lost, and perhaps last, poem (in *this* world anyway). One week later he told Marriott: "Since I wrote to you about 'Leonainie' I have read it many times & have it by heart, & on comparing it with the other poems by Poe which I have it seems to me to be in many respects the *most perfect* of all. The rhythm is most exquisite, and the form of verse different from any other I can call to mind in the double triplets of rhymes in each verse, carried on throughout by simple, natural and forcible expressions while the last verse seems to me the very finest in any of his poems." Wallace reprinted the poem for Marriott at the end of the November 2 letter:

Leonainie

Leonainie, angels named her, and they took the light
Of the laughing stars and framed her, in a smile of white,
And they made her hair of gloomy midnight, and her eyes of bloomy
Moonshine, and they brought her to me in a solemn night.
In a solemn night of summer, when my heart of gloom
Blossomed up to greet the comer, like a rose in bloom.
All foreboding that distressed me, I forgot as joy caressed me,
Lying joy that caught and pressed me, in the arms of doom.

Only spake the little lisper in the angel tongue,
Yet I, listening, heard the whisper; "songs are only sung
Here below that they may grieve you, tales are told you to deceive you,
So must Leonainie leave you, while her love is young."
Then God smiled, and it was morning, matchless and supreme,
Heaven's glory seemed adorning earth with its esteem,
Every heart but mine seemed gifted with a voice of prayer and lifted,
When my Leonainie drifted from me like a dream.

In response to Marriott's uncritical acceptance of Wallace's conviction that Poe had written from beyond the grave, Wallace wrote: "Your letter about the 'Poems from the Inner Life' very much pleased as it shows you are open to conviction. I therefore send for your acceptance a copy of my little book—'Miracles & Modern Spiritualism.' " Wallace's spiritualistic leanings are apparent in this and the following letter of December 19, 1903:

> Such teachings as these are in my opinion worth all the poems he wrote during life. The "Farewell to Earth" is such a favourite of mine that I know it by heart, & use it as an opiate if I lay awake. It really contains the essence of modern spiritualistic teaching, and such lines as—"Where the golden line of duty / Like a living pathway lies" strikes a higher note than anything in Poe's earthy poems.

With little evidence to go on, however, on the first day of 1904 Wallace noted that he still needed a scene and a motive for the poem: "I presume Poe was never in California, but I shall be glad to know if, at anytime, shortly before his death, he is known to have travelled anywhere in an almost penniless condition, where such an incident as his paying for a night's board & lodging with a poem *might* have occurred." Undaunted by a lack of evidentiary support, however, and in his usual eagerness to get into print with an exciting new find, on January 6 Wallace told Marriott: "I think I can see when Leonainie was probably written & I shall now send it with a few preliminary remarks to the Editor of the *Fortnightly,* & its publication may possibly lead to its origin being traced in America." Growing bolder by the day, on January 10 Wallace announced that the poem would be published and that "taking all the circumstances into consideration . . . I have come to the conclusion that this was the very last thing Poe wrote, & it was *probably* written only a few days before his death."

Five days later Wallace was in print with the poem and, as usual, found himself embroiled in controversy. Apparently someone identified the poem as a fake, written by one James Whitcomb Riley, but Wallace spin-doctored this attribution, setting a standard of proof he had not held for himself: "Till we have the alleged *proof* that Riley wrote 'Leonainie,' it seems to me quite as probable that *he* found it, and on the suggestion of a friend made use of it to gain a reputation" (February 8). But then Wallace received a letter from a Mr. Law (reprinted in the February 8 letter to Marriott), implicating Riley as the perpetrator of the hoax, cajoled by friends who told him that if he could write like Poe he could achieve enough fame to establish himself as a poet of high caliber. Riley, speculates Wallace, then wrote "Leonainie," submitted it, and "after it had run the gauntlet of Poe critics and been pronounced genuine if not canonical Riley proved the authorship. This drew attention to his own works, and he has never since lacked for praise and pudding."

On February 15, Wallace received more bad news, this from the "Librarian of the London Library," who "obtained a copy of Riley's 'Armazindy'—which contains 'Leonainie' & has sent it to me. The publishers say that this vol. 'contains some of Mr. Riley's latest and best work including 'Armazindy' & the famous Poe Poem.' " Despite the overwhelming evidence that "Leonainie" was a well-perpetrated hoax, Wallace was unable to recant. The remainder of the February 15 letter is a critique of Riley's other poetry, with Wallace's analysis that Riley did not have the skill to write "Leonainie," and his conclusion that the real hoax is that Riley *found* the Poe poem and pretended to have written it! On March 1, still obsessed with the problem, Wallace told Marriott: "I have now looked through 4 vols. of Riley & can find no sign of his being able to write *Leonainie* with all its defects." Wallace then conducted a careful line-by-line analysis and comparison of "Leonainie" with Riley's other poetry, and drew up a final summary of the whole affair: "The more I consider the matter the more I am convinced he did *not* compose the poem. It looks to me very much as if he really got hold of the poem in the form I have it or nearly,—that to cover himself from exact copying he made the alterations in words, which he might think would make it more like his own work, and the alteration in the arrangement of lines &c. so that it might be accepted as a bad imitation of Poe."

The entire incident encapsulates Wallace's heretic personality—his eagerness to investigate unusual claims, his thorough, almost obsessive analysis of a subject, his willingness to make a serious commitment to a position early in the absence of substantial evidence, and his resolution, regardless of contrary evidence, to maintain his original position (even using contradictory evidence in his favor). When he was right, as in his discovery of natural selection, these traits worked in his favor. But when he was wrong, as is most likely the case here as in his investigations of spiritualism, Wallace's heretic personality brought down upon him the scorn and ridicule of scientists, skeptics, and more conservative personalities.

The "Anti-Body"

From his letters one gets a sense of Wallace's feeling of being an outcast—a rebel of sorts—even among his closest colleagues. He turned down many honorary degrees and only reluctantly agreed to be admitted as a Fellow of the Royal Society after Sir W. T. Thiselton-Dyer invited him three times. After receiving the Order of Merit in 1908, the highest honor ever conferred on him, Wallace wrote to his close friend, Mrs. Fisher, about the number of awards he was being offered:

> Is it not awful—two more now! I should think very few men have had three such honours within six months! I have never felt myself worthy of the Copley medal—and as to the Order of Merit—to be given to a red-hot Radical, Land nationaliser, Socialist, Anti-Militarist, etc., etc., etc., is quite astounding and unintelligible![21]

From the tone of such letters, it is obvious that Wallace was both aware of and proud of his autonomous stand on certain issues that his more conservative scientific colleagues would not vouchsafe to take. In a November 4, 1905, letter to Raphael Meldola, after the reviewers of his autobiographical *My Life* took note of his "faddish" interests in the fringes of science and spiritualism, he recalled: "Yesterday I got a notice in a paper called *'Reviews'* with a very fair bit . . . he says—'For on many subjects Mr. Wallace is an antibody. He is anti-vaccination, anti-state endowment of education, anti-land-laws, and so on. To compensate, he is pro-spiritualism, and pro-phrenology, so that he carries, as cargo, about as large a dead weight of fancies and fallacies as it is possible to float withal." A more conservative scientist might blanch at such a description, but not a heretic personality. Three days later, again to Mrs. Fisher, Wallace actually boasted of his new title:

> The reviewers are generally very fair about the fads except a few. The Review invents a new word for me—I am an "anti-body"; but the Outlook is the richest: I am the one man who believes in spiritualism, phrenology, anti-vaccination, and the centrality of the earth in the universe, whose life is worth writing. Then it points out a few things I am capable of believing, but which everybody else knows to be fallacies, and compares me to Sir I. Newton writing on the prophets! Yet of course he praises my biology up to the skies—there I am wise—everywhere else I am a kind of weak, babyish idiot! It is really delightful![22]

Wallace's anti-body attitudes, which at times led him to greatness, occasionally took him down paths that led to dead ends. An unparalleled observer in some fields, he was almost blind in others.

Qualms of Doubt

Like all species in nature, human variation and individual differences are the norm, not the exception. There is a wide range of variability in all behaviors and beliefs. Not everyone is equally tempted by heretical ideas. Heretic personalities are significantly more tempted by unorthodox ideas, and, further, are less willing to analyze such claims with the same critical scrutiny as they might apply to other belief systems, or their more conservative colleagues might view these heresies.

It is true that Wallace's belief in spiritualism had, in his mind, a certain amount of scientific evidentiary support behind it. But the sense one gets from reading the vast correspondence and literature Wallace produced on the subject is that his beliefs were driven as much by his emotions as his rationality. That is, the proximate cause of his belief in spiritualism is the evidence he believed existed for it; the ultimate cause—the deeper substrate underlying this proximate cause—is a personality suited for seeking and finding such evidence to be viable. As in his involvement with the flat-earthers and Poe's last poem, Wallace's traits for radicalness overwhelmed his states of caution. Consider the following letter, written by Wallace in 1894 to a friend, on the death of his beloved sister, Frances:

> Death makes us feel, in a way nothing else can do, the mystery of the universe. Last autumn I lost my sister, and she was the only relative I have been with at the last. For the moment it seemed unnatural and incredible that the living self, with its special idiosyncrasies you havé known so long can have left the body, still more unnatural that it should (as so many now believe) have utterly ceased to exist and become nothingness. With all my belief in and knowledge of Spiritualism, I have, however, occasional qualms of doubt, the remnant of my original deeply ingrained scepticism; but my reason goes to support the psychical and spiritualistic phenomena in telling me that there must be a hereafter for us all.[23]

Such commentary in Wallace's writings is bountiful, and he received much support over many years and from around the world from both the lay public and his fellow scientists. A typical letter in a folio of Wallace's correspondence on matters spiritual is from Professor Theo D. A. Cockerell, in Las Cruces, New Mexico, who had corresponded with Wallace several times on biological matters. On September 24, 1893, Cockerell wrote Wallace regarding the death of his wife in childbirth:

> When a man speaks of his "better half,"—it is usually a form of speech, but I feel as if my better half was indeed taken away from me, and scarcely know how the other half can work to any purpose until they are reunited. The more

> one thinks, the more one sees every reason to hope and indeed believe that the present separation is only temporary. I am sure you will agree to this. The outlook in every way should make one cheerful, but it is impossible to be so philosophical as to ignore the present, which is hard enough. I find so few people who seem to have any clear notion about immortality. To me, it seems simply axiomatic just like the infinity of time and space—although in each case the conception eludes our mental capacity.[24]

Wallace responded with his usual sensitivity, as he did to all inquirers in such otherworldly matters, and by so doing he added the weight of scientific credibility to such beliefs. While there were many heretic personalities in Wallace's time to whom the public might and did turn for comfort in such matters, Wallace was among the most important because of his stature as a scientist.

Those with heretic personalities—scientists and nonscientists alike—must be more cautious than most, for while their boldness may lead them to extraordinary success in one field, it may occasionally turn to temerity and lead them down the road to deception and self-deception in others. The rub in science is to find the right balance between being so open to heretical ideas that it becomes difficult to separate sense from nonsense; and so closed to heretical ideas that it becomes difficult to abandon the status quo. Heretic personalities, so numerous among the various pseudosciences, need to temper their beliefs with a little caution. Skeptics, so numerous among the various sciences, need to moderate their skepticism with a little boldness. Where the heretic meets the skeptic a creative scientist will emerge. Finding that exquisite balance was always difficult for Wallace, and became ever more so in the final stage of his life as he entered the confusing and contentious world of social and political causes.

11

The Last Great Victorian

Trying to capture the essence of a man who had so many adventurous explorations, diverse interests, intellectual pursuits, and activist causes, all played out over nine decades, mirrors the risk the taxonomist faces in attempting to classify a wide range of varieties into a few species. When should we lump Wallace's various interests and projects together into one taxon, and when should we split them apart? Which were created ex nihilo in his mind and which were the product of descent with modification from previous intellectual ancestors? And how much did his ideas change through time? We have seen how Wallace synthesized the various themata of his research into a unified scientistic theory of everything, but rarely does the ontogeny of one's final work recapitulate the phylogeny of one's life work. Ideological genetic drift coupled to countless intellectual mutations shape the morphology of one's personal worldviews. The process was no different for Wallace, but the extra decades of his life afforded him the time to impose additional order on the seeming chaos of his intellectual development. And in his life, as for the organisms he so assiduously studied, it was a struggle for existence throughout.

Economic Struggles

In the final decades of his life Wallace's finances waxed and waned. Like his father, Wallace was a dreadful businessman. Although his entrepreneurial independence was admirable and afforded him (barely) the time to do science, his progress as an investor was fitful at best, and he often found himself sliding back down into the financial wave's trough. At one point, for example, he

Figure 11-1 Wallace's home at Nutwood Cottage, Godalming, 1881. Though he struggled financially most of his life, his prolific writings, coupled with a government pension procured for him by Darwin and others in the scientific community, allowed him to lead a comfortable middle-class life. (From *My Life,* 1905, v. II, 103)

cashed in his blue chips for speculative stock "whose fluctuations in value I was quite unable to comprehend" and promptly forfeited his portfolio, whose "loss was so great as to be almost ruin." He also invested in a lead mine, but lost all when the price of lead in England collapsed with the discovery of lead and silver mines in Nevada. "The result of all this was that by 1880 a large part of the money I had earned at the risk of health and life was irrecoverably lost."[1] Wallace's autobiography, in addition to containing much of his personal correspondence, is riddled with unpretentious honesty about his blunders and failures, embarrassments and blemishes, which cannot be seen as false modesty or self-effacing irony. He was all too human in his capacity to err, and he was not ashamed to admit it.

Wallace's economic struggles would haunt him much of his life. He was a prolific writer, in part, because of financial need, and this led him to write for popular magazines in addition to his scientific writing. But bad timing, awkward social skills, and sometimes pure tactlessness also contributed to his frequent bouts of unemployment. In 1864, for instance, the *Royal Geographical Society* advertised for an assistant secretary. Unfortunately for Wallace, his old friend Henry Walter Bates also applied for the job and got it on his administrative experience, which Wallace lacked. In 1869 the government announced plans to open a branch of the South Kensington Museum at Bethnal Green, and Wallace was so certain he would get the job as curator (he had letters of recommendation from Darwin, Lyell, Hooker, and Huxley) that he moved his home to within walking distance of the museum. Government comptrollers, however, decided that they could not afford another curator, so the administration dissolved the position before interviews began. In 1878 Epping Forest was declared an "open space" for which a superintendent was needed to oversee the use and development of the 5,560 acres. Wallace, in a letter to his friend the influential botanist Joseph Hooker, claimed: "It is a post which would exactly suit me, and which I believe I am fitted for. I have long been seeking some employment which would bring me in some fixed income while still allowing me some leisure for literary work."[2] Once again Darwin, Lyell, Hooker, and Huxley recommended him for the job. Before he was hired he explained to Hooker, "What I think should be done in order to deal best with the forest, and take full advantage of the grand opportunity now afforded, I have written an article in the forthcoming 'Fortnightly Review,' which I sincerely trust will meet with your approval."[3] The article was entitled "Epping Forest, and How to Deal with It," in which Wallace insisted that the forest be left in a pure and unsullied state. Unfortunately, he voiced his conservationism too strongly, too soon, and, worst of all, in print. Merchants, real estate brokers, and other businessmen who wanted to finance skating rinks, golf courses, hotels, and the like, responded swiftly and harshly, demanding that a more flexible man be hired. Wallace once again lost out.[4]

Over his long life of ninety years, regular work eluded him, so that the teaching post at Leicester Collegiate School Wallace held when he was twenty-one years old was his last formal position of regular employment. "I had now to depend almost entirely on the little my books brought me in, together with a few lectures, reviews, and other articles." Nevertheless, and with his usual positive spin on matters disastrous, toward the end of his life he noted that the lack of regular work was "really for the best, since it left me free to do literary work."[5] And literary work he did, in spades, producing tens of thousands of words a year in numerous genres both professional and popular, through essays, articles, reviews, letters, and especially books—twenty-two all told.

Wallace's financial situation was aided in 1881 when, thanks to the political machinations of Darwin, Huxley, and others well connected in British science and politics, Wallace was awarded a pension of 200 pounds a year for life, directly approved by Prime Minister Gladstone and justified by Wallace's scientific and geographical exploratory contributions to the British Empire during the height of her imperialistic expansiveness. True to form, however, even this act of seeming goodwill did not come about without a struggle.

When Darwin sought support for Wallace from the newly knighted Sir Joseph Hooker, who well knew Wallace's contributions to science, the latter recalled with disdain Wallace's involvement in the spiritualist movement, and in particular the BAAS incident when the Barrett paper on psychic phenomena was read in the biology section. Hooker complained to Darwin:

> I have well considered the pros and cons of the proposal to enlist sympathy in the matter of a pension for Wallace, and greatly doubt its advisability.
>
> Wallace has lost caste terribly, not only for his adhesion to Spiritualism, but by the fact of his having deliberately and against the whole voice of the committee of his section of the British Association, brought about a discussion on Spiritualism at one of its sectional meetings, when he was President of that section.
>
> This he is said to have done in an underhanded manner and I well remember the indignation it gave rise to in the British Council, and amongst the members at large. . . .
>
> I think that under these circumstances it would be very difficult to ask one's friends to sign an application to Govt. for a pension. Added to which Govt. should in fairness be informed that the candidate is a public and leading Spiritualist![6]

Despite his own misgivings about and disgust over Wallace's endorsement of and involvement in spiritualism, Darwin, who had supported Lankester against Wallace in the Slade trial, nevertheless pressed Hooker to sign the petition, and even wrote Prime Minister Gladstone directly, and in the end had his way. Once again Darwin's and Wallace's paths met at a congenial juncture and they remained close to the end of Darwin's life in 1882.

The First Darwinian Brings Darwinism to America

At Darwin's funeral Wallace shared pallbearer duties with James Russell Lowell, of the Lowell Institute of Boston, who invited him to come to America to be a speaker in their prestigious lecture series. Several years later, at a youthful age sixty-three, Wallace once again sailed west for the Americas. His opening lecture was on November 2, 1886, and the Boston *Transcript* reported a favorable response: "The first Darwinian, Wallace, did not leave a leg for anti-Darwinism to stand on when he got through his first Lowell lecture last evening. Mr. Wallace, though not an orator, is likely to become a favourite as a lecturer, his manner is so genuinely modest and straight-forward."[7]

On completion of his obligatory lectures Wallace fanned out across the American landscape, both cultural and geographic. He was well known enough on that side of the Atlantic to have garnered meetings with such luminaries as Oliver Wendell Holmes, Henry George, William James, and even President Grover Cleveland at the White House. After several months

Figure 11-2 Charles Darwin in 1882 at age seventy-three in the final year of his life. Wallace served as a pallbearer at Darwin's funeral. (From F. Darwin, 1887, frontispiece)

and many thousands of miles he reached California, where he visited his older brother John in San Francisco. It must have been quite a reunion, as the two had not seen each other for forty years. While in northern California he hiked around the dramatic glacier-cut walls of Yosemite Valley, toured the famous redwood forests with the renowned naturalist John Muir, and even got a tour of the newly built Stanford University by Leland Stanford himself. He also delivered what was apparently a wildly successful lecture on "If a Man Die, Shall He Live Again?," bringing his full focus on the question of reincarnation as it relates to spiritualism, although here his answer to the title question was negative.[8]

Like Alexis de Tocqueville's descriptions of the American democratic experience, Wallace's correspondences are riddled with the observations of a foreigner in a foreign land. In a letter to his friend Raphael Meldola written from Boston, for example, Wallace commented that American hotels "certainly far surpass us in hotel arrangement," but he was even more impressed with the Museum of Comparative Anatomy at Harvard, also known as the Agassiz Museum: "Alex Agassiz took me round. It *is* a museum! The only one worthy of the name I have ever seen," particularly, Wallace concluded, because "no architect [was] allowed to interfere—except to carry out Agassiz' own design."[9] Seven months later, on June 19, 1887, from Stockton, California, he wrote Meldola again, describing how the American landscape was even more impressive than Agassiz's museum:

> I have crossed this mighty continent from Plymouth Rock to the Golden Gate. I have crossed the Alleghenies & the Rockies & the Sierra Nevadas. I have wondered on the mighty prairie where the bones of the now almost extinct buffalo lie in heaps upon the prairie. I have gazed upon mighty Niagara, and on the liquid torrent of [water] that "mighty Missouri rolls down to the sea." I have looked up with aching eyes and breaking back at the Washington Monument and at the huge precipices of the Yosemite, and lastly & most recently—I have wondered for days in the glorious pines forests which grow the majestic Sequoias,—the one thing that has more impressed and satisfied me than any thing else I have seen in America. Amid all the exaggerations of guidebooks & popular writers, they remain one of the living wonders of the world, perhaps more than anything else to a lover of nature, worth a journey across America to see.[10]

It is a sight only a naturalist would place above all others the American continent and culture has to offer, but one that continues to invoke awe.

Wallace's American journals, which are still unpublished and at the Linnean Society, are filled with observations of American nature and culture. His entry for Wednesday, April 6, 1887, is typical:

> Talk with Judge Holman about Irish in America. He has known them for 50 years. Near him in Indiana is a township half Irish half Germans both Catholics settled 40 years. The Germans have increased—the Irish diminished, by emigration further west & other causes. Many of the Irish became public men of eminence & many took a good position. They cultivated their farms as well as the Germans & showed equal industry. On the whole Judge Holman is of opinion, that, considering the low class of Irishman who came over & their usual extreme poverty as compared with the Germans & other emigrants it can not be said that they are at all inferior in industry & in success in life.[11]

Many entries are entertaining, such as this one he penned on the train headed west from Kansas City in May 1887: "In train a lady chewing gum—

saw her at intervals for an hour her jaws going all the time like those of a cow ruminating."[12] Other entries are not so amusing, such as this disturbing entry of Wednesday, May 4, 1887, from Sioux City, Iowa, the morning after his previous evening's lecture: "Morning with Mr. Talbot to see pork curing establishment—kill 1000 hogs a day—hogs walk up to top of building, hung up by one leg slide along to man who cuts throat, drop into tank of boiling water, into machine which takes off most of hair, the[n] along counter where other men finish scraping, then cut up, entrails pour into tanks where lot of men clean them—fat out to make lard, sausages, hams, salted pork, blood & refuse all passed through steam heater dryers & form a dry fertilizer—whole place dark, confined passages, steep ladders, all wet with water brine, blood etc. very sickening."[13]

Sightseeing and social commentary aside, the primary focus of this trip was the popularization of Darwinism, as clearly denoted in the title of his lecture—"The Darwinian Theory"—for which he typically earned a tidy sum of fifty dollars per evening for an hour and a half lecture. His audiences ranged from a hundred to three hundred people and he used the new visual aid technology of lantern slides.[14] The lectures helped him organize his thoughts and notes for what would become his definitive statement on evolutionary theory, entitled simply *Darwinism.* After ten months in America he returned to England and promptly began work on the project. In less than two years he generated 494 typeset pages, written in a style accessible to the lay public.

Figure 11-3 When Wallace traveled to America in 1886 on a lecture tour he was stunned by the stark beauty of the American West. He was particularly struck by the gateway to the Garden of the Gods, framing Pike's Peak. (From *My Life,* 1905, v. II, 180)

Darwinism was published in May 1889, but it was more than a recapitulation of the *Origin,* as Wallace explained: "A weakness in Darwin's work has been that he based his theory, primarily, on the evidence of variation in domesticated animals and cultivated plants. I have endeavoured to secure a firm foundation for the theory in the variation of organisms in a state of nature."[15]

This was an exaggeration, to be sure, since the *Origin* does contain copious observations from nature. What Wallace most certainly meant, as evidenced from his opening paragraph in the 1858 Ternate paper, is how he used domesticated animals to a different purpose than Darwin. Darwin used them as an example of variations being selected artificially by humans as an analogue to how nature operates (for which he did provide numerous examples from nature). Wallace demonstrated how varieties in nature are different from varieties in domestication, and that while the latter may return to their original condition (indicating the stability of species), the former do not. And that is a critical point of departure for Wallace from Darwin. As he noted in the third paragraph of the Ternate paper, written before the *Origin,* "It will be observed that this argument rests entirely on the assumption, that *varieties* occurring in a state of nature are in all respects analogous to or even identical with those of domestic animals, and are governed by the same laws as regards their permanence or further variation. But it is the object of the present paper to show that this assumption is altogether false, that there is a general principle in nature which will cause many *varieties* to survive the parent species, and to give rise to successive variations departing further and further from the original type, and which also produces, in domesticated animals, the tendency of varieties to return to the parent form."[16] As always, Wallace had to cut his own path through the thicket of evolutionary theory.

The Common Man's Scientist

Wallace's writings were as diverse as they were plenteous. While scientific themes dominated his total production, he was not diffident about tackling social issues, political controversies, and, of course, matters spiritual. Among his books, for example, he wrote *Miracles and Modern Spiritualism* (1875), *Land Nationalization* (1882), *Bad Times* (1885), *The Wonderful Century* (1898), *Studies Scientific and Social* (1900), *Man's Place in the Universe* (1903), *Is Mars Habitable?* (1907), and in the ninety-first and final year of his life (1913), steadfast and indefatigable to the end, he wrote *Social Environment and Moral Progress* and *The Revolt of Democracy.*

In the last two decades of his life Wallace's voluminous writings, coupled to his vast diversity of interests, brought him some degree of world renown. Scholars and scientists, as well as general well-wishers from all over the world (particularly America, where he even had a fan club), wrote him regularly.

Most of the nonscientist correspondents were unknown to Wallace and do not appear in biographical dictionaries. For example, one R. H. Arnot, on October 3, 1898, wrote him: "On behalf of my associates of the 'Alfred Russel Wallace Scientific Club of Rochester' [New York], I thank you most cordially for the honor which you have most graciously accorded us. [Wallace had agreed to the use of his name and sent them a signed photograph.] To bear such a distinguished name as yours is indeed an incentive to determined work in the field of science. For we Americans have been taught to reverence the name of Alfred Russel Wallace."[17]

Because of his foray into spiritualism, his belief in an afterlife, his theory of the existence of a higher intelligence than man, as well as his support of fringe movements such as land nationalization, anti-vaccination, and women's rights, Wallace became a folk hero, often using science and scientific reasoning to defend and support many popular movements and beliefs. Eveleen Myers, for example, on the death of her husband who had zealously read about Wallace's researches into the afterlife, wrote him from Albuquerque, New Mexico, on February 7, 1901: "My dear husband held you in such reverence and such admiration that to feel your sympathizing friendship is such a comfort to me."[18] Caroline A. Foley, appreciative of Wallace's support of a nascent women's movement in a magazine interview, wrote him on December 4, 1893: "I trust you will not feel put out if, as an individual woman and by a private letter, I venture to offer you homage and thanks for your published utterance respecting women which I had read in the *Daily Chronicle* of today. At this time of day, it is true our prospects are no larger than what they were and you as their champion resemble happiness as characterized by Goethe."[19] Four days later Wallace received a letter from Frances Willard, who called his interview "the highwater-mark of the woman's movement thus far."[20] Another letter, dated June 26, 1910, from Lilian Whiting in Paris, is nothing short of adulatory:

> For many years it has been the dream of my life, or the prayer of my life, or both, to meet you. Perhaps it is not a prayer to be granted in this present state of life, and I am not impatient regarding it.—
>
> "I can wait Heaven's perfect hour
> Through the innumerable years"
>
> as our Emerson says. Of course my name is unknown to you and this little printed data I enclose merely to explain the better that in many of my books I have allowed myself the privilege of enriching them by quoting from you.[21]

Reluctant Recognition

Having outlived most of his nineteenth-century colleagues to become one of the "last great Victorians," Wallace received numerous awards, medals, honorary degrees, lecture invitations, and the like, but the essence of the man that allowed him to survive for long stretches alone in the tropics returned again in his final years. He was alone in a crowded culture and cherished his solitude most: "Really the greatest kindness my friends can do me is to leave me in peaceful obscurity, for I have lived so secluded a life that I am more and more disinclined to crowds of any kind. I had to submit to it in America, but then I felt exceptionally well, whereas now I am altogether weak and seedy and not at all up to fatigue or excitement."[22]

This was no coquettish ploy, as evidenced by the fact that many of these accolades Wallace flatly refused. In a confidential letter of January 17, 1893, Alfred told his colleague W. T. Thiselton-Dyer, who had been working to get Wallace admitted into the Royal Society, "I cannot understand why you or anyone should care about my being an F.R.S., because I have really done so little of what is usually considered scientific work to deserve it. I have for many years felt almost ashamed of the amount of reputation & honour that has been awarded me. I can understand the general public thinking too highly of me, because I know that I have the power of clear exposition, and, I think, also, of logical reasoning. But all the work I have done is more or less amateurish & founded almost wholly on other men's observations; and I always feel myself dreadfully inferior to men like Sir J. Hooker, Huxley, Flower, & scores of younger men who have extensive knowledge of whole departments of biology of which I am totally ignorant."[23] His protestation to the contrary, Wallace was offered an invitation from the Royal Society to become a Fellow and, characteristically, he refused. The Royal Society, in turn, refused his refusal, and in 1893 he was made a Fellow of the Royal Society.

Even the famed Darwin–Wallace medal awarded to him on the fiftieth anniversary of the July 1, 1858, Linnean Society joint presentation of their papers was only reluctantly accepted. He told his friend Raphael Meldola, in a P.S. of a letter marked "*Very Private & Quite confidential!*": "I suppose I have to thank either yourself or Poulton for this quite 'outrageous' attempt to put me on a level with Darwin! If I live through it, I shall have something to say on this point!! A.R.W."[24] More important, in these later years Wallace shifted his emphasis from science to history of science, and in one extensive essay in particular he reviewed the history of his own science with admirable perspective.

The Historian of Science

As the nineteenth century came to a close, publishers scrambled to come out with collections of commentaries by leading lights from varied fields. Wallace was tapped to reflect on the theory of evolution. The article, entitled simply

"Evolution," was originally published as the first in a series of articles in New York–based *The Sun* under the overall title "The Passing Century." It was republished the following year in *The Progress of the Century,* as the lead essay in an edited collection looking back at the achievements of a number of sciences and related subjects, including evolution, chemistry, archaeology, astronomy, philosophy, medicine, surgery, electricity, physics, war, naval ships, literature, engineering, and religion.[25]

As a mark of Wallace's clear exposition and ability to communicate complex ideas clearly, the first handwritten draft of the manuscript, in the holdings of the British Museum of Natural History, was remarkably close to a Mozartian final draft. By this time, at age seventy-seven, after fifteen books and six hundred journal articles, Wallace had mastered the craft of writing. There were almost no corrections, deletions, additions, or changes made in the original manuscript. The few changes in the published version are mostly grammatical, though it is interesting that Wallace always capitalized "Evolution" but not "creator," whereas the opposite occurs in the published manuscript. This is consistent with other first-draft, handwritten manuscripts of his (as well as his letters), though it is possible the publisher might have found the capitalization of the former too honorific in contrast with the latter. The manuscript is divided into six subsections: "The Nature and Limits of Evolution, The Rise and Progress of the Idea of Evolution, The First Real Step Towards Evolution, Evolution of the Earth's Crust, Organic Evolution: Its Laws and Causes, and The Theory of 'Natural Selection.' " In it we see two Wallaces:

1. Wallace the scientist. Wallace provides a summary of the state of the field of evolutionary biology at the turn of the century, although he is no neutral observer. He emphasizes the role of natural selection (at the exclusion of other mechanisms of change, particularly sexual selection) to account for the origin of structures and functions of organisms, although he does not emphasize the limitations of natural selection.
2. Wallace as historian of science. Although not hagiographic, there is a bit of Whiggish historiography in his depiction of a progressive march through time toward the triumph of evolutionary theory with his own and Darwin's achievement. As usual, however, Wallace is self-effacing and praises Darwin to the hilt.

Wallace's writing style is emblematic of the pre-twentieth-century paragraph-length sentence, in which he fits numerous thoughts, qualifications, and extrapolations, all tied together with various grammatical tools. As a historical essay it is a classic, despite its prepublication flaws. His opening paragraph is meant to grab the reader's attention through a bold claim, though there is every indication that Wallace really believed about the theory of evolution that "while upon the greatest problems of the mode of origin of the various forms of life—long considered insoluble—it throws so clear a light that to many biologists it seems to afford as complete a solution, in principle, as we can ever expect to reach." Yet he tempers his claim with a touch of philosophical realism by noting that the theory is only a partial explanation "for no complete explanation is possible to finite intelligence."[26] For Wallace, this seeming contradiction between a problem being both soluble and insoluble was resolved through his belief that there is an intelligence higher than finite for whom a complete explanation is presumably known.

Wallace proceeds to recount the many misconceptions about evolution, and in the third paragraph defines his subject: "Evolution . . . implies that all things in the universe as we see them have arisen from other things which preceded them by a process of modification, under the actions of those all-pervading but mysterious agencies known to us as 'natural forces' or more generally the 'laws of nature.' " Interestingly, without overtly stating it, Wallace identifies the etiology of the word evolve—meaning to "unfold"—originally used to describe the embryological process of an organism unfolding into a more mature state from a less mature one. "More particularly, the term evolution implies that the process is an 'unrolling' or an 'unfolding,' derived probably from the way in which leaves and flowers are usually rolled up or crumpled up in the bud and grow into their perfect form by unrolling or unfolding." But he notes that this is too general a usage for the term and therefore "it must not be taken as universally applicable, since in the material world there are other modes of orderly change under natural laws to which the terms development or evolution are equally applicable." And evolution does have its limitations because "even if it is essentially a true and complete theory of the universe, can only explain the existing conditions of nature by showing that it has been derived from some preexisting condition through the action of known forces and laws." As we have seen, Wallace believed that there are other forces and laws in existence pertaining to matters spiritual, as well as structures and functions involving humans that cannot be explained by material evolution and natural selection.

Like most historians of evolutionary thought, Wallace begins with such ancient Greeks as Thales, Anaximander, Anaxagoras, Empedocles, and especially Lucretius and his great work *On the Nature of Things* in which he laid out a complete reductionistic and materialistic philosophy of science, starting with the assumption that all events have causes, that things cannot have come from nothing, that the first particles would have been solid and of a finite size, and that over the course of time these were built up into larger particles, now made of parts, and so on up to humans. "Lucretius was an absolute materialist, for though he did not deny the existence of Gods he refused them any share in the construction of the universe, which he again and again urges, arose by chance after infinite time, by the random motions and collisions and entanglements of the infinity of atoms." So far so good, until Lucretius speculates wildly on the origin of the first human infants: "For much heat and moisture would then abound in the fields; and therefore wherever a suitable spot offered, wombs would grow attached to the earth by roots; and when the warmth of the infants, flying the wet and craving the air, had opened these in the fullness of time, nature would turn to that spot the pores of the earth and constrain it to yield from its opened veins a liquid most like to milk." To this Wallace offers a mild rebuke and understated assessment of his field: "The fact that this mode of origin commended itself to one of the brightest intellects of the 1st century B.C., enlightened by the best thought of the Grecian philosophers may enable us the better to appreciate the immense advance made by modern Evolutionists."

Wallace next makes an important distinction between two types of change, both embodied by the word *evolution*, but with radically different implications. The type of evolution discovered by Tycho Brahe, Johannes Kepler, and Isaac Newton is that of constant and unvarying change as represented in the clockwork universe of planetary orbits and described by the theory of universal gravitation. But, Wallace notes, "all this implied no law of development, and it was long thought that the solar system was fixed and unchangeable—that some altogether unknown or miraculous agency must have set it going, and that it had itself no principle of change or decay but might continue as it now is to all eternity." He then properly credits Laplace as the pinnacle of this mode, whose Nebular Hypothesis was "the first attempt ever made to explain the origin of the solar system under the influence of the known laws of motion, gravitation and heat, acting upon an altogether different antecedent condition of things—a true process of Evolution." By this usage Wallace takes "evolution" to mean the inevitable unfolding of a system according to fixed laws. There is no history in this system. Planets whirl around suns, moons around planets, all determined (predetermined actually) by unvarying laws of nature.

Wallace then shows how this type of evolutionary change was applied to earth history, dividing the historical sequence from the first modern evolutionary thinkers to Darwin and himself into two lineages—geological and biological. For the former he begins with the Scottish geologist James Hutton, who presented the principle of uniformitarianism, but Wallace notes that despite the efforts of Hutton and Playfair, it was Charles Lyell who put the concept on the scientific map. Not limiting himself to Great Britain, however, Wallace turns to the continent to review the work of Cuvier, who he claims "never appeals to known causes, but again and again assumes forces to be at work for which no evidence is adduced and which are totally at variance with what we see in the world today." In contrast to Cuvier, Wallace demonstrates that it was the work of Lyell in applying the "principles of Evolution to the later phases of the earth's history" that "prepared the way for the acceptance of the still more novel and startling application of the same principles to the entire organic world." The principle of uniformitarianism, discovered by Hutton, reached fruition in Lyell's *Principles of Geology,* demonstrating, according to Wallace, that "not only have all the chief modifications during an almost unimaginable period of time been clearly depicted, but they have in almost every case been shown to be the inevitable results of real and comparatively well-known causes, such as we now see at work around us."

Wallace then brings the reader up date on geological science since Lyell's death, answering the criticisms of uniformitarianism that had been hurled against the doctrine in recent decades. Critics, he notes, "alleged that it is unphilosophical to take the limited range of causes we now see in action, as a measure of those which have acted during all past geological time. But neither Lyell nor his followers make any such assumption. They merely say, we do not find any *proof* of greater or more violent causes in action in past times, and we *do* find many indications that the great natural forces then in action—seas and rivers, sun and cloud, rain and hail, frost and snow, as well as the very texture and constituents of the older rocks, and the mode in which the organisms of each age are preserved in them—must have been in their general nature and magnitude very much as they are now." He addresses additional objections, then concludes: "Lyell's doctrine is simply that of real against imaginary causes, and he only denies catastrophes and more violent agencies in early times, because there is no clear evidence of their actual existence, and also because known causes are quite competent to explain all geological phenomena. It must be remembered, too, that uniformitarians have never limited the natural forces of past geological periods to the precise limits of which we have had experience during the historical period. What they maintain is, that forces of the same *nature* and of the same *order of magnitude,* are adequate to have brought about the evolution of the crust of the earth as we now find it."

Despite the application of the principle of uniformitarianism, neither Hutton nor Lyell, Wallace explains, was able to discover natural selection. Thus, it was the naturalists, not the geologists, who made this important step, and Buffon and Lamarck are the objects of his pre-Darwinian focus. Interestingly, Wallace argues that Buffon's discovery of the homologous nature of mammalian skeletons (e.g., the similarity in the bones of the arm of a man with those of the leg of the horse), coupled to his thorough classification of both the plant and animal kingdoms, should have made Buffon, not Darwin (and by implication himself), the discoverer of natural selection. What held Buffon back? The lack of religious toleration, Wallace argues, backing his claim by noting that "to save himself from the ecclesiastical authorities he [Buffon] at once adds this saving clause [to his analysis of the degeneration of apes from men]:—'But no! It is certain, from revelation, that all animals have alike been favoured with the grace of an act of direct creation, and that the first pair of every species issued full formed from the hands of the creator.' "

Here Wallace has likely misread the historical record. We should probably take Buffon at his word. Such a mistake, however, is understandable when Wallace quotes such passages as this from Buffon: "If we once admit that there are families of plants and animals, so that the ass may be of the family of the horse, and that the one may only differ from the other by degeneration from a common ancestor, we might be driven to admit that the ape is of the family of man, that he is but a degenerate man, and that he and man have had a common ancestor." At first this sounds like an "anticipation" of evolution, and in fact, such passages are easy to find in numerous pre-Darwinian authors. But Buffon, Lyell, and many others were expounding the accepted belief that the forces of nature were acting to preserve created kinds through the elimination of deviants, and not acting to change them into new and different kinds. This essentialistic belief in the fixity of species was a vital part of the understanding of nature since the time of Aristotle. What Darwin and Wallace did was to turn this mechanism of species preservation into one of species modification.

Giving a nod to Goethe on his way to the nineteenth century, Wallace briefly discusses "his views of the metamorphosis of plants," noting Goethe's mistakes, and then picks up with his march in time and moving on to Lamarck, "the first systematic evolutionist," and summarizes the French naturalist's belief in the inheritance of acquired characteristics. Of course, Wallace points out that "no direct evidence of this has ever been found, while there is a good deal of evidence showing that it does not occur," citing many examples, such as "the feathers of the peacock, the poison in the serpents fangs, the hard shells of nuts, the prickly covering of many fruits, the varied armour of the turtle, porcupine, crocodile, and many others." He then concludes: "For these reasons Lamarck's views gained few converts; and although some of his arguments have been upheld in recent years the fatal objections to his general principle as a means of explaining the evolution of organic forms, has never been overcome."

Wallace then tips his hat to one of his intellectual mentors, Herbert Spencer, who strongly influenced him in his attitude toward a number of social issues, but whose theory of evolution was not up to scientific snuff. Although Wallace says of Spencer's *First Principles* that it was "a coherent exposition of philosophy, coordinating and explaining all human knowledge of the universe into one great system of evolution everywhere conforming to the same general principles, must be held to be one of the greatest intellectual achievements of the Nineteenth Century," he assesses its specific value rather less flatteringly: "It left, however, the exact method of evolution of organisms untouched, and thus failed to account for those complex adaptations and appearances of design in the various species of animals and plants, which have always been the stronghold of those who advocated special creation." Wallace then credits Darwin for meeting "this difficulty" in 1859, modestly excluding himself as a contributor to the solution.

In contrast to the Laplacean clockwork machine, Wallace identifies the Darwinian historical system with all the concomitant uniqueness that comes with unpredictable elements in a nonlinear system. This is the distinction between the experimental sciences and the historical sciences and why physics can never be the model of science for evolutionary biology or any other historical science. This type of evolution—historically contingent, not physically necessary—is best embodied in the theory of natural selection, which Wallace masterfully summarizes in a brief historical account, acknowledging the anonymous author of *The Vestiges of Creation* (Robert Chambers). But, for Wallace, this lengthy historical sequence was merely paving the way for Darwin, who "produced a work which at once satisfied many thinkers that the long-desired clue had been discovered" and "will probably take its place, in the opinion of future generations, as the crowning achievement of the Nineteenth Century."

With the history lesson complete, Wallace moves to the structure of evolutionary theory: "The first group of facts consists of the great powers of increase of all organisms, and the circumstances that, notwithstanding this great yearly increase, the actual population of each species remains stationary, there being no permanent increase." From these two observations the deduction follows that there is a struggle for existence: "Individuals of the same species struggle together for food, for light, for moisture; they struggle also against other species having the same wants; they struggle against every kind of enemy, from parasitic worms and insects up to carnivorous animals; and there is a continual struggle with the forces of nature—frosts, rains, droughts, floods, and tempests." What determines who survives this struggle? Once again he credits his elder colleague with the answer: "Darwin calls this process of extermination one of 'natural selection,'—that is, by this process nature weeds out the weak, the unhealthy, the unadapted, the imperfect in any way." Noting that it was Spencer who called this process "survival of the fittest," Wallace places the emphasis here on negative selection (selection against the least fit individuals) as opposed to positive selection (selection for the most fit individuals): "The struggle is so severe, so incessant, that the smallest defect in any sense, organ, and physical weakness, any imperfection in constitution will almost certainly, at one time or another, be fatal." The most fit individuals are the result of the negative selection process: "This continual weeding out of the less fit, in every generation, and with exceptional severity in recurring adverse seasons, will produce two distinct effects, which require to be clearly distinguished. The first is the preservation of each species in the highest state of adaptation to the conditions of its existence; and therefore, so long as those conditions remain unchanged, the effect of natural selection is to keep each well-adapted species also unchanged. The second effect is produced whenever the conditions vary, when, taking advantage of the variations continually occurring in all well adapted and therefore populous species, the same process will slowly but surely bring about complete adaptation to the new conditions."

To reinforce his emphasis on negative selection as the driving force behind evolution, Wallace hammers home his position one final time: "Suffice it to say here that this theory of Natural Selection—meaning the elimination of the least fit and therefore the ultimate 'survival of the fittest'—has furnished a rational and precise explanation of the means of adaptation of all existing organisms to their conditions, and therefore of their transformation from the series of distinct but allied species which occupied the earth at some preceding epoch. In this sense it has actually demonstrated the 'origin of species' from other distinct species, and, by carrying back this process step by step into earlier and earlier geological times, we are able mentally to follow out the evolution of all forms of life from one or few primordial forms. Natural Selection has thus supplied that motive power of change and adaptation that was wanting in all earlier attempts at explanation, and this has led to its very general acceptance both by naturalists and by the great majority of thinkers and men of science."

With this historical retrospective of the intellectual trajectory that led to his scientific work, Wallace set about writing his autobiography and testing the limits of his remarkable memory and prodigious recordkeeping habits over the decades that afforded him ample resources from which to draw.

"The Most Important Ideas I Have Given the World"

In 1905 Wallace's two-volume autobiography, *My Life,* was published, followed by an edited one-volume version released in 1908.[27] An ambitious recapitulation of his first eighty-two years, Wallace's humble beginnings and modest lifestyle are seen as clear determinants of his lifelong feeling that he never quite belonged with the great Victorian scientists and scholars of his time. "Thanks for your remarks on what an autobiography ought to be. But

Figure 11-4 Alfred Russel Wallace at age seventy-nine, shortly after the publication of an edited compilation of his past works, *Studies, Scientific and Social,* and his essay-length history of evolutionary thought through his own time, titled simply "Evolution." (From *My Life,* 1905, v. I, frontispiece)

I am afraid I shall fall dreadfully short," he told his friend Mrs. Fisher on April 17, 1904. "I seem to remember nothing but ordinary facts and incidents of no interest to anyone but my own family. I do not feel myself that anything has much influenced my character or abilities, such as they are." The gentleman protests too much, however, since he then confesses that "lots of things have given me opportunities, and those I can state. Also other things have directed me into certain lines, but I can't dilate on these; and really, with the exception of Darwin and Sir Charles Lyell, I have come into close relations with hardly any eminent men. All my doings and surroundings have been commonplace!"[28]

Figure 11-5 In his final years Wallace still enjoyed his communion with nature in the form of walks and picnics in the woods (top) and playtime with grandchildren (bottom). (Courtesy of Alfred John Russel Wallace and Richard Russel Wallace)

On the contrary, the autobiography is filled with distant lands, numerous eminent scientists and statesmen he came to know, and an assortment of influences on his personality and events that led to his achievements. It is a remarkable work that reveals Wallace's prodigious memory and towering intellect that, while perhaps a modicum below the most eminent scientists of his age, allowed him to excel in every field he entered, to significantly change many of them, and to lead or create several new areas of research. In the second volume's final chapter he even lists chronologically what he considers his most important "ideas, or suggestions, or solutions of biological problems, which I have been the first to put forth."[29] They include:

1. His 1858 Ternate paper that triggered the Darwinian revolution that "has been so fully recognized by Darwin himself and by naturalists generally that I need say no more about it here."
2. His 1864 paper "The Origin of Human Races and the Antiquity of Man Deduced from the Theory of 'Natural Selection' " that he called "the most original and important part of which was that in which I showed that so soon as man's intellect and physical structure led him to use fire, to make tools, to grow food, to domesticate animals, to use clothing, and build houses, the action of natural selection was diverted from his body to his mind, and thenceforth his physical form remained stable while his mental faculties improved." Although he complained that "owing to its having been published in one of my less known works, 'Contributions to the Theory of Natural Selection,' it seems to be comparatively little known," it was, in fact, the most cited article he ever wrote, outdistancing even his 1858 Ternate paper.
3. His 1867 "solution of the cause of the gay, and even gaudy colours of many caterpillars."
4. His 1868 theory of birds' nests, connecting female bird coloration to nest structure, which he characteristically stated in the form of a law: "When both sexes of birds are conspicuously coloured, the nest conceals the sitting bird; but when the male is conspicuously coloured and the nest is open to view, the female is plainly coloured and inconspicuous."
5. A class of contributions lumped under the general heading of animal coloration, such as " 'recognition colours,' which are of importance in affording means for the young to find their parents, the sexes each other, and strayed individuals of returning to the group or flock to which they belong." Even more important is his identification of "the use of these special markings or colours during the process of the development of new species adapted to slightly different conditions, by checking intercrossing between them while in process of development." This, he believed, falsified "Darwin's theory of brilliant male coloration or marking being due to female choice," or sexual selection, further reinforcing his commitment to natural selection as the only materialistic force for evolutionary change.
6. "The general permanence of oceanic and continental areas" that formed the basis of the study of biogeography.

7. The causes of glacial epochs (ice ages), including geographical changes that lead to snow and ice producing cumulative effects that cause global temperatures to decrease by a few degrees.
8. Creating the science of island biogeography, especially as presented in his 1880 *Island Life*.
9. His 1881 hypothesis that mouth gestures were a "factor in the origin of language" in that "every motion of the jaws, lips, and tongue, together with inward or outward breathing, and especially the mute or liquid consonants ending words which serve to indicate abrupt or continuous motion, have corresponding meanings in so many cases as to show a fundamental connection." For this idea he even found a supporter in Prime Minister Gladstone, who "informed me that there were many thousands of illustrations of my ideas in Homer."
10. "The gradual improvement of the race by natural process," by which he meant natural selection, of course, that can only come about "under a social system which gives equal opportunities of culture, training, leisure, and happiness to every individual." This was certainly his most ambitious attempt to link his science to his politics, presented in numerous journals and magazines and republished in his volume *Studies Scientific and Social.* He considered this to be "by far the most important of the new ideas I have given to the world."

The autobiography comes to a close with a narrative account of spiritual "predictions fulfilled," most of which were spectacularly uninteresting examples of selective memory. But then, as he so thoughtfully reflects, "every one who reaches my age enjoys 'retrospection,' but that kind of general looking back to the past is very different from the detailed Retrospection I have had to make in searching out the many long-forgotten incidents and details of my very varied life as here recorded."[30]

"As Twenty Years Is to One Week": The Origins of Natural Selection

Wallace's insistence that his contributions had been "almost entirely overlooked" was more a reflection of his modesty (or perhaps one of his rare bouts of low self-esteem) than it was of how he was really perceived by his colleagues and the general public at large. In addition to the numerous awards already heaped on him, one of the most significant was presented to him in 1908 when he was honored with the first Darwin–Wallace medal at the fiftieth anniversary of the 1858 Linnean Society meeting where his and Darwin's papers were jointly read. This led *The Popular Science Monthly* to invite him to write a piece on "The Origin of the Theory of Natural Selection," to clear up the historical confusion that the 1908 celebration rekindled.

It had become apparent to Wallace that there was much misunderstanding of what actually happened in the years leading up to 1858. "Since the death of Darwin in 1882, I have found myself in the somewhat unusual position of receiving credit and praise from popular writers under a complete misapprehension of what my share in Darwin's work really amounted to," he began, then continued by citing the even more egregious claim that he had scooped Darwin, then stepped aside. "It has been stated (not unfrequently) in the daily and weekly press, that Darwin and myself discovered 'natural selection' simultaneously, while a more daring few have declared that I was *the first* to discover it, and that I gave way to Darwin!"[31]

In this article we see Wallace's generosity in offering more of the share of the credit to Darwin (whom he refers to as "my honored friend and teacher"), while at the same time firmly reestablishing what he did and did not do. "The idea came to me, as it had come to Darwin, in a sudden flash of insight: it was thought out in a few hours—was written down with such a sketch of its various applications and developments as occurred to me at the moment,—then copied on thin letter-paper and sent off to Darwin—all within one week. *I* was then (as often since) the young man in a hurry; *he,* the painstaking and patient student, seeking ever the full demonstration of the truth that he had discovered, rather than to achieve immediate personal fame." The paper also contains a certain amount of the obligatory modesty that is usually elicited when one is being so honored, such as when Wallace states that the share of the credit should be allocated "proportional to the time we had each bestowed upon it . . . that is to say, as twenty years is to one week."[32]

Wallace did discover and describe natural selection all in the course of a week in late February 1858, but his four years in the Amazonian tropical rain forest and another four (to that time) in the Malay Archipelago hardly represent one week to Darwin's twenty years (in fact, Darwin was five years on his voyage, Wallace a total of twelve in two voyages). It is true, however, that had Darwin published, "after ten years—fifteen years—or even eighteen years" instead of the twenty following the opening of his notebook in 1838, Wallace "should have had no part in it whatever, and he would have been recognized as the sole and undisputed discoverer of 'natural selection.' "[33] The fact is, however, Darwin waited twenty years, and would have likely waited longer had Wallace not triggered Darwin's productive burst.

Greatness In Spite of Himself

That same year Wallace was also presented with the Order of Merit from King Edward, the highest award of distinction from the Crown for science, literature, and the arts.

Wallace was, in fact, sought after by one and all for his views on all manner

DARWIN-WALLACE MEDAL.
1st July, 1908.

Figure 11-6 The Darwin–Wallace Medal presented to Wallace in 1908 on the fiftieth anniversary of the 1858 meeting of the Linnean Society of London when the Darwin–Wallace papers were jointly read into the record. Neither Darwin nor Wallace was present at the original meeting. (From *The Darwin–Wallace Celebration Held on Thursday, 1st July, 1908,* published by the Linnean Society of London, 1908, 2)

of scientific and social issues. In 1909, for example, he gave an interview to Ernest Rann for *The Pall Mall Magazine,* in which he explained his working habits at this late stage of his life: "I am always at work. As a rule I manage two steady hours every morning. In the afternoon I take a quiet doze, or content myself with watching the harbour, which you can see from my window there; and in the evening I am ready for another spell of writing or study." Wallace also confessed a preference for the philosophy behind vegetarianism, even if he could not consistently apply it: "I am afraid that on another subject I am still misunderstood. That is vegetarianism, in which I thoroughly believe; but although it may appear inconsistent, I am a meat-eater myself, as I have found that meat-eating, in the way I eat it, is, with a diet regulated in other ways, a remedy for a troublesome complaint from which I suffered for many years, and might be suffering now, if I had not changed my mode of living. You cannot alter the habits of mankind in a single generation. Vegetarianism is a reform which will come, but it must come gradually, when people have learned that there are other foods than those to which they have been accustomed. You cannot force the pace; if you try to do so, it simply gives a set-back to the movement."[34]

Wallace's fame as scientist and social activist never waned, as he told Rann of his weekly requests for his autograph, especially from America, for which "I always send it, particularly if a stamp is included for return." Wallace also emphasized his continued support for and acknowledgment of the founder of modern evolutionary biology when he told his interviewer: "What I want to do is to state once more the essential truth of Darwinism, and its relation to the world of life. I am busily engaged at this moment in preparing a lecture which I am to deliver in London before the Royal Institution. We of this generation seem to have forgotten the fundamental truths set forth by Spencer in his 'First Principles,' as well as by Darwin. There is a tendency to belittle, if not to ignore, their work; but Darwin, I tell you, will stand secure in the coming ages against all criticism." Rann ended his profile of Wallace with this beautiful and fitting tribute to the man who bridged two epochs:

> It was the impatience of the old warrior stirred afresh at the prospect of conflict. My last memory of him is as he paused at his study door to bid me farewell—a venerable figure crowned with white, standing, as it were, midway between the centuries. Behind him lies the one that he has done so much to mould and alter and convince; before him that mysterious future on which he gazes with unshadowed faith in the ultimate triumph of his views.[35]

His energies barely waning, in 1907 Wallace took on the world-renowned American astronomer Percival Lowell, whose 1906 book, *Mars and Its Canals,* presented his controversial idea that the "canali" first observed on Mars by the Italian astronomer Schiaparelli (who meant by the attribution "channels" and never presumed that they were intelligently designed "canals") were constructed by a dying race attempting to quench its final thirst by diverting the waters of the Martian poles to the parched cities around the planet. Because Wallace had already committed himself several years earlier to the position that humans are unique in the cosmos, he could not let such apparent contradictory evidence be presented without a challenge. "The one great feature of Mars which led Mr. Lowell to adopt the view of its being inhabited by a race of highly intelligent beings . . . is that of the so-called 'canals'—their straightness, their enormous length, their great abundance, and their extension over the planet's whole surface from one polar snow-cap to the other. . . . The very immensity of this system, and its constant growth and extension during fifteen years of persistent observations, have so completely taken possession of his mind, that, after a very hasty glance at analogous facts and possibilities, he has declared them to be 'non-natura,'—therefore to be works of art—therefore to necessitate the presence of highly intelligent beings who have designed and constructed them." In his book-length critique, *Is Mars Habitable?,* Wallace set to work dismantling Lowell's theory by first arguing that the canals were possibly "produced by the contraction of heated outward crust upon a cold, and therefore non-contracting interior," then moved to demonstrating the illogic of assuming that they could have been made by intelligent Martians:

Figure 11-7 Wallace in his garden at Broadstone standing next to a fully blooming king's-spear plant in 1905, the year his two-volume autobiography, *My Life,* was published. (Courtesy of Richard Milner and Gareth Nelson)

> The innumerable difficulties which it raises have been either ignored, or brushed aside on the flimsiest evidence. As examples, he never even discussed the totally inadequate water-supply for such world-wide irrigation, or the extreme irrationality of constructing so vast a canal-system the waste from which, by evaporation, when exposed to such desert conditions as he himself describes, would use up ten times the probable supply. . . . The mere attempt to use open canals for such a purpose shows complete ignorance and stupidity in these alleged very superior beings, while it is certain that, long before half of them [the canals] were completed their failure to be of any use would have led any rational beings to cease constructing them.[36]

By now there was virtually no subject beyond the purview of this latent Renaissance polymath. Wallace continued to write steadily and work in his garden on botanical breeding experiments through the final years of his life. He corresponded with such social and political luminaries as Asquith, Lloyd

Figure 11-8 Wallace in his greenhouse, tending his plants. (Courtesy of Richard Milner and Gareth Nelson)

George, and Sir Edward Gray, and wrote three more books, *The World of Life, Social Environment and Moral Progress,* and *The Revolt of Democracy.* Neither his mind nor his pen ceased activity, but his body finally gave out in his sleep on November 7, 1913, after ninety years and ten months.

Alfred Russel Wallace's life spanned the reigns of King George IV, King William IV, Queen Victoria, King Edward VIII, and King George V, and was one of relative peace; he was born two years after Napoleon's death and died only ten months before the outbreak of the First World War. In Wallace's youth the study of nature had all the properties of an avocation—a hobby for the curious with leisure time on their hands. In his dotage the field of biology had erupted into a full-blown science, with multiple subdisciplines and specialties, two of which—evolutionary biology and biogeography—were founded on the work of two men, Charles Robert Darwin and Alfred Russel Wallace, their names forever linked in the annals of science history.

Although Wallace was buried in the little cemetery at Broadstone, where he had lived since 1902, a memorial plaque was affixed to the floor at Westminster Abbey two years after his death, next to Darwin's and near those of

Figure 11-9 Alfred Russel Wallace in 1913 at age ninety, the last of the last great Victorians in the final year of his life. (From *The Revolt of Democracy,* 1913, frontispiece)

Lister, Hooker, Herschel, Kelvin, and Lyell, five of the most eminent scientists of the age. Honored simultaneously with Hooker and Lord Lister, Wallace's greatness was put into perspective by the Dean of the Abbey, who noted: "As is so often observable in true greatness, there was in them an entire absence of that vanity and self-advertisement which are not infrequent with smaller minds. It is the little men who push themselves into prominence through dread of being overlooked. It is the great men who work for the work's sake without regard to recognition, and who, as we might say, achieve greatness in spite of themselves."[37] It was a fitting tribute to a man whose greatness was owed to the fact that he was not just a scientist, but a persistent and fearless heretic scientist.

12

The Life of Wallace and the Nature of History

On so many different levels Alfred Russel Wallace was one of the most revolutionary thinkers in the history of evolutionary thought. Unmatched as a keen observer of nature, and prolific in collecting and writing, Wallace combined both talents in his discovery of a mechanism for evolutionary change that he proselytized for the rest of his life. Though eclipsed by Darwin in both his and our time, he made significant contributions to many branches of the biological sciences in the nineteenth century, and much of his pioneering work in the field of biogeography and evolutionary theory is still relevant today. With his theory on the origin of species, including and especially humans, Wallace rejected scriptural arguments for separate creation by divine intervention. His empirical research on the mechanism of natural selection provided evidence in support of the quantitative connection of humans with the apes and lower animals, proving our descent with modification from these common ancestors. His exploration of several deep and overarching themata allowed him to tie together his seemingly disparate and apparently unconnected scientific and social interests into a coherent worldview.

Further, Wallace was exceptional in rejecting the standard racial dogma of his age, with his arguments for the equal intellectual capacities of "savages" and Europeans. Investigator of unusual phenomena and champion of unpopular causes, Wallace had the courage and strength to defend his position (once he was convinced of its validity through scientific analysis), despite frequently finding himself on the heretical fringe. He achieved a level of fame in his own time that, while virtually nonexistent today, made him a world-class figure in the sciences and a recognized authority in the general social sphere. Finally, his decency as a caring human being transcended the boundaries of a competitive scientific community in which, for Wallace, friend and foe alike were treated with the utmost respect, even over the issue of priority—certainly one that would test the mettle of the most spirited of individuals. Alfred Russel Wallace was an event-making man, who found a fork in the historical road and altered the path as he strove down it.

Having praised Wallace, however, let us put him into a proper sociohistorical context. We would do well to strike a balance between the internalist and externalist interpretations of history, as I have tried to do in the Historical Matrix Model. In this analysis, a significant historical figure is both product and producer of the age in which he lives. Such is the case for Wallace, who made contributions that lived well beyond his own time, while simultaneously being deeply influenced by his culture. With twenty-first-century hindsight, we can critique Wallace's theories with ease, and find his selectionistic arguments with regard to humans, while logically consistent, deeply flawed. His knowledge of native peoples, while extensive in the sense of lasting several years, was based on limited ethnographic study, since his primary focus was the natural history of insects, birds, and mammals, and only secondarily human evolution. His involvement in phrenology and spiritualism, the movement of which, after over a century of serious investigation, has been found lacking in any empirical evidence, was a bust. And, most relevant to his hyperselectionism and his theoretical break with Darwin, Wallace was unaware of possible solutions to the problem of how natural selection could account for a relatively large brain, along with a number of other seemingly inexplicable anomalies in human evolution.

These criticisms are largely anachronistic and in no way lessen Wallace's impact as a seminal thinker, particularly if we understand the disparate forces, internal and external, that combined to shape his thoughts and ideologies over a long lifetime. As Gunnar Myrdal has observed about the difficulty of teasing out the effects of social forces on individual thought and action: "Cultural influences have set up the assumptions about the mind, the body, and the universe with which we begin; pose the questions we ask; influence the facts we seek; determine the interpretation we give these facts; and direct our reaction to these interpretations and conclusions. But there must be still other countless errors of the same sort that no living man can yet detect, because of the fog within which our type of Western culture envelops us."[1] A goal of the historian is to look to the past to discover what cultural influences have operated on an individual's ideas and behaviors, and how those forces determined the questions we ask about and interpretations we make of that thinker. To the extent that this has been accomplished in examining Alfred Russel Wallace, then this psychobiographical approach has done its job as a heuristic tool in the research kit of the historical scientist.

The Nature of Historical Change

This book, on one level, has been one long argument on the nature of historical change as part of an attempt to reach a deeper understanding of the general process of history itself. As we have seen in this particular historical sequence, which may have applications to others, history is an abundant and complex set of relationships. To recall Darwin's and Wallace's branching bush metaphor, the farther out on a limb that a step from the past is located, the more complex the pathways leading to it. Countless twigs are connected by innumerable branches to untold trunks somewhere in the past, so that the present is contingent on the past in a way that determines or necessitates the future.

Contingency—Necessity was one of the themata we explored in Wallace's work. Historians and philosophers have been cognizant for millennia of this basic tension between what may not be at all and what cannot be otherwise, between the particular and the universal, between history and nature, between contingency and necessity. But such synonyms can take us only so far (and may lead to problems of meaning and emphasis). Precise definitions are needed to formulate a model of change. Thus, in this analysis contingency will be taken to mean: *a conjuncture of events occurring without design;* and necessity to mean *constraining circumstances compelling a certain course of action.* Contingencies are the sometimes small, apparently insignificant, and usually unexpected events of life—the kingdom hangs in the balance awaiting the horseshoe nail. Necessities are the large and powerful laws of nature and trends of history—once the kingdom has collapsed, the arrival of 100,000 horseshoe nails will not save it. Leaving either contingency or necessity out of the formula, however, is to ignore an important component in the development of historical sequences—their interactions. The past is constructed by an interaction of contingencies and necessities, therefore it might be useful to combine the two into one term that expresses this interrelationship—*contingent–necessity*—taken to mean: *a conjuncture of events compelling a certain course of action by constraining prior conditions.*[2]

Contingency and necessity are not mutually exclusive properties of nature; rather, they vary in the amount of their respective influence and at what point their influence is greatest in the historical sequence. Necessities such as laws of nature (gravity), economic forces (supply and demand), demographic trends (birth and death rates), geographical currents (immigrations and emigrations), and political ideologies (egalitarianism) exert a governing force on the individuals falling within their sphere of influence. Often necessities are so powerful they override all other factors, including human freedom. Contingencies, however, exercise influence sometimes despite the necessities involved. At the same time, contingencies reshape future necessities. But how and when do contingencies matter? And how do they interact with necessities?

One can find in history meaningful interactions between contingencies and necessities, as they vary over time, in what I call the *model of contingent–necessity: In the development of any historical sequence the role of contingencies in the construction of necessities is accentuated in the early stages and attenuated in the later.* There are six corollaries that encompass various aspects of the model:

Corollary 1: The earlier in the development of any historical sequence the more chaotic the actions of the individual elements of that sequence; and the less predictable the future actions and necessities.

Corollary 2: The later in the development of any historical sequence the more ordered the actions of the individual elements of that sequence and the more predictable the future actions and necessities.

Corollary 3: The actions of the individual elements of any historical sequence are generally postdictable, but not specifically predictable, as regulated by Corollaries 1 and 2.

Corollary 4: Change in historical sequences from chaotic to ordered is common, gradual, followed by relative stasis, and tends to occur at points where poorly established necessities give way to dominant ones so that a contingency will have little effect in altering the direction of the sequence.

Corollary 5: Change in historical sequences from ordered to chaotic is rare, sudden, followed by relative nonstasis, and tends to occur at points where previously well-established necessities have been challenged by others so that a contingency may push the sequence in one direction or the other.

Corollary 6: Between origin and bifurcation, sequences self-organize through the interaction of contingencies and necessities in a feedback loop driven by the rate of information exchange.

At the beginning of any historical sequence, actions of the individual elements tend to be chaotic, unpredictable, and have a powerful influence in the future development of that system. But as the system gradually develops and the pathways slowly become more worn, out of chaos comes order, as the individual elements sort themselves into their allotted positions, as dictated solely by what came before.

The forward movement of these historical pathways, which may begin as single-track trails and eventually develop into interstate highways, is only adequately understood when we look back to see the road from which we came. The view forward, beyond even the closest of chronological landmarks, becomes befogging to our intellectual senses. We may know from where we have come, but just where are we going? The problem is made no clearer with epistemological lenses, because the difficulty is not in the seeing but in what is observed. Future histories, beyond a scant few days, are not and never will approach the predictability achieved in some areas of the physical sciences. The reason is contingency, in interaction with necessity.

Contingency and Necessity in Wallace

The *Contingency–Necessity* theme in both Wallace's work and in history is beautifully illustrated in the 1908 paper he wrote for the fiftieth anniversary celebration of the Darwin–Wallace 1858 joint papers at the Linnean Society of London, entitled "The Origin of the Theory of Natural Selection." It is interesting to note Wallace's recognition of the role of both contingencies and necessities in the development of scientific discoveries as he explores "a curious series of correspondences, both in mind and in environment, which led Darwin and myself . . . to reach identically the same theory." This series is a recounting of the contingencies of thought and culture leading to the necessity of discovering natural selection. The concatentation of these contingencies struck the necessity of discovering natural selection, in Wallace's words, like "friction upon the specially-prepared match."[3]

After first clarifying that he and Darwin *independently,* not simultaneously, discovered natural selection ("the idea occurred to Darwin in October, 1838, nearly twenty years earlier than to myself in February, 1858"), Wallace identifies the role of contingency in scientific discovery when he notes: "It was really a singular piece of good luck that gave to me any share whatever in the discovery."[4] He then turns to an analysis that shows how a number of contingencies in the lives of both men led to the necessary discovery of natural selection, including:

1. Being "ardent beetle-hunters, [a] group of organisms that so impresses the collector by the almost infinite number of its specific forms, the endless modifications of structure, shape, color and surface-markings that distinguish them from each other, and their innumerable adaptations to diverse environments."
2. Having "an intense interest in the mere variety of living things . . . which are soon found to differ in several distinct characteristics."
3. A "superficial and almost child-like interest in the outward forms of living things, which, though often despised as unscientific, happened to be the only one which would lead us towards a solution of the problem of species."
4. Both "were of a speculative turn of mind [and] constantly led to think upon the 'why' and the 'how' of all this wonderful variety in nature."
5. "Then, a little later (and with both of us almost accidentally) we became travellers, collectors and observers, in some of the richest and most interesting portions of the earth" (Darwin's five-year global circumnavigation and Wallace's four years in the Amazon and eight in the Malay Archipelago).

"Thence-forward our interest in the great mystery of *how* species came into existence was intensified."

6. Both men on their voyages and in their home lives enjoyed "a large amount of solitude . . . which, at the most impressionable period of our lives, gave us ample time for reflection on the phenomena we were daily observing."
7. Both men carefully read Lyell's *Principles of Geology* and Malthus's *Principles of Population,* the latter "at the critical period when our minds were freshly stored with a considerable body of personal observation and reflection bearing upon the problem to be solved," that acted on both like "that of friction upon the specially-prepared match, producing that flash of insight which led us immediately to the simple but universal law of the 'survival of the fittest.' "[5]

All of these contingencies created necessities (what Wallace calls "the combination of certain mental faculties and external conditions") that drove the two naturalists down parallel paths that became cut ever deeper until they finally crossed in the spring of 1858. This historical tension between what might be and what must be—the contingent and the necessary—for an eighty-five-year-old Wallace reflecting back on a life of science, explains why it was Darwin and himself who finished first and "a very bad second," in the "truly Olympian race" to discover the mechanism of evolutionary change; and why it was not the "philosophical biologists, from Buffon and Erasmus Darwin to Richard Owen and Robert Chambers." For Wallace, the explanation is simple. These "great biological thinkers and workers" were on different paths at different times that made it impossible for them to "hit upon what is really so very simple a solution of the great problem." An adequate explanation of a historical development requires a healthy balance of the internal and the external, individual thought and collective culture, or "the combination of certain mental faculties and external conditions that led Darwin and myself to an identical conception."[6]

Finally, Wallace applies his model to the larger picture of the development of ideas in general, and comes to the conclusion that "no one deserves either praise or blame for the ideas that come to him, but only for the actions resulting therefrom." Wallace is suggesting that the vagaries and nuances of our life and thoughts lead us down certain paths toward conclusions that can be reached only by way of that particular road. Wallace and Darwin shared nearly parallel paths for a time (which later diverged on other issues), and Wallace acknowledges the role of such historical contingencies and necessities in the larger scale of the discovery of scientific ideas: "They come to us—we hardly know how or whence, and once they have got possession of us we can not reject or change them at will." Wallace also addresses the even larger role of human freedom within historical trends by explaining that it is not the development of ideas but in the "actions which result from our ideas" that individuals have the most say in their historical context. Here we catch a glimpse of Wallace, the hard-working, common man who made a most uncommon discovery: "It is only by patient thought and work, that new ideas, if good and true, become adopted and utilized; while, if untrue or if not adequately presented to the world, they are rejected or forgotten."[7] Such is the nature of science and history.

Counterfactuals and "What if?" History

Another path into the murky past that can aid us in discerning cause-and-effect relationships in history is what is known in the trade as counterfactual history, or in popular circles "What if?" history, as in its humorous extreme "What if Napoleon had the atom bomb?" Well, what if he did? He no doubt would have dropped it on Blücher before he arrived at Waterloo to rescue Wellington. We know this is ridiculous, of course, but hyperbole does not equal superfluity. In moderation we can play this game to great historical insight, and the process even has a technical name: counterfactual conditionals.

In logic, conditionals are statements in the form "if *p,* then *q,*" as in a more realistic counterfactual for Waterloo where "if Blücher arrives in time to reinforce Wellington's troops then Napoleon loses," where *q* depends on *p* (and *p* is the antecedent since it comes before *q*). Counterfactual conditionals alter the factual nature of *p,* where *p'* is counter to the facts, thus altering its conditional element *q* into *q'*. Counterfactual conditionals are said to be modal in nature; that is, changing the antecedent changes the modality of the causal relationship between *p* and *q* from necessary (what had to be) to contingent (what might have been). Change *p* to *p'* and instead of *q* you may get *q',* as in "if Blücher does not arrive in time, Napoleon may win." In other words, the modal nature of the relationship between *p* and *q* changes from necessary to contingent in counterfactual conditionals.

Counterfactual modal thinking is prevalent in works of history, as historians try to understand causal relationships by considering what might have happened in a different replay of the historical sequence. Histories of the American Civil War, for example, are filled with counterfactual conditionals whereby the South might have gained independence from the North at various contingent stages of the war. The Battle of Antietam is a case study in counterfactual history.[8] In preparing to defend his territory against invading Confederates, General George B. McClellan caught a break when one of his soldiers stumbled onto Robert E. Lee's battle plans in the infamous Special Orders 191, wrapped in cigar paper and accidentally dropped in an open field. With Lee's plans in hand, the impossibly refractory and interminably sluggish

McClellan was able to fight Lee to a draw, thwarting his invasion plans. In the factual time line the conditional sequence is "if *p* (McClellan has Lee's plans), then *q* (the invasion is turned back)." From this, additional conditional series arise where: if *q,* then *r* (the war continues), *s* (England does not recognize the South as a sovereign nation), *t* (the northern blockade continues), *u* (the South's diminishing resources hinders their war effort), *v* (Lee is defeated at Gettysburg), *w* (Grant becomes the head northern general), *x* (Sherman destroys everything in his path from Atlanta to Savannah), *y* (Lee surrenders to avoid the utter destruction of the South), and *z* (America remains a single nation).

In a counterfactual conditional in which McClellan does not get Lee's plans, one possible outcome is that he is dealt a major defeat, the invasion continues until the South earns the recognition from England as a sovereign nation, bringing the British navy to bear on the Northern blockade (in order to retain trading channels), thereby allowing the South to replenish her rapidly depleting resources and carry on the war until the North finally gives up. Here *p'* (no plans to McClellan) leads to *q'* (Lee wins), with the modal cascading consequences of *r'* through *z',* and America ever after is divided into two nations, changing almost everything that has happened since. Whether I am right about this counterfactual is not relevant here. More than just an example of the modal nature of counterfactuals, we see how they help us think about cause-and-effect relationships in historical sequences. We do not want to just know how the Civil War unfolded, we want to know *why.* Why questions are deeper than how questions, and require an appropriately deeper level of analysis.[9]

Counterfactual modal thinking is also common in science once you look for it. Scientists study systems of interacting elements—astronomers examine the movement of planets, stars, and galaxies, biologists record the complex web of an ecosystem, and psychologists observe the interaction of people in a crowd. The component parts of these systems can be labeled, as in the conditional "if *p,* then *q.*" If a star shows a certain type of wobble, then it means it has a planetary body orbiting it. If a chemical is introduced into an ecosystem, then certain animals will disappear. If a crowd is of a certain size, then an individual will most likely comply. These are all real examples of conditional statements for which counterfactual conditionals help us test hypotheses. The astronomer knows that stellar wobble means a planetary body is present because of the counterfactual observation that when such wobble is not present in other nearby objects, no other body is present. The biologist knows about the relationship between introduced chemicals and local extinction because of the counterfactual observation that when the chemical is removed the affected species return. The psychologist understands the correlation between group size and compliance because of the counterfactual observation that when group sizes are smaller less compliance occurs.

In some sciences counterfactual conditionals are directly testable in the form of experimental and control group comparisons, tested definitively with statistical tools. In other sciences the counterfactuals must be inferred, as in the search for extra-solar-system planets, where none have ever been directly observed. The use of inference in the physical and biological sciences is commonplace, so we should not disparage its application in the historical sciences, including human history. An astronomer or biologist setting up a conditional string of components labeled "if *p,* then *q, r, s, t, u, v, w, x, y, z*" uses the counterfactual conditional "if p' then $q', r', s', t', u', v', w', x', y', z'$" no differently than does the historian. Newton's famous formula $F = MA$ (force equals mass times acceleration) is, in its essence, the conditional statement "if a certain force is applied then an object will move as a function of its mass and acceleration," or, counterfactually, "an object's acceleration is dependent on its mass and the force applied to it."

Unfortunately history has not found its Newton and no such simple formulas exist for historians (or any of the social sciences for that matter). Still, a good start may be found comparing conditionals and counterfactual conditionals in historical time lines. That is, we can conduct thought experiments in the "if *p* then *q*" mode, informed by real historical examples of similar events to see how they unfolded in the "if p' then q''" condition. This is what I call the comparative method in historical science. What I am proposing is that when we compare, say, one culture or one time period to another to see how and why they differ, we are also rerunning the time line in a counterfactual experiment. We cannot literally rerun the time line, of course, but we can approximate it. I will show how that can be done in the Epilogue.

What if Wallace Had Discovered Natural Selection First?

Counterfactual thinking can also help us better understand the role Wallace played in the history of evolutionary thought. In fact, Wallace's own sequence of contingent events that paralleled Darwin's and that led the two of them to independently discover natural selection is a form of counterfactual modal thinking. Implicit in each step is that *if* it had not happened, *then* natural selection might not have been discovered. *If* Wallace had not been an "ardent beetle hunter," *then* he might not have seen the importance of "innumerable adaptations to diverse environments." *If* Wallace had not taken his voyages of discovery, *then* he would not have had "ample time for reflection on the phenomena." *If* Wallace had not read Lyell and Malthus, *then* that "flash of insight which led us immediately to the simple but universal law of the 'survival of the fittest' " might not have happened.

Historian C.F.A. Pantin employs counterfactual reasoning in his analysis of Wallace's role in the history of evolutionary thought, when he expands Wallace's list of parallel contingencies between him and Darwin. Pantin begins: "Both took the variation of domesticated animals and plants as their starting-point. Both noted the power of domestic selection. Both noted the common assumptions that the variations of breeds met a natural limit, and that domestic varieties left to a state of nature tended to revert to type."[10] Pantin then applies the counterfactual, altering the historical contingencies to reconstruct a different necessity: "Had there been no *Origin of Species* and had the Darwinian essays of 1842 and 1844 never come to light it would have been Wallace's Linnean essay rather than Darwin's that would have been given pride of place in the history of the theory of evolution."[11]

Perhaps, but one's counterfactual must be within reason. Here one might as well assume Darwin never lived and imagine a Wallacean revolution in biology. This would require a lot more from Wallace than he gave. Would he have completed his "plan," described to Darwin on September 27, 1857? Would he have been able to marshal together the enormous amount of material that Darwin did to support the grand conclusion of the origin of species by means of natural selection? Would his arguments, even if similar to Darwin's, been received in the same manner within the scientific community, and then beyond? These are counterfactual questions worth considering and trying to test. Pantin apologizes for his method in noting that "historical phantasy is amongst the most idle of occupations," but he concludes nonetheless that "had it been Charles Darwin who had died at the age of 39 . . . we should have had the interesting spectacle of Alfred Wallace with the complete skeleton of a theory of evolution."[12] But would Wallace have fleshed out that skeleton with the musculature of a complete scientific research program? I have my doubts.

Counterfactual conditionals are not historical fantasizing and certainly are not idle, but if we are going to play the game we might as well make it instructive. Let's imagine making some smaller contingent changes, perhaps giving Wallace the confidence, scientific training, and professional contacts (not to mention financial independence) that Darwin had, and wonder if he would have discovered the idea of natural selection sooner than Darwin. As we have seen, both internal and external forces played a role, some more, some less, in the shaping of Wallace's theories. Change any one of them and the outcome would certainly have been different, particularly earlier in Wallace's life. Have Wallace raised in an upper-class family, or have him, as a young man, read Adam Smith instead of Robert Owen, or in his twenties send him to Russia instead of South America, or have him miss the séance in 1865, or, or, or, and the pattern of change, while equally determined, would likely have been considerably different. But different how? Would he have become another Darwin, or (at least) been the architect of his own revolution?

The interactive nature of contingency and necessity, coupled to Wallace's own thoughts on the subject, can shed some light on this counterfactual thought experiment and show us why this would likely never have happened. It was, in fact, the very *lack* of these social conditions that created the independence of thought in Wallace and compelled him to go into the sciences in the first place. If these counterfactual changes were made, he may have never gone into science at all, as Wallace himself keenly reflects in his autobiography:

> Had my father been a moderately rich man and had supplied me with a good wardrobe and ample pocket-money; had my brother obtained a partnership in some firm in a populous town or city, or had established himself in his profession, I might never have turned to nature as the solace and enjoyment of my solitary hours, my whole life would have been differently shaped, and though I should, no doubt, have given some attention to science, it seems very unlikely that I should have ever undertaken what at that time seemed rather a wild scheme, a journey to the almost unknown forests of the Amazon in order to observe nature and make a living by collecting. All this may have been pure chance, as I long thought it was, but of late year I am more inclined to Hamlet's belief when he said—
>
> "There's a divinity that shapes our ends,
> Rough-hew them how we will."[13]

History is cumulative in this contingent sense, and all works of history, in a way, are detailed narratives of the connections between ideas, people, and events, all influenced by culture, folded into chronological sequences. A history offers us a glance at the richly detailed and highly convoluted linkages of the past. In his own autobiographical history Wallace would seem to agree: "I have good reasons for the belief that we are surrounded by a host of unseen friends and relatives who have gone before us, and who have certain limited powers of influencing, and even, in particular cases, almost of determining, the actions of living persons, and may thus in a great variety of indirect ways modify the circumstances and character of any one or more individuals in whom they are specially interested. Sometimes they only aid in the formation of character; sometimes they also lead to action which gives scope for the use of what might have been dormant or unused faculties (as, I think, has occurred in my own case)."[14]

Wallace, sensitive to the importance of both contingency and necessity in historical systems, would appear to see such an interaction not only in the history of science and in his own life, but in the history of life itself, as the laws of nature are guided by, and help guide, the events of evolution that, though apparently chaotic and labyrinthian, actually contain a plan to be discovered, as he explains in a passage penned as the concluding paragraph to *Island Life:*

> Not only does the marvellous structure of each organised being involve the whole past history of the earth, but such apparently unimportant facts as the presence of certain types of plants or animals in one island rather than in another, are now shown to be dependent on the long series of past geological changes—on those marvellous astronomical revolutions which cause a periodic variation of terrestrial climates—on the apparently fortuitous action of storms and currents in the conveyance of germs—and on the endlessly varied actions and reactions of organized beings on each other. And although these various causes are far too complex in their combined action to enable us to follow them out in the case of any one species, yet their broad results are clearly recognisable; and we are thus encouraged to study more completely every detail and every anomaly in the distribution of living things, in the firm conviction that by so doing we shall obtain a fuller and clearer insight into the course of nature, and with increased confidence that the "mighty maze" of Being we see everywhere around us is "not without a plan."[15]

Bound Together

On one level Wallace's entire working career was dedicated to a scientific understanding of the cause of physical, biological, and social phenomena. He would not settle for (and by temperament was not comfortable with) leaving the apparent chaos of nature unaccounted for without at least a rational search for order—unraveling the maze, discovering the plan. In the physical world, his analysis of the structure of the earth, solar system, and cosmos in *Man's Place in the Universe* found order in chaos by invoking a higher intelligence, necessary because of contingency—change one small component in the early evolution of the universe, and life would have been radically different or, more likely, nonexistent. Wallace's replay of the time line of intelligent human life would necessitate the contingencies of evolution to occur in exactly the sequence they did in the original sequence: "In order to produce a world that should be precisely adapted in every detail for the orderly development of organic life culminating in man, such a vast and complex universe as that which we know exists around us, may have been absolutely required."[16] That requirement was, for Wallace, met by an all-pervading higher intelligence.

In the biological world, of course, his own and Darwin's theory of natural selection as the driving mechanism of evolutionary change in the history of life nearly completely orders the chaotic diversity of nature, as he indicates in a passage from a letter to his childhood friend Thomas Sims, from the island of Timor, on March 15, 1861: "It is the vast *chaos* of facts, which are explicable and fall into beautiful order on the one theory [natural selection], which are inexplicable and remain a *chaos* on the other [the critics of Darwin], which I think must ultimately force Darwin's view on any and every reflecting mind." For Wallace, the wholesale collection of the facts of nature produces mere chaos without a unifying and ordering principle to tie them together: "The human mind cannot go on for ever accumulating facts which remain unconnected and without any mutual bearing and bound together by no law."[17] That is the deepness and insight of Wallace's wisdom.

EPILOGUE

Psychobiography and the Science of History

Late in his life Albert Einstein reflected back on a long career in science and concluded: "Science is the attempt to make the chaotic diversity of our sense-experience correspond to a logically-uniform system of thought."[1] This description of science was behind the development and general goal of the Historical Matrix Model, as well as the application of the Five Factor personality scale and the use of modern psychological research and theories on a historical figure, presented in the Prologue and used as heuristic tools to help get our minds around a complex array of data from the life, work, and culture of Alfred Russel Wallace. A device was needed to help sort through the chaotic diversity of historical data on a man who lived a very long and complex life in order to create a logically uniform system to explain his thoughts and actions. The HMM is one such device. "The traditional view of research amounts to studying the relation between one independent variable and one dependent variable," explain Kerlinger and Pedhazur in *Multiple Regression in Behavioral Research* regarding the limitations of such a simple research design. But, they argue, while "one can hardly say that the traditional view is invalid, one can say that in the behavioral sciences it is obsolescent, even obsolete. One simply cannot understand and explain phenomena in this way because of the complex interaction of independent variables as they impinge on dependent variables."[2]

The Historical Matrix Model, however, is only a model and makes no pretense of representing reality. And while more complex than previous historical interpretations of Wallace, it is, like all models, relatively simple compared to the actual messiness of life and history. Does this mean we should not construct models? Must we abandon the scientific method because of its inability either to completely postdict human history or predict human actions? Of course not. What we are discussing is a matter of degree of accuracy in the representation of reality. Too few variables and the model oversimplifies, leaving us with a cardboard characterization of our subject that leads to more myth than materiality. Too many variables and the model overcomplicates, leaving us confused and befuddled in a fog of facts without structure. Somewhere in between lies a workable medium where the social scientist can construct a model representing a subject, and know that enough of the most important variables to adequately explain the phenomenon in question have been included, without leaving out or adding too many of those variables whose roles were relatively insignificant. Modeling is what scientists, both experimental and historical, do. It is not possible to reconstruct nature exactly in a laboratory, any more than it is possible to write a history "as it really happened," because it would be nearly infinite in detail. A moderate approximation is a worthy goal. In his 1947 classic, *Multiple Factor Analysis* (a pioneering work in the development of the factorial matrix), L. L. Thurston notes the limitation of science in reconstructing nature, while at the same time emphasizing that the construction of models is our only option if we want to do science:

> It is the faith of all science that an unlimited number of phenomena can be comprehended in terms of a limited number of concepts or ideal constructs. Without this faith no science could ever have any motivation. To deny this faith is to affirm the primary chaos of nature and the consequent futility of scientific effort. The constructs in terms of which natural phenomena are comprehended are man-made inventions. To discover a scientific law is merely to discover that a man-made scheme serves to unify, and thereby to simplify, comprehension of a certain class of natural phenomena.[3]

The methodologies employed in this biography have been presented with a certain assumption that many historians may not recognize as valid. That is, the HMM, the Five Factor personality profile, and the other psychological and social variables employed in this biography, patterned after the rigidly controlled methodologies of experimental psychologists, assume that history, like psychology, is a scientific enterprise. As such, historians are scientists attempting to identify, understand, and explain past human behavior. Past attempts at psychohistory have been met with considerable scorn by both historians and psychologists, and in many cases with good reason. Much of early psychohistory was based on the deeply flawed psychological theories of such psychologists as Freud and Erikson. Clearly, one's psychohistory, or psychobiography, can be no better than the psychological theory on which it is based.[4] Lawrence Stone, for example, quipped sardonically: "I just do not think that such things as the extermination of six million Jews can be explained by the alleged fact that Hitler's mother was killed by treatment given her by a Jewish doctor in an attempt to cure her cancer of the breast; or that Luther's defiance of the Roman church can be explained by the brutal way he was treated by his father or by his chronic constipation."[5] Agreed, and anyone who would suggest that a single variable produces any type of complex human or social behavior is not doing good psychology or history.

The critics of psychohistory and psychobiography have presented a number of important methodological hurdles that must be met in order for history to be practiced as a psychological science. William Runyan has outlined three separate positions, "each partially overlapping in time and each still alive in some quarters and in contention with the others," including:

> (1) a naive overemphasis on the role of prominent individuals in influencing and representing the course of historical events; (2) a rejection of the study of individuals, in favor of larger structures, whether in the form of modes of production, class relationships, demographic and ecological processes, or quantitative studies of social groups; and (3) a search for ways of reintegrating individuals and their psychological processes into analyses of their reciprocal causal relationships with broader economic, demographic, and institutional forces.[6]

Clearly balance between positions is called for here, and when Runyan writes that "there is a certain primacy in the historical perspective, in that it is more encompassing than the social scientific perspective," and that "historical inquiry . . . must attend not only to the ordered, structured, and lawlike aspects of human and social reality, but also to the disorderly, the particular, the idiosyncratic, the transient, and the random,"[7] he is saying that we need to attend to both contingencies and necessities, which is done in the *model of contingent–necessity* presented in the last chapter.

Given the vigorous debate among historians over the validity and usefulness of psychohistory, it is understood that attempting to adopt the scientific paradigm to psychobiography in particular as I have tried to do in this work, or more generally to psychohistory in the context of modern scientific paradigms, an adequate defense of this program is needed, which, first and foremost, requires semantic precision.

The Science of History

Science is a specific way of thinking and acting common to most members of a scientific group, as a tool to understand information about the past or present. More formally, I define science as *a set of methods to describe and interpret observed or inferred phenomena, past or present, aimed at building a testable body of knowledge open to rejection or confirmation.* Methods include hunches, guesses, ideas, hypotheses, theories, and paradigms; as well as background research, data collection and organization, colleague collaboration and communication, experiments, correlation of findings, statistical analyses, manuscript preparation, conference presentations, and paper and book publications.

There are two major methodologies in the sciences—experimental and historical. Experimental scientists (e.g., physicists, geneticists, experimental psychologists) constitute what most people think of when they think of scientists in the laboratory with their particle accelerators, fruit flies, and rats. But historical scientists (e.g., cosmologists, paleontologists, archaeologists) are no less rigorous in their methods to describe and interpret past phenomena, and they share the same goal as experimental scientists of building a testable body of knowledge open to rejection or confirmation. Unfortunately, a hierarchical order exists in the academy, as well as in the general public, in two orthogonal directions: (1) experimental sciences are more rigorous than historical sciences; (2) physical sciences are harder than biological sciences, which themselves are harder than social sciences. If anything, this hierarchical ranking is precisely backwards. The hardest problems of all to solve are social and historical because, in the first instance, they have so many variables, and in the second case much inference is required. But, in any case, such hierarchies discolor our perceptions of how science is done and the sooner we can get past them the deeper will be our understanding of the nature of the scientific enterprise.

One common element within both the experimental and historical sciences, as well as within the physical, biological, and social sciences is that they all operate within defined paradigms, as originally described by Thomas Kuhn in 1962 as a way of thinking that defines the "normal science" of an age, founded on "past scientific achievements . . . that some particular scientific community acknowledges for a time as supplying the foundation for its further practice."[8] Kuhn's concept of the paradigm has achieved nearly cult status in both elite and populist circles, but he has been challenged time and again for his multiple usages of the term without semantic clarification.[9] His 1977 expanded meaning of "all shared group commitments, all components of what I now wish to call the disciplinary matrix," still fails to give the reader a sense of just what Kuhn means by paradigm.[10]

Because of this lack of clarity, and based on the earlier definition of science, I define a paradigm as *a framework shared by most members of a scientific community, to describe and interpret observed or inferred phenomena, past or present, aimed at building a testable body of knowledge open to rejection or confirmation.* The modifier "shared by most" is included to allow for competing paradigms to coexist, compete with, and sometimes displace old paradigms, and to show that a paradigm may exist even if all scientists working in the field do not accept it. The other component of science that makes it different from all other paradigms is that it has a self-correcting feature that operates, after a fashion, like natural selection. Science, like nature, preserves the gains and eradicates the mistakes. When paradigms shift (e.g., during scientific revolutions), scientists do not necessarily abandon the entire paradigm any more than a new species is begun from scratch. Rather, what remains useful in the paradigm is retained, as new features are added and new interpretations given, just as in homologous features of organisms the basic structures remain the same while new changes are constructed around it. Thus, I define a *paradigm shift* as *a new cognitive framework, shared by a minority in the early stages and a majority in the later, that significantly changes the description and interpretation of observed or inferred phenomena, past or present, aimed at improving the testable body of knowledge open to rejection or confirmation.*[11]

History can be, and on many levels already is, a science. Practicing historians have developed a set of methods in their historical analyses that attempt to *describe and interpret past phenomena aimed at building a testable body of knowledge open to rejection or confirmation.* Their methods are learned in graduate training and beyond, while an organized historical work becomes universally valid within the community of historical scientists through the testing of hypotheses and theories by examining historical data. Through this process, historical facts are discovered and described; and facts in historical sciences may be as reliable as facts in the experimental sciences, if by *fact* we mean *a conclusion confirmed to such an extent that it would be reasonable to offer provisional agreement.* Historian James Kloppenberg has argued as such in his description of "pragmatic hermeneutics," when he notes that historical facts, hypotheses, and interpretations "can be checked against all the available evidence and subjected to the most rigorous critical tests" and "if they are verified provisionally, they stand" and if "disproved, new interpretations must be advanced and subjected to similar testing."[12]

Despite the very powerful and convincing arguments of the historical relativists,[13] deconstructionists,[14] and epistemological anarchists[15] against a science of history, using a broadened definition of science and a definition of facts as provisional conclusions, it can be argued that history is a science. The distinction is not between the "hard" and "soft" sciences, or between the physical, biological, and social sciences, but between the experimental and historical sciences. All the sciences contain both experimental and historical fields—for example, cosmology and geology in the physical sciences, evolutionary biology and paleontology in the biological sciences, and archaeology and human history in the social sciences. The inability to observe past events or set up controlled experiments is no obstacle to a sound science of paleontology or geology, so why should it be for a sound science of human history? The key is the ability to test one's hypothesis, as paleontologist Stephen Jay Gould notes: "We cannot see a past event directly, but science is usually based on inference, not unvarnished observation. The firm requirement for all science—whether stereotypical or historical—lies in secure testability, not direct observation. History's richness drives us to different methods of testing, but testability is our criterion as well. We work with our strength of rich and diverse data recording the consequences of past events; we do not bewail our inability to see the past directly."[16]

Kloppenberg argues as much in his analysis of critics of historical science when he notes that beyond the "noble dream" of pure and unsullied objectivity "and the nightmare of complete relativism lies the terrain of pragmatic truth, which provides us with hypotheses, provisional synthesis, imaginative but warranted interpretations, which then provide the basis for continuing inquiry and experimentation." As historical scientists, Kloppenberg insists in his closing remark, "we cannot aspire to more than a pragmatic hermeneutics that relies on the methods of science and the interpretation of meanings. But we should not aspire to less."[17]

The various methods employed in this biography aim to bring together the rich and diverse data on a historical subject, formulate a hypothesis about what happened, and then test this hypothesis in the historical record. The ultimate test comes when other members of the community of historical scientists investigate and test the hypothesis themselves (by examining the evidence and checking it against other data from this time period or subject), and then decide—provisionally—whether this interpretation might stand or fall. If corroboration and confirmation by the historical community is received, then this claim will become a provisional historical fact until new evidence is presented, or a more comprehensive or adequate interpretation is proffered for the same evidence. Such is the social nature of all science.

Based on this social and pragmatic definition of science—both experimental and historical—we can construct a definition of what all historical scientists are studying: *History is a product of the discovery and description of some past physical, biological, or human-action phenomena.* (The phrase "product of the discovery and description" is useful as it recognizes the objective nature of knowledge and data to be "discovered," while simultaneously acknowledging the subjective nature of "description," since the facts never just speak for themselves, and the scientific enterprise is fundamentally a social one.)

In contrasting the experimental and historical sciences it is assumed that the historical sciences are nonexperimental in nature. We cannot rerun the past, alter a variable here or there, and then observe the effects. This does not mean, however, that we cannot make causal inferences from what has already transpired, or that we cannot use the methods of experimentation, as we saw in the last chapter on counterfactual history. In reality, the experimental sciences have many of the same problems of interpretation as do the historical sciences, as statistician Hubert Blalock points out in *Causal Inferences in Nonexperimental Research:* "Reality, or at least our perception of reality, admittedly consists of ongoing processes. No two events are ever exactly repeated, nor does any object or organism remain precisely the same from one moment to the next."[18] Experimental scientists deal with this problem by constructing a model to interpret the confusing array of nature's variables, as Blalock explains: "In developing these models the scientist temporarily forgets about the real world. Instead, he may think in terms of discrete 'somethings,' or systems, made up of other kinds of somethings (subsystems, elements) which have fixed properties."[19] So both experimental and historical scientists manufacture models to represent "reality," but neither one is superior to the other in the final quest for understanding causality. Similarly, philosopher of science Philipp Frank argues that causal laws are really just practical assumptions made by the scientist for the purpose of data interpretation. The actual linkage between two variables, whether observed in the laboratory or through the historical record, is always a product of the discovery and description of a cause-and-effect relationship.[20] Blalock concludes: "The dilemma of the scientist is to select models that are at the same time simple enough to permit him to think with the aid of the model but also sufficiently realistic that the simplifications required do not lead to predictions that are highly inaccurate. Put simply, the basic dilemma faced in all sciences is that of how much to oversimplify reality."[21]

Since history is, in part, human action writ past, the observations of scientists studying present human actions may be insightful. The experimental psychologist Hans Eysenck, for example, has made a useful distinction between *description* and *explanation* in the study of human behavior: "The main difference between description and explanation . . . is essentially one of breadth and latitude; description is essentially of individual phenomena, explanation is in terms of laws derived from large numbers of individual phenomena and applicable to literally infinite numbers of further individual events."[22] We may then, using Eysenck's definition, speak of specific historical observations as *descriptions,* and attempts at interpreting them in a meaningful way as *explanations.* "In arriving at such laws and generalizations it often becomes necessary also to invent or discover certain concepts which are of a peculiar abstract nature." But are these historical explanations more abstract or less viable than those in the physical or biological sciences? No, says

Eysenck: "Newton's gravitational force was such a concept; Pavlov's 'conditioning' is such another. These terms do not refer to actual observable objects or events, but to hypothetical constructs which make thinking about observable events easier and which may enter into our equations describing and 'explaining' the behavior of objects, animals, or human beings."[23] For psychologists Clarence Brown and Edwin Ghiselli, "Why is it so?" is the ultimate question to be asked about any human action, and "Explanation is the fundamental method through which we discover the answer to this type of question." For the social scientist, they argue, "explanation proceeds to the discovery of higher-order meanings by means of the manipulation of concepts" where theories and interpretations form the basis of these higher-order meanings.[24]

One important question asked in this book is this: Why did Wallace believe it was not possible for evolutionary mechanisms to account for the advanced development of the human mind? There is no reason that this question cannot be studied and answered in a scientific manner using the best available tools of description and explanation. The goal in this biography, then, is to address this and other questions about Wallace in the most rigorous, scientific manner possible, teasing out the factors and variables that mattered most, using the factorial matrix design to construct a test of this relative influence, and analyzing the historical data to reach a deduction from the test. The conclusions stand or fall based on the evidence and interpretation. That is the efficiency and elegance of science. Like all scientists, historians want to understand causality, in this case causes of past effects. It so happens that because of its set of methods that includes rigorous testing, careful examination of the data, corroboration among other members of the community, and the agreement that all conclusions are provisional, science is the best cultural tradition we have for understanding the cause of things, present and past.

Historical Science in Practice

How can the normal types of evidence that historians use—documents, letters, memos, photographs, and the like—be employed in a more scientific fashion? We might consider four higher-order levels of analysis that have been and are being used by other historical scientists.

1. Convergence of Evidence. In August 1996, NASA announced that it had discovered life on Mars. The evidence was the Allan Hills 84001 rock believed to have been ejected out of Mars by a meteor impact millions of years ago, which subsequently fell into an earth-crossing orbit and struck our planet. On the panel of NASA experts was paleobiologist William Schopf, a historical scientist specializing in ancient life. Schopf was skeptical of NASA's claim because, he said, the four "lines of evidence" claimed to support the find did not converge toward a single conclusion. Instead, they pointed to several possible conclusions.

Schopf's "lines of evidence" analysis reflects a method defined by the nineteenth-century philosopher of science, William Whewell, as a *consilience of inductions* (discussed in Chapter 8). To prove a theory, Whewell believed, one must have more than one induction, or a single generalization drawn from specific facts. One must have multiple inductions that converge on one another, independently but in conjunction. Whewell said that if these inductions "jump together," it strengthens the plausibility of a theory. "Accordingly the cases in which inductions from classes of facts altogether different have thus jumped together, belong only to the best established theories which the history of science contains. And, as I shall have occasion to refer to this particular feature in their evidence, I will take the liberty of describing it by a particular phrase; and will term it the Consilience of Inductions."[25]

The theory of evolution is confirmed by the fact that so many independent lines of evidence converge to a single conclusion. Independent sets of data from geology, paleontology, botany, zoology, herpetology, entomology, biogeography, comparative anatomy, physiology, and many other sciences each point to the conclusion that life has evolved. This is a consilience of inductions, a convergence of evidence. Creationists demand "just one fossil transitional form" that shows evolution. But evolution is not proved through a single fossil. It is proved through a convergence of fossils, and many other lines of evidence, such as DNA sequence comparisons across species. For creationists to disprove evolution they would need to unravel all these independent lines of evidence and find a rival theory that can explain them better than evolution. Similarly, in analyzing the claims of the Holocaust deniers I used the consilience approach to answer their challenge to present "just one proof" of the Holocaust, demonstrating that the Holocaust is not proved through a single piece of evidence.[26] It is proved through a convergence of evidence from letters, memos, orders, bills, speeches, memoirs, confessions, photographs, blueprints, and physical evidence. No one of these amounts to something we think of as "the Holocaust," but they "jump together" to that conclusion.

2. *Model Building.* Scientists studying the history of the universe, the earth, or life construct models in order to make generalizations about data and test specific hypotheses that are a part of the model. There is no reason why historians studying human history should not do the same. In fact, the physicist and Nobel laureate Murray Gell-Mann is doing just that at his Santa Fe Institute in New Mexico. Gell-Mann won the Nobel prize for constructing a model of the subatomic world (quarks, etc.), and now he and his colleagues

have constructed a model of complexity: *complex adaptive systems.* Applying the model to various historical events, Gell-Mann explains that "in each one a complex adaptive system acquires information about its environment and its own interaction with that environment, identifying regularities in that information, condensing those regularities into a kind of 'schema' or model, and acting in the real world on the basis of that schema. In each case, there are various competing schemata, and the results of the action in the real world feed back to influence the competition among those schemata."[27]

Science itself can be modeled as a complex adaptive system, says Gell-Mann, "in which the schemata are theories, giving predictions for cases that have not been observed before. There is a tendency for theories that give successful predictions to assume a dominant position. . . . Older, less successful theories may be retained as approximations for use in restricted sets of circumstances."[28] Scientists at the Institute and elsewhere are computer modeling everything from the stock market to weather systems, from the evolution of languages to the prehistory of the American Southwest.[29] Is model building science? Do those simulation models in Gell-Mann's computer represent something in the real world? They do, to the extent that any model does, and when the model leads to ways of testing hypotheses it becomes a vital part of the scientific process. One way that model building can become a useful method for historical sciences is when it is used in the comparative method.

3. The Comparative Method. Punctuated equilibrium, as a model for the history of life, at its core is Ernst Mayr's theory of allopatric speciation, which describes how modern species change, applied to the fossil record.[30] Like Mayr, Niles Eldredge and Stephen Jay Gould took a modern theory of change and applied it to history.[31] Similarly, Luigi Luca Cavalli-Sforza and his colleagues compare data from fifty years of research in population genetics, geography, ecology, archaeology, physical anthropology, and linguistics to trace the evolution of the human races.[32] Using both the consilience and the comparative methods led them to conclude that "the major stereotypes, all based on skin color, hair color and form, and facial traits, reflect superficial differences that are not confirmed by deeper analysis with more reliable genetic traits and whose origin dates from recent evolution mostly under the effect of climate and perhaps sexual selection."[33] They discovered, for example, that Australian Aborigines are genetically more closely related to southeast Asians than African blacks, which makes sense from an evolutionary perspective considering the migration pattern from Africa to Asia to Australia.

A superb piece of scientific history applying the comparative method and using evolutionary thinking can be found in Jared Diamond's book *Guns, Germs, and Steel,* in which he explains the differential rates of development between civilizations around the globe over the past 13,000 years.[34] Why, Diamond asks, did Europeans colonize the Americas and Australia, rather than Native Americans and Australian Aborigines colonize Europe? Diamond rejects the theory that there are inherited differences in abilities between the races that have prevented some from developing as fast as others. In its stead he proposes a biogeographical, environmental theory having to do with the availability of domesticated grains and animals to trigger the development of farming, metallurgy, writing, nonfood-producing specialists, a large population, a military and government bureaucracy, and other components of Western cultures. Australian Aborigines could not strap a plow to or mount the back of a kangaroo like Europeans could the ox and horse. Indigenous wild grains that could be domesticated were few in number and located only in certain regions of the globe—those regions that saw the rise of the first civilizations. The East–West axis of the Euro-Asia continent lent itself to diffusion of knowledge and ideas much better than the North–South axis of the Americas. Through constant interactions with domesticated animals and other peoples, these Euroasians developed immunities to numerous diseases that, when brought by them to Australia and the Americas along with their guns, produced a genocide unprecedented in history. "History followed different courses for different peoples because of differences among peoples' environments," says Diamond, "not because of biological differences among peoples themselves."[35]

How does he prove this theory of history? Diamond, a physiologist and evolutionary biologist, applies the sciences of genetics, molecular biology, behavior ecology, and biogeography to understand the development of crops and domesticated animals and the effects they had on the people who used them. He applies models from molecular biology and epidemiology to understand the spread of human diseases between peoples of the past. He uses genetics, linguistics, and archaeology to explore the origins and diffusion of languages, writing, technology, and political organizations around the globe. Unfortunately few historians are up to the task for the simple reason that they know so little about any of these fields. It takes a scientist to synthesize such diverse scientific knowledge, and Diamond, as readers of his previous books and articles know, is just such a synthesizer. And as a scientist Diamond looks for ways to test his ideas and to falsify his hypothesis. After over three hundred pages of data, comparisons, and analyses to test his historical theory, Diamond concludes his book by arguing that the long-held classification of history as a nonscientific subject of the humanities is due for a change. He sees the historical sciences as similar to the experimental sciences in four critical ways: methodology, causation, prediction, and complexity.

Historians can and do use similar *methodologies* as other sciences. For example, astronomers infer stellar evolution by observing the "natural experiments" of red giants, white dwarfs, pulsars, and black holes—all stars in various stages of their history. Darwin used the same method in inferring the evolution of coral reefs. The different types of coral reefs, he determined, were actually all the same but at different stages of development. Historians can also use natural experiments. If some races were biologically inferior, for example, how do you explain the fact that modern Australian Aborigines can learn, in less than a generation, to fly planes, operate computers, and do anything that any European inhabitant of Australia can do? Similarly, when European farmers were transplanted to Greenland they went extinct because their environment, not their genes, prevented them from being successful farmers. Native Americans first encountered horses and guns when Europeans brought them over in the sixteenth century. It was not long before Native Americans were using both to defend themselves against the invading hordes from the east.

Historical and experimental scientists both look for proximate and ultimate *causes.* A biologist who studies the phenomenon of mammals increasing in size as the climate gets colder wants to understand both the physics of heat transference as a function of the surface-to-volume ratio, as well as the deeper cause of how natural selection would have favored this phenomenon in evolution. Likewise, historians are interested not only in the particular history of a people or a region, they are also curious to know the universal principles or causes behind these events. Speculative, or universal history, operates at this deeper level.

Most people think that the major dividing point between history and science is *prediction.* How can historians make predictions? Diamond shows precisely how his theory can be falsified by making predictions. If historians discover that Native Americans had an elaborate writing system and advanced metallurgy, yet never initially developed a correspondingly complex system of farming and domesticated animals, his theory would be doomed.

Finally, history is *complex,* with seemingly countless independent variables. But so too is most of biology, which is surely in anyone's pantheon of sciences. "Thus, the difficulties historians face in establishing cause-and-effect relations in the history of human societies are broadly similar to difficulties facing astronomers, climatologists, ecologists, evolutionary biologists, geologists, and paleontologists," Diamond concludes. "To varying degrees, each of these fields is plagued by the impossibility of performing replicated, controlled experimental interventions, the complexity arising from enormous numbers of variables, the resulting uniqueness of each system, the consequent impossibility of formulating universal laws, and the difficulties of predicting emergent properties and future behavior."[36] Nevertheless, we must try. We may have to work a lot harder than physicists and astronomers in isolating our variables and testing them, but test them we must.

4. Historical Hypothesis Testing. The historical sciences are rooted in the rich array of data from the past that, while nonreplicable in the laboratory sense, are nevertheless valid as sources of information for piecing together specific events and testing general hypotheses. Based on data from the past the historian tentatively constructs a hypothesis, then checks that against "new" data uncovered from the historical record. As an example of this, I once had the opportunity to dig up a dinosaur with Jack Horner, Curator of Paleontology at the Museum of the Rockies in Bozeman, Montana. In his book *Digging Dinosaurs,* Horner reflected on the historical process in two stages of the famous dig in which he exposed the first dinosaur eggs ever found in North America. The initial stage was "getting the fossils out of the ground; the second was to look at the fossils, study them, make hypotheses based on what we saw and try to prove or disprove them."[37] The first phase of unsheathing the bones from the overlying and surrounding stone is back-breaking work. As you move from jackhammers and pickaxes to dental tools and small brushes, however, the historical interpretation accelerates as a function of the rate of bone unearthed. "Paleontology is not an experimental science; it's an historical science," Horner explained. "This means that paleontologists are seldom able to test their hypotheses by laboratory experiments, but they can still test them."[38] How?

When I arrived at Horner's camp I expected to find the busy director of a fully sponsored dig barking out orders to his staff. I was surprised to come upon a patient historical scientist, sitting cross-legged before a cervical vertebrate from a 140-million-year-old *Apatosaurus* (formerly known as *Brontosaurus*), wondering what to make of it. Soon a reporter from a local paper arrived (apparently a common occurrence, as no one took notice) inquiring of Horner what this discovery meant for the history of dinosaurs. Did it change any of his theories? Where was the head? Was there more than one body at this site? And so on. Horner's answers were consistent with those of the cautious scientist: "I don't know yet." "Beats me." "We need more evidence." "We'll have to wait and see."

The dig was historical science at its best. After two long days of exposing nothing but solid rock and my own ineptness at seeing bone within stone, one of the preparators pointed out that the rock I was about to toss away was a piece of bone that appeared to be part of a rib. *If* it was a rib, *then* the bone should retain its riblike shape as more of the overburden was chipped away. This it did for about a foot, until it suddenly flared to the right. Was it a rib or something else? Jack moved in to check. "It could be part of the pelvis," he suggested. *If* it was part of the pelvis, *then* it should also flare out to the left when more was uncovered. Sure enough, Jack's prediction was verified by further empirical evidence.

This process in science is called the *hypothetico-deductive* method, where one forms a hypothesis based on existing data, deduces a prediction from the hypothesis, then tests the prediction against further data. For example, in 1981 Horner discovered a site in Montana that contained approximately thirty million fossil fragments of approximately 10,000 *Maiasaurs.* Horner and his team extrapolated this estimate from selected exposed areas in a bed 1.25 miles by .25 miles. The hypothesizing began with a question: "What could such a deposit represent?"[39] There was no evidence that predators had chewed the bones, yet many were broken in half, lengthwise. Further, the bones were all arranged from east to west—the long dimension of the bone deposit. Small bones had been separated from bigger bones, and there were no bones of baby *Maiasaurs,* just those of individuals between nine and twenty-three feet long. What would cause the bones to splinter lengthwise? Why would the small bones be separated from the big bones? Was this one giant herd, all killed at the same time, or was it a dying ground over many years?

An early hypothesis of a mud flow burying the herd alive was rejected because "it didn't make sense that even the most powerful flow of mud could break bones lengthwise . . . nor did it make sense that a herd of living animals buried in mud would end up with all their skeletons disarticulated." Horner constructed another hypothesis: "It seemed that there had to be a twofold event, the dinosaurs dying in one incident and the bones being swept away in another." Since there was a layer of volcanic ash 1.5 feet above the bone bed, volcanic activity was implicated in the death of the herd. From this hypothesis he deduced that only fossil bones split lengthwise, therefore the damage to the bones came long after the dying event, which might have been a volcanic eruption, especially since volcanoes "were a dime a dozen in the Rockies back in the late Cretaceous." This hypothesis-deduction process led to his conclusion: "A herd of *Maiasaura* were killed by the gases, smoke and ash of a volcanic eruption. And if a huge eruption killed them all at once, then it might have also killed everything else around," including scavengers or predators. Then perhaps there was a flood, maybe from a breached lake, carrying the rotting bodies downstream, separating the big from the small bones (which are lighter) and giving them a uniform orientation. "Finally the ash, being light, would have risen to the top in this slurry, as it settled, just as the bones sank to the bottom."[40]

A paleontological dig is a good example of hypothetico-deductive reasoning and how a historical science can make predictions based on initial data that are then verified or rejected by later evidence. The digging up of history, whether bones or letters, is the experimental procedure of the historical scientist putting a hypothesis to the test. The best example I have seen of this process, and one that uses all three methods of consiliences, models, and comparisons to derive testable hypotheses, can be found in Frank Sulloway's *Born to Rebel.* We have already seen the strength of its methodology in understanding the Darwinian revolution, and why Darwin and Wallace were more likely than most of their colleagues to lead and support this intellectual revolution. Figure E-1 shows the birth-order effect for receptivity to evolutionary theory, comparing social class to birth order. Despite numerous books by historians of science claiming that social class was the determining factor in the Darwinian revolution, the data show that it was birth order, not social class, that was the strongest variable predicting who supported the revolution and who opposed it. The only way to know this is through hypothesis testing. Anecdotes alone tell us nothing.

Sulloway, of course, is not claiming that birth order is the only factor

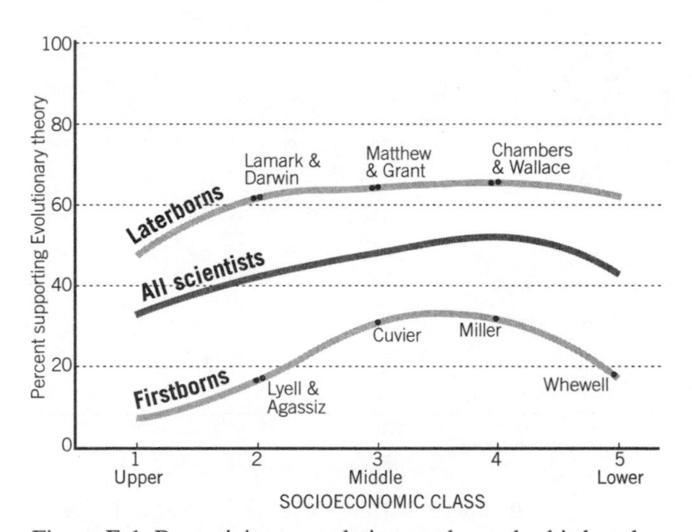

Figure E-1 Receptivity to evolutionary theory by birth order and social class. Why were Wallace and Darwin more likely than most of their colleagues to lead and support this intellectual revolution? Historian of science Frank Sulloway has tested numerous variables in the lives of prominent figures who supported or opposed evolutionary theory, and his data here show the birth-order effect for receptivity to evolutionary theory compared to social class. Despite numerous books by historians of science claiming that social class was the determining factor in the Darwinian revolution, the data show that it was birth order, not social class, that was the strongest variable predicting who supported the revolution and who opposed it. The only way to know this is through hypothesis testing. (From Sulloway, 1996, 237)

involved in determining receptivity. Birth order is a predisposing variable that is part of a suite of variables within the dynamics of a family that sets the stage for numerous other variables, such as age, sex, personality, and social class, to influence receptivity. Sulloway's theory is a dynamic and interactive one, capturing the contingencies and complexities of human history. In Figure E-2, for example, he shows that the strength of the birth-order effect depends on the radicalness of the theory. Not all theories are equally radical. In general, the more radical the idea, the more laterborns are attracted to it; the more conservative the idea, the more firstborns are attracted to it. The theory of evolution is one of the more radical and thus more susceptible to birth-order effects.

In support of his theory Sulloway uses the *consilience* method, showing that the effects he found in one scientific revolution can be found in dozens of scientific revolutions, as well as in thirty-two political revolutions, the Prot-

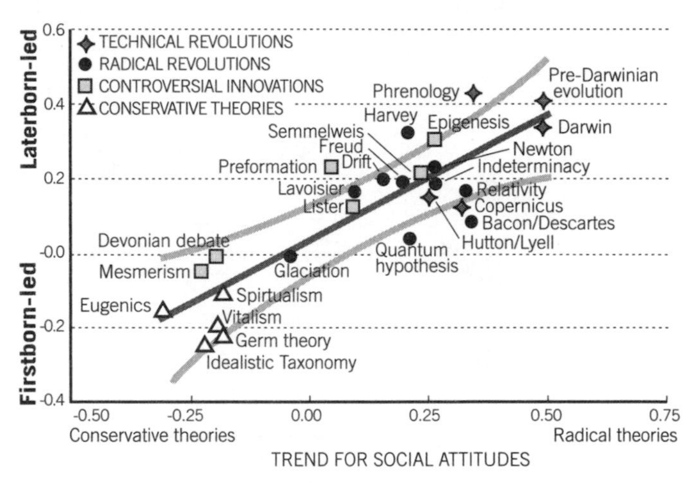

Figure E-2 Birth-order trends in science as related to the religious and political implications of the idea, computed from over 19,000 ratings made by expert historians. Birth order is not the only factor involved in determining receptivity to radical ideas. Birth order is a predisposing variable that is part of a suite of variables within the dynamics of a family that sets the stage for numerous other variables, such as age, sex, personality, and social class, to influence receptivity. Here we see that the strength of the birth-order effect depends on the radicalness of the theory. Not all theories are equally radical. In general, the more radical the idea, the more laterborns are attracted to it; and the more conservative the idea, the more firstborns are attracted to it. The theory of evolution is one of the more radical and thus more susceptible to birth-order effects. (From Sulloway, 1996, 332)

estant Reformation, sixty-one reform movements in American history, and U.S. Supreme Court voting. (Firstborns tend to vote in a conservative direction, laterborns tend to vote in a liberal direction; and, as further confirmation, Republican presidents tend to nominate firstborn justices, and Democratic presidents tend to nominate laterborn justices.) He has also applied the *comparative* method, analyzing 196 previously published birth-order studies involving 120,800 subjects, as well as hundreds of other animal and human studies to show that this is what siblings of many species do. Finally, Sulloway has built a working *model* that can be applied to other fields and revolutions, and can be tested for its continued veracity. This is a splendid example of how a science of history can be conducted.[41]

To return to where we began, I defined history as the product of the discovery and description of past events. The description can take the form of either narration or analysis, as long as one's methods of discovery are as scientifically rigorous as possible. Along with psychology, sociology, and anthropology, history adds to our understanding of human behavior by providing data of what people have actually done. But history is more than this. Humans are storytelling animals and history is our story. Since history is an interaction of both contingencies and necessities, it involves the study of both unique one-time events and universal repetitive trends: time's arrow and time's cycle. For both, but especially the latter, the social sciences are an integral part to finding the repetitive signals in the background noise, and as such the study of history can benefit greatly from the social sciences. The problem, Sulloway explains, is that "historians are attempting to write about humans and their behavior without knowing the best science available on these difficult questions. I think that is a big mistake. To really understand history historians need to test their claims. Unfortunately historians never take any courses in hypothesis testing."[42] Historians can still write great stories, but they must be stories grounded in facts, not fiction; in science, not anecdotes. The historian can act as a psychologist and sociologist of the past by employing statistical and other scientific methodologies to estimate the probabilities of events, causes, and our confidence in them. As the mathematician Jacob Bronowski observed: "History is neither determined nor random. At any moment, it moves forward into an area whose general shape is known, but whose boundaries are uncertain in a calculable way."[43]

Let us now move forward into the past with the best science available, so that the uncertainties of the historical boundaries will be made as clear as our faculties allow. It will be hard work, but as Darwin liked to say of the key to success, "It's dogged is as does it."

NOTES

References to unpublished letters, documents, and manuscripts are given in the following abbreviations, with more details on the archives found in Appendix I: Wallace Archival Sources.

AJRW: The private collection of Wallace's two grandsons, Alfred John Russel Wallace and Richard Russel Wallace. Letters are designated by letter number, corresponding to a catalogue of the collection.

BL: British Library, Department of Manuscripts. Wallace letters and manuscripts are catalogued by volume (v.) and folio (f.) number.

BMNH: British Museum (Natural History). No referencing designation.

CCD: *Correspondence of Charles Darwin.* F. Burkhardt and S. Smith (Eds.), Cambridge University Press.

DAR: Darwin Archives, Cambridge University Library. The "DAR" designation is followed by a volume and folio page number.

Hope: Hope Entomological Collection of Oxford University Museum. No referencing designation.

ICL: Imperial College London. Letters are listed by folio page number.

Kew: Kew Royal Botanic Gardens. Letters are listed by volume and letter number.

LS: Linnean Society of London. Wallace collection has manuscript numbers, as well as journal or notebook volume and entry numbers.

RES: Royal Entomological Society. No referencing designation.

RGS: Royal Geographical Society. No referencing designation.

RS: Royal Society of London. No referencing designation.

UCL: University College London. Letters listed by volume numbers only.

UL: University of London. Letters listed by MSS number.

WI: Wellcome Institute. No referencing designation.

ZSL: Zoological Society of London. No referencing designation.

References to Wallace's books are designated by a "WB" and the number, and his published articles by a "W" and the number, referencing Appendix II: Wallace's Published Works. Additional details such as pagination are provided here. For example, his 1905 autobiography *My Life* is WB17; his 1855 paper "On the Law which has Regulated the Introduction of New Species" is W21. Note references here will provide volume and page number as needed. For secondary sources, last name, date, and page number are provided, with the full bibliographic reference found in the Bibliography.

Preface. Genesis and Revelation

1. Hardison, 1988, 1.
2. Poulton, 1923.

Prologue. The Psychology of Biography

1. Poulton. 1913. Obituary for Alfred Russel Wallace. *Zoologist.* November.

2. Ibid. An oil painting on canvas by J. W. Beaufort was made in 1923 and can be seen on the cover of this book. It is a majestic portrait that portrays the "last of the great Victorians." The original is at the

Natural History Museum in London, NHM Picture Library identification TO4283/B.

3. In F. Darwin, 1889, II: 121. For the elevation of the quote to a dictum see Shermer, 2001a.

4. Marchant, 1916, 63.

5. Kinsey, 1948, 639.

6. Hook, 1943, 155–57.

7. Shermer, 1988.

8. Keppel, 1973, 169–70. Keppel also noted: "A great deal of the research in the behavioral sciences consists of the identification of variables contributing to a given phenomenon. Quite typically, an experiment may be designed to focus attention upon a single factor. The factorial experiment is probably most effective at the *reconstructive* stage of a science, where investigators begin to approximate the 'real' world by manipulating a number of independent variables simultaneously."

9. Ibid., 178–79.

10. Ibid., 181–82.

11. An attempt to construe history as a science requires precise and semantically clear definitions of key terms. In mathematics, a *matrix* is "a rectangular array of quantities or symbols" (*OED,* V1, 1744); and in science a *model* is "a representation in three dimensions of some projected or existing structure . . . showing the proportions and arrangement of its component parts" (*OED,* V1, 1827). Based on these general usages, and Keppel's discussion, the following definition is given: *The Historical Matrix Model is a three-dimensional rectangular factorial of historical variables showing the interactive and additive nature of its component parts.*

12. Sulloway, 1996, 22–23.

13. Nickerson, 1998.

14. Shermer, 2001b, 218–19.

15. Shermer, in press.

16. Ibid. See also Sulloway, 1987. Sulloway notes Gerald Holton's important contributions to understanding the role of such themata in the history of science: "Gerald Holton has argued that all science is inspired by such bipolar 'themata,' which transcend the strictly empirical character of science by giving a primary role to human imagination." See Holton, 1988.

17. Charles Smith, http://www.wku.edu/smithch/home.htm

18. Ibid.

19. Ibid.

20. Ibid. The original count was made by Charles Smith, who tallied up "those seven books and articles of his that have been cited the most times in the technical literature over the last fifty years or so (based on information gleaned from *Science Citation Index, Social Sciences Citation Index,* and *Arts and Humanities Citation Index*)." I double-checked the figures.

21. W1, reprinted in WB17, I: 201–4.

22. In Marchant, 1916, 65–67.

23. W715.

24. W27.

25. W2, xii.

26. W632.

27. WB16, 73.

28. WB5.

29. W561.

30. George, 1964, x.

31. *Oxford English Dictionary,* I, 1294; Guilford, 1959, 5–6.

32. Guilford, 1959, 5–6.

33. For discussions of and data on the Five Factor model of personality see Digman, 1990; Costa and McRae, 1992; Goldberg, 1993.

34. The historians of science and Wallace experts who took the Five Factor personality inventory to evaluate Wallace's personality were Janet Browne, Gina Douglas, Michael Ghiselin, David Hull, John Marsden, Richard Milner, James Moore, Michael Shermer, Charles Smith, and Frank Sulloway. The complete data set appears in table on p. 331.

35. Sulloway, 1990.

36. Sulloway, 1996.

37. Sulloway, 1990, 15.

38. Ibid., 1.

39. Ibid.

40. Ibid., 6.

41. Sulloway, 1996.

42. Sulloway, 1990, 10.

43. Sulloway, 1990, 11.

44. Turner and Helms, 1987, 175.

45. Adams and Phillips, 1972.

46. Kidwell, 1981.

Dimension	Alpha	Mean	Percentile	Raters	Variables
Conscientiousness	.68	6.74	84th	10	8
Agreeableness	.63	6.68	90th	10	8
Openness to experience	.75	7.37	86th	10	10
Extroversion	.43	6.10	58th	10	7
Neuroticism	.45	3.21	22nd	10	7

Mean reliability = .59.

47. Markus, 1981.
48. Hilton, 1967.
49. Nisbett, 1968.
50. See also Bank and Kahn, 1982; Dunn and Kendrick, 1982; Koch, 1956; Sutton-Smith and Rosenberg, 1970.
51. Sulloway, 1990, 19.
52. Sulloway, 1996, 73.
53. Personal correspondence, January 25, 1991.
54. Ibid.
55. Sulloway, 1996, 21.
56. W1, reprinted in WB17, I, 201–4.

Chapter 1. Uncertain Beginnings

1. WB12, 447.
2. Hope Entomological Collection of Oxford University Museum. No referencing designation. John G. Wilson (2000, 3) "obtained a copy of Wallace's baptism record in the Baptismal Register of the little church at Llanbadoc, not far from the Wallace family home, where the entry reads Feb. 16th 1823. Alfred Russel son of Thomas Vere & Mary Ann Wallace, Llanbadock gentleman."
3. WB17, I, 2–4.
4. Ibid., 8
5. Ibid., 15.
6. Ibid., 17–18.
7. Ibid., 21.
8. Ibid., 20–21.
9. Ibid., 61.
10. For more on Wallace's early life in Hertford, see Green, 1995.
11. WB17, II 416.
12. Olson, 1986, 1.
13. Quoted in Olson, 1986, 120.
14. See Jacob, 1976.
15. Ibid.
16. Ibid.
17. Quoted in ibid., 156.
18. Ibid., 161.
19. Ibid., 107–42.
20. Burnet, 1691.
21. Ibid., 16.
22. Ibid., 21.
23. Ibid., 89.
24. Ibid., 23, 54.
25. Ibid., 6. For a discussion of Burnet in the context of the history of geology, and of the broader subject of time's arrow and time's cycle, see Gould, 1987. For more on the cultural influences on pre-Darwinian science see Tillyard, 1944; Greene, 1959; Thomas, 1971; Hill, 1980; Cohen, 1985; Olson, 1987.
26. WB17, II, 417.
27. Reader, 1964, 7.
28. Houghton, 1957, 1.
29. Reader, 1964, 27.
30. Ibid., 58.
31. Crouzet, 1982.
32. WB17, I, 87.
33. Ibid., 88.
34. Ibid. For the influence of the Mechanics' Institutes on Wallace, see Eaton, 1986, and Morgan, 1978.
35. Ibid., 88–89.
36. Owen, 1813, 1.
37. Ibid.
38. WB17, I, 104.
39. Stephens and Roderick, 1972, 350.
40. Inkster, 1976, 284.
41. Shapin and Barnes, 1977, 35–36.
42. Laurent, 1984, 592.
43. Sheets-Pyenson, 1985, 562.
44. Shapin and Barnes, 1977, 33.
45. *Edinburgh Review,* 17: 168–69.
46. WB17, I, 130.
47. Ibid., 223–24.
48. Ibid., 224.

49. Ibid., 78, 227.
50. Ibid., 87–88.
51. Ibid., 228.
52. See Shapin's and Barnes's discussion, 1977, 59–64.
53. Letter of January 15, 1840, AJRW, l. 3.
54. Letter of April 25, 1859, AJRW, l. 44.
55. WB17, I, 106.
56. Ibid., 102–3.
57. Ibid., 105.
58. Ibid., 201–2.
59. Ibid., 202.
60. Herschel, 1830, 14.
61. Ibid., 92.
62. WB17, I, 232.
63. Chambers, 1844, 222.
64. Ibid., 152.
65. In Desmond and Moore, 1991, 341.
66. WB17, I, 254.

Chapter 2. The Evolution of a Naturalist

1. In Marchant, 1916, 26.
2. Darwin, 1839.
3. Humboldt, 1818.
4. WB17, I, 232.
5. Edwards, 1847, 29.
6. WB17, I, 264.
7. WB2, 2.
8. Bates, 1863, 5.
9. Ibid., 265.
10. Ibid., 266.
11. Ibid., 268.
12. Ibid., 269.
13. WB2, 1–6.
14. Ibid., 157.
15. W4, 74–75.
16. WB2, 19–25. The use of hot-air balloons to glide across the rain-forest canopy became a common method of naturalists over a century later in the 1980s.
17. W5, 156–57.
18. WB2, 29.
19. Ibid., 22.
20. Bates, 1851, 3144.
21. *Annals and Magazine of Natural History,* 1850, 2d. ser. 5.
22. WB17, I, 277–78.
23. WB2, 222–23.
24. In WB17, I, 270.
25. WB2, 56.
26. Ibid., 75.
27. Ibid., 85.
28. Ibid., 121.
29. Ibid., 280–90.
30. Ibid.
31. Ibid., 290–96.
32. Ibid., 300–309.
33. Ibid., 322.
34. Ibid., 326–27.
35. Ibid., 327–28.
36. W8, 3641–43.
37. WB2, 400–401.
38. Hooker, 1854, 62.
39. Spruce, 1855.
40. W6–14.
41. W9, 109.
42. Ibid., 110.
43. W14, 257–58.

Chapter 3. Breaching the Walls of the Species Citadel

1. WB17, I, 323–24.
2. RGS, 1853.
3. WB27, I, 327.
4. Ibid., 354.
5. LS, ms. 140a.
6. WB3, 3.
7. W21.
8. WB27, I, 355.
9. W21. All quotes from the "On the Law" paper are from this reference.
10. Ibid.
11. Ibid.
12. Ibid.
13. Ibid.
14. Ibid.
15. CCD, 6, 514.
16. DAR, Vol. 98, no. 8 of 15 letters received by Darwin from Blyth in Calcutta, December 8, 1855. Barbara Beddall (1972, 158) concludes: "If Blyth's letter went off as expected, Darwin could have received it as early as late February 1856. But whether he received it before or after the meeting with Lyell in April, he knew from it that more than one other person had grasped the significance of Wallace's paper."
17. In Wilson, 1970, xxxix.
18. In Wilson, 1970, xliii–xlvii.
19. Dated September 5, 1857. CCD, 6, 446. In fact, Beddall (1968, 292) correctly (I think) remarks: "Why did Darwin send

this statement to Gray, who responded that it was "grievously hypothetical" rather than to Wallace (who was hardly at this time, however, a scientific peer), who would have understood? Was it Wallace rather than Chambers of whom he was afraid? If so, a recent outline of his views, including the important addition on divergence, mailed to the eminent American botanist might protect his ideas. And, whatever his intentions may have been, this indeed was the result."

20. CCD, 6, 387. Huxley too was so taken with Wallace's work that he later declared that "no enumeration of the influences at work" in the dissemination of the theory of natural selection "would be complete without the mention of his [Wallace's] powerful essay." See Darwin, F. 1887, v. II, 185.

21. WB17, I, 355.

22. See, for example, Poulton, 1903; Jordan, 1905; de Vries, 1906; Stresemann, 1919; Dobzhansky, 1937; Goldschmidt, 1940; Mayr, 1942, 1957, 1963, 1982; Gould, 1982. Jack Horner argues that the entire Linnean classification system should be scrapped and replaced with one that recognizes the fluidity of evolution: "I think the Linnean system is the worst thing we've hung onto from pre-Darwin; I think it has created most of the biological misunderstandings that we have, certainly in the general public. The general public's understanding of biology is based on the Linnean system that everything fits in a box. That's why we can't explain evolution, why we can't define it, because we can never find the transitional taxon. We talk about things like archeopteryx, this transitional taxon between reptiles and birds, but at the class level, and things don't evolve at the class level. They don't even evolve at the species level. They evolve at the individual level, whether you call it a population or whatever you want to call it, it is still at the individual level. And so here we are talking about a transitional organism between classes. That's bullshit. There's no such thing" (in Shermer, 1998).

23. In Mayr, 1982, 223.

24. In Gould, 1982, 206.

25. Ibid.

26. Mayr, 1982, 273. Mayr notes that the " 'actual vs. potential' distinction is unnecessary since 'reproductively isolated' refers to the possession of isolating mechanisms, and it is irrelevant for species status whether or not they are challenged at a given moment." Mayr offers these "more descriptive" definitions: "A species is a reproductive community of populations (reproductively isolated from others) that occupies a specific niche in nature." And: "Species are the real units of evolution, as the temporary incarnation of harmonious, well-integrated gene complexities" (1963). I was one of those students who memorized Mayr's definition.

27. In Lovejoy, 1942, 24.

28. Mayr, 1982, 304.

29. Osborn, 1929, 46.

30. Sarton, 1960, 176.

31. Osborn, 50.

32. Ibid., 51.

33. Ibid., 52.

34. Mayr, 1982, 304.

35. Sarton, 1960, 535.

36. Lovejoy, 1942, 58–59.

37. In Sarton, 1960, 534.

38. In Minkoff, 1983, 45.

39. In Mayr, 1982, 306.

40. Stannard, 1978, 430.

41. In Steneck, 1976, 109.

42. In Mayr, 1982, 256.

43. Ibid., 334.

44. Ibid.

45. Ibid.

46. In Ritterbush, 1964, 110.

47. Ibid., 111.

48. In Goerke, 1973, 96.

49. Ibid., 102.

50. In Blunt, 1971, 248.

51. In Ritterbush, 1964, 113.

52. Ibid., 114.

53. See Fellows and Milliken, 1972, 40–50.

54. Ibid.

55. Buffon, 1756, VI, 59–60.

56. Buffon, 1749, Vol. III, *Le Pigeon.*

57. Buffon, 1749, Vol. I, 38.

58. Ibid., 36.

59. Ibid., 1749, Vol II, 10.

60. Ibid., 14.

61. Ibid., 1753, Vol. IV, 381.

62. Ibid.

63. Ibid., 383.

64. Ibid.

65. Buffon, 1755, Vol. VI, 59.

66. Buffon, 1749, Vol. I, 43.

67. Richards, 1987, 560–74.
68. Ibid., 574–93.
69. Mayr, 1982, 403.

Chapter 4. The Mystery of Mysteries Solved

1. Letter in BMNH.
2. Letter in BMNH.
3. W37.
4. W40.
5. Ibid.
6. In Marchant, 1916, 54.
7. Letter dated January 8, 1858, in Marchant, 1916, 55.
8. For a comparison of Wallace's 1855 and 1858 papers, see Pantin, 1959a, 1959b, 1960.
9. WB17, I, 396.
10. WB14, 140.
11. W44, 53–62.
12. Ibid., 53.
13. Ibid., 54.
14. Ibid., 55.
15. Ibid., 55–56.
16. Ibid., 56.
17. Ibid., 57.
18. Ibid., 58–59.
19. List of insect species collected in WB17, I, 360, from a letter to Bates.
20. Ibid., 60–61.
21. Ibid., 61–62.
22. In F. Darwin, 1892, 197.
23. Darwin, 1859, 1–2.
24. Darwin, 1892, 43.
25. For a further discussion of this distinction see Mayr, 1982.
26. See, for example, Bowler, 1976. For a thorough summary of "two decades of debate over natural selection" between Darwin and Wallace, see Malcolm Kottler's essay in Kohn, 1985, 367–432.
27. Darwin, 1892, 197.
28. Darwin and Wallace, 1858.
29. Ibid.
30. Ibid.
31. In Marchant, 1916, 109.
32. CCD, 7, 240.
33. In Marchant, 1916, 113.
34. Ibid., 114.
35. W54.
36. Ibid.
37. Ibid.
38. LS, v. II, 39.
39. LS, v. II, 21.
40. LS, v. II, 17.
41. In Marchant, 1916, 67.

Chapter 5. A Gentlemanly Arrangement

1. Medawar, 1984, 2–3.
2. Evolutionary biologist Ernst Mayr argues that Kuhn's general model of how science changes fits *no* particular scientific revolution, but especially not the Darwinian revolution (Mayr, 1982, 1988; Shermer, 2000).
3. Brackman, 1980.
4. AJRW, 1.40. This is the private collection of Wallace's two grandsons, Alfred John Russel Wallace and Richard Russel Wallace. Letters are designated by letter number, corresponding to a catalogue of the collection.
5. Brackman, 1980, 78. For reviews of Brackman, see Himmelfarb, 1980, and Browne, 1981. See also Bernstein, 1982.
6. Brooks, 1984, 258.
7. Ibid., 261–63.
8. Ibid., 257.
9. McKinney, 1972, 139.
10. Ibid., 141.
11. See Beddell, 1968 and 1988, for a detailed history of the development of Wallace's ideas.
12. Smith, C., 1999.
13. Personal communication, June 21, 2001. For the full development of Smith's ideas on the development of Wallace's theory see Smith, C., 1999. "Alfred Russel Wallace on Evolution: A Change of Mind?" Presented February 26 at the Symposium on the History of Medicine and Science at the University of Southern Mississippi, Hattiesburg, http://www.wku.edu/~smithch/WALLTALK.htm.
14. Smith, C. 1992. "Alfred Russel Wallace on Spiritualism, Man, and Evolution: An Analytical Essay." Published on: http://www.wku.edu/~smithch/ARWPAMPH.htm.
15. Darwin, 1859, 488.
16. Letter reprinted in WB17, I, 367–68.
17. F. Darwin, 1887, xviii–xix.
18. In Darwin and Wallace, 1858, 58. See also Gardiner, 1995.

19. CCD, 7, 94–95.

20. Ibid., 117.

21. Ibid., 129.

22. CCD, 7, 166. Richard Milner called my attention to this letter, in the possession of Quentin Keynes and on display at the Huntington Library as part of a Darwin exhibit organized by Alan Jutzi. The original letter is published with permission of Quentin Keynes.

23. Frank Sulloway recognized in Wallace's October 6, 1858, letter the implications for the rules of how priority was established in the nineteenth century and before, and why the actions of Darwin, Lyell, and Hooker, and Wallace's amicable and conciliatory attitude toward them, was fully consistent with those rules. (Personal communication, June 21, 2001.) See Sulloway, 1996, 100–104, for a discussion on birth order differences in the response to priority disputes. See also Merton, 1973.

24. Sulloway, 1996, 101.

25. CCD, 7, 240–41.

26. CCD, 7, 323. Darwin wrote to Wallace on August 9, 1859: "I received your letter & memoir on the 7th & will forward it tomorrow to Linn. Soc[y.] But you will be aware that there is no meeting till beginning of November. Your paper seems to me *admirable* in matter, style & reasoning; & I thank you for allowing me to read it. Had I read it some months ago I sh[d.] have profited by it for my forthcoming volume."

27. DAR, 47:145 (Darwin Archives, Cambridge University Library, catalogue and number). The deliberate cutting up of letters, manuscripts, notes, and various forms of correspondence by Darwin was his regular, rather disjointed method of organizing his major publishing projects. When one requests original manuscript and papers of Darwin's at the Cambridge Library, for example, one receives a box filled with clippings, snippets, notes cribbed on the backs of envelopes, and the like. Darwin collected these and labeled them as to their source, date of receipt, the chapter into which they would fit, and so on. The letter clipping from Wallace to Darwin, labeled by Darwin, fits this pattern.

28. AJRW, l. 41. Emphasis in original.

29. F. Darwin, 1887, 68.

30. Von Neumann and Morgerstern, 1947. See also Dawkins, 1976.

31. Trivers, 1971.

32. Axelrod and Hamilton, 1981.

33. Marchant, 1916, 113.

34. Ibid., 131.

35. CCD, 7, 117–19.

36. Personal communication, August 1990.

37. Hitching, 1982, 196.

38. Ibid., 196–97.

39. Eisley quoted in Hitching, 1982, 199.

40. Ibid.

41. Lyell, 1835, II, 159.

42. Ibid., 162.

43. W44, 60.

44. In Pantin, 1959a, 75.

45. In Gould, 1985a, 336.

46. Discussed by Pantin, 1959a, 73.

47. DAR:106, 107.

48. W401.

49. W572.

50. WB17, II, 16–22.

51. Bell, 1859, viii–ix.

52. In F. Darwin, 1887, ii, 126.

53. In Marchant, 1916, 57.

Chapter 6. Scientific Heresy and Ideological Murder

1. WB17, I, 386.

2. See Appendix II for a complete listing of all of Wallace's published works.

3. WB17, I, 404.

4. Ibid., 405–6.

5. WB3; for citations of Wallace's works, see Prologue or http://www.wku.edu/smithch/index1.htm

6. WB17, I, 409.

7. Ibid., 410. For the traditions of Victorian marriage see Himmelfarb, 1986.

8. Ibid.

9. Ibid.

10. Ibid., 412.

11. Ibid.

12. W125.

13. WB17, I, 415.

14. Ibid., 420.

15. W148.

16. Ibid., 391–92.

17. Ibid., 394.

18. April 28, 1869; in Lyell, 1881, v. ii, 442.

19. In Marchant, 1916, 197.
20. Ibid.
21. Ibid., 199.
22. Ibid., 206.
23. Ibid., 200.
24. Ibid., 210.
25. Hope.
26. In Marchant, 1916, 450.
27. Gould, 1980a, 43–53.
28. Kottler, 1974, 162–63.
29. Ibid., 163. In her 1964 biography of Wallace Wilma George makes the same argument in noting that spiritualism was the sole cause for the post-1865 shift in Wallace's view.
30. Kottler, 1974, 189.
31. Ibid.
32. Ibid., 191.
33. It would seem that Kottler has slightly attenuated his enthusiasm for the monocausal argument, as in a 1985 article (in Kohn), he states that Wallace was "motivated primarily, if not exclusively, by his underlying belief in Spiritualism" (421), and in asking the question "Why did Wallace become such a strict adaptationist?" he answers it by stating: "Unfortunately I do not have a good answer to this good question" (426).
34. Schwartz, 1984a, 285.
35. W745.
36. Schwartz, 1984a, 285.
37. Ibid., 288.
38. Durant, 1979, 32.
39. McKinney, 1972, 138. McKinney has reversed his metaphor, as it was Mr. Hyde, not Dr. Jekyll, who was the deviant. Dr. Jekyll was the "gentle" character, Mr. Hyde the "vicious" side of the split personality title figure in R. L. Stevenson's *The Strange Case of Dr. Jekyll and Mr. Hyde.*
40. Ibid., 30.
41. Ibid., 34.
42. Ibid., 47–48.
43. Ibid., 53.
44. Popper, 1968.
45. W148.
46. W155.
47. W167, 332–71.
48. Ibid., f.n.
49. Ibid. All quotes are from W167.
50. WB12, 469–75.
51. Ibid., 478.
52. W392.
53. W745, 618–22.

Chapter 7. A Scientist Among the Spiritualists

1. Winter 1998. For more on the relationship between science and culture in Victorian England, see Cannon, 1978. For a discussion of the relationship between science and pseudoscience in nineteenth-century America, see Wrobel, 1987.
2. Cooter, 1984, 17.
3. WB17, I, 234.
4. Ibid., 235.
5. Ibid.
6. Ibid., 235–36.
7. Ibid., 25–26.
8. Ellenberger, 1970, 84.
9. See Winter, 1998, and Shapin, 1994, for excellent discussions of the social history and nature of science.
10. Hacking, 1988, 435–37.
11. W119, 10.
12. Ibid., iii.
13. Ibid., 1.
14. Ibid., 2.
15. Ibid., 3. This brings to mind Arthur C. Clarke's famous quip that "any sufficiently advanced technology is indistinguishable from magic." See also Gillispie, 1960, for a discussion of the "edge of objectivity" in science.
16. Ibid., 4. Wallace's observation is true so far as it goes, and a reasonable argument on its surface, but he could not have known that similar arguments would be made by cranks and those on the fringe throughout the next century, to both bolster their own confidence in their claims, as well as justify them to a skeptical public.
17. W176.
18. W119, 7.
19. Ibid., 9.
20. In Marchant, 1916, 423.
21. In WB17, II, 336–37.
22. F. Darwin, 1887, v. ii, 364–65, letter from Charles Darwin to unnamed correspondent, January 18, 1874.
23. Ibid., Charles Darwin letter to Thomas Huxley, January 29, 1874.
24. Ibid., January 18, 1874.
25. Unpublished letter from George Darwin to Thomas Huxley, January 1874, Thomas Huxley Papers, Darwin Library, American Philosophical Society, Philadelphia.
26. Huxley, 1900, 1, 419–20.

27. W193.
28. W488.
29. W572.
30. The alternative—that the spirit world is real and that Frances had witnessed a genuine spiritual miracle—is not appropriately addressed here and I refer readers to my book *Why People Believe Weird Things.*
31. Wallace's unpublished American journal is in the archives of the Linnean Society of London.
32. Reported in *The Times* (London), October 3, 1876.
33. The best account of the Slade trial is by historian of science Richard Milner, in three sources: Milner, 1990, 1994, 1996. Milner based his account on primary documents never published before, including a fascinating discussion of Charles Darwin's indirect involvement in the trial. The irony of Lankester's role in this case was that he fully accepted the Piltdown Man fraud and went to his grave convinced of its genuineness.
34. Reported in *The Times* (London) on September 13, 1876.
35. The letter by Lankester was published in *The Times* (London) on September 16, 1876.
36. Report on the trial in *The Times* (London), "A Spirit Medium," appeared on October 3, 1876.
37. The letter by Donkin was published in *The Times* (London) on September 16, 1876.
38. Letter to *The Times* (London) from A. Lane Fox, F.R.S., President of the Anthropological Institute, September 16, 1876.
39. Letter to *The Times* (London) from Wallace, September 19, 1876.
40. Letters from A. Joy and George Joan in *The Times* (London) on September 19, 1876.
41. Letters to *The Times* (London) from J. Park Harrison, Horatio B. Donkin, and Edward W. Cox, September 20, 1876.
42. Letter to *The Times* (London) from W. F. Barrett, September 22, 1876.
43. Reported in *The Times* (London), October 11, 1876.
44. Ibid., October 23, 1876.
45. Ibid.
46. Ibid., October 30, 1876.
47. Ibid.
48. Ibid., November 1, 1876.
49. Quoted in Milner, 1990, 34.
50. Ibid., 29.
51. Ibid., 38.
52. Letter from Charles Darwin to George Romanes, June 4, 1877, 63, in Romanes, 1896.
53. In WB17, II, 336.
54. W245, 630–57.
55. Letter in the Hope Entomological Collection of Oxford University Museum, no referencing designation given other than under Wallace's name.
56. *Chambers' Encyclopaedia,* 1892, v. ix, 645–49.
57. WB5, vii–viii.
58. W643.
59. Letter in the British Museum archives, under Wallace, v. 46439, f. 3.
60. Ibid., ff. 35–46.

Chapter 8. Heretical Thoughts

1. Merz, 1896–1914, 1–2.
2. Randall, 1926; Olson, 1991, 3.
3. DiGregorio, 1984, 131.
4. In Kohn, 1985, 367–68.
5. Herschel, 1830; Whewell, 1837, 1840; Mill, 1843.
6. Quoted in Hull, 1973, who provides an excellent discussion of this debate and its effects on the reception of evolutionary theory in the nineteenth century.
7. Whewell, 1840, 230.
8. Darwin, 1892.
9. Quoted in Hull, 1973.
10. In F. Darwin, 1889, II: 115.
11. Ibid., II: 121.
12. In F. Darwin, 1903, II: 323.
13. Darwin, 1892, 98.
14. Huxley, 1896, 72.
15. Marchant, 1916, 63.
16. Mayr, 1982, 506–9.
17. Mayr, 1982, 479–80; 1988, 215–20.
18. Sulloway, 1979, 1982a, 1982b.
19. Adopted and paraphrased from Mayr, 1982, 501.
20. See, for example, Gould, 1980a. Charles Smith has noted this fact in both his 1991 anthology and in his 1999 paper.
21. W27, 30–31.
22. W136, 47.
23. After a thorough review of the literature in his paper on the debate over natural selection and why Wallace shifted from non-

adaptationism to adaptationism, Malcolm Kottler confessed, "I do not have a good answer to this good question" (1985, 426).

24. W94, clxv–clxvi.
25. WB12, 137.
26. W167.
27. WB12, vii–viii.
28. All letters from which quotes are taken are at the Hope Entomological Collections of Oxford University Museum. There are approximately 200 letters in two boxes between Wallace and primarily E. B. Poulton, dated 1886–1913, and Raphael Meldola, dated 1879–1910.
29. Ibid.
30. Ibid.
31. Letter to Sir Joseph Dalton Hooker, dated November 10, 1905, in the archives at Kew Royal Botanic Gardens.
32. Letter to A. Smith Woodward, dated, April 21, 1907, in the archives at University College London.
33. Ibid.
34. Letter to Sir W. T. Thiselton-Dyer, dated February 8, 1911, in the archives at the Linnean Society of London.
35. Ibid.
36. Scarpelli, 1985, 1.
37. Ibid., 19–20.
38. W639, 38.
39. W430, 167.
40. Letter to Gladstone in the archives at the British Museum, v. 44526, f. 252.
41. Letter to Galton in UCL.
42. Ibid.
43. Stocking, 1987, 67.
44. Ibid.
45. Ibid., 96.
46. Marchant, 1916, 181–82.
47. W94, 168.
48. Ibid., 173.
49. Ibid., 174.
50. Ibid., 177–78.
51. Ibid., 179–80.
52. Ibid., 185–86.
53. In Marchant, 1916, 277.
54. W94, 179.
55. In Marchant, 1916, 55.
56. W114, 671.
57. Ibid., 672.
58. WB17, II, 198–99.
59. Ibid., 395.
60. Ibid.
61. W94, 178.
62. Merz, 1896–1914, 280–81.

Chapter 9. Heretical Culture

1. Kuhn, 1977, 65.
2. See Sulloway, 1996.
3. Kuhn, 1977, discusses the importance of social commitments in science. See Shermer, 2001b, for a discussion on how such commitments are altered.
4. Mayr, 1982, 832.
5. Quoted in Mayr, 1982, 834.
6. Aquinas, 1952, v. 19, 13.
7. Paley, 1802, 436.
8. Voltaire, 1759, 238.
9. Hume, 1953, 169.
10. Ibid.
11. Arthur O. Lovejoy's 1942 *The Great Chain of Being* presents a complete history of this teleological purposefulness, particularly as it relates to the study of the natural world.
12. See Stocking, 1987, 142, for a detailed history and cultural background to the development of these ideas.
13. Ibid., 185.
14. W148, 185.
15. WB12, 476–77.
16. Ibid.
17. See Krieger, 1977; Stern, 1973; Collingwood, 1956.
18. W148, 391.
19. W678, 17.
20. WB16, 6.
21. WB20, xxxviii.
22. Teilhard de Chardin, 1955; Dyson, 1979; Barrow and Tipler, 1986.
23. Moore, 1979, 280.
24. Mivart, 1873, 1876, 1882, 1889.
25. Henslow, 1871, 1873.
26. Originally attributed to Spencer in the Darwin archives at Cambridge, its correct attribution was made by Kim Ziel. See Introduction.
27. DAR
28. Lyell, 1863.
29. WB17, 63.
30. In F. Darwin, 1887, II, 194.
31. Ibid., 196.
32. Ibid., 199.
33. WB17, I, 342–43.
34. W120.

35. Ibid., 221.
36. LS, *Malay Archipelago Journal.*
37. Darwin, 1871, 618–19.
38. LS, *Malay Archipelago Journal.*
39. WB2, 331–32.
40. LS, *Malay Archipelago Journal,* April 6, 1857.
41. WB2, 344.
42. Ibid., 360.
43. Ibid., 332.
44. WB17, I, 45.
45. To Sydney Cockerell, letter dated August 21, 1904, BM, v. 46442, ff. 16–17.
46. Ibid., letter dated August 23, 1904, BM, v. 46442, ff. 16–17.
47. Ibid., letter dated December 17, 1905, BM, v. 46442, ff. 16–17.
48. W94, clxx.
49. In Marchant, 150.
50. Letter dated November 12, 1873, in archives at UL, MSS. 791, no. 89.
51. W609, 22.
52. In Marchant, 1916, 78.
53. Spencer, 1893, I, 31.
54. Spencer, 1904, 634–38.
55. Spencer, 1862, 48.
56. In WB17, II, 217.
57. Spencer, 1842, "Letter I," 15 June.
58. W94, clxii-clxiii.
59. Ibid.
60. W148, 455–56.
61. Ibid.
62. WB17, II, 79.
63. BM, v. 38, 835, ff. 61.
64. W94, 183–85.
65. W159, W160.
66. W233.
67. Hyder, 1907, vi–viii.
68. W227.
69. Document in the archives of ZSL.
70. Document in BM, v. 46440, ff. 140–41.
71. Ibid., ff. 162–63.
72. Ibid.

Chapter 10. Heretic Personality

1. Marchant, 1916, 451.
2. Quoted in Marchant, 1916, 451.
3. W401.
4. W505.
5. W572.
6. W733.
7. W745.
8. Wallace description of his personality as "shy" can be found in WB17, I, 127–28. Survey results collected from the expert raters throughout February 2001.
9. Ibid.
10. Ibid.
11. WB17, II, 365–76.
12. Ibid.
13. Ibid.
14. Ibid.
15. Letter in the archives of the Royal Society, no referencing designation given other than under Wallace's name.
16. Quoted in WB17, II, 368–69.
17. Letter in the Hope Entomological Collection of Oxford University Museum, no referencing designation given other than under Wallace's name.
18. Letter in the archives of the Royal Society, no referencing designation given other than under Wallace's name.
19. In WB17, II, 372.
20. W747, 5.
21. In Marchant, 1916, 447.
22. Letter in the Hope Entomological Collection of Oxford University Museum, no referencing designation given other than under Wallace's name.
23. British Library, Department of Manuscripts, catalogued volume #46436, folio #299.
24. Letter dated September 24, 1893, in BMNH, v. 46436, ff. 299.

Chapter 11. The Last Great Victorian

1. WB17, II, 258–61.
2. Letter dated August 27, 1878, in Kew, v. 104.
3. Letter dated October 24, 1878, in Kew, v. 104.
4. W295.
5. WB17, II, 264.
6. Unpublished letter from Joseph Hooker to Charles Darwin, December 18, 1879, in DAR 104: 136–37.
7. WB17, II, 295.
8. W408.
9. Letter dated November 14, 1886 in Hope.
10. Hope.

11. LS.
12. Ibid.
13. Ibid.
14. Ibid.
15. WB12, 294.
16. W44.
17. Letter in BMNH, v. 46441, f. 17.
18. Ibid., v. 46441, f. 222.
19. Ibid., v. 46441, f. 128–29. The interview referenced in the letter is W499.
20. Ibid., v. 46441, f. 130.
21. Ibid., v. 46439, ff. 379–80.
22. Marchant, 1916, 443.
23. LS, ms. 140a.
24. Hope, June 23, 1908.
25. The original handwritten essay is housed in the BMNH and, according to the archives, it was never published. There are no printer's marks or proofs on the original autograph manuscript, and this is the only extant copy of the original, written in Wallace's hand on his stationery, and dated 1900 (Volume 46433, FF 108–61). In fact, it was not only published in *The Sun,* it was republished in *The Progress of the Century* in 1901. See W609.
26. See W609 for the published edition of the essay. All quotes taken from the original handwritten manuscript, which I copied in pencil by hand (photocopying not allowed).
27. WB17.
28. Marchant, 1916, 449.
29. WB17, 383–86.
30. Ibid., 400.
31. RES, 397 (Royal Entomological Society. No referencing designation given. Number designates page in the article).
32. Ibid.
33. Ibid.
34. W698.
35. Ibid.
36. WB18.
37. Marchant, 1916, 474. The prominent paleontologist Henry Fairfield Osborn also wrote an obituary for Wallace, in Osborn, 1913. See also Prance, 1999.

Chapter 12. The Life of Wallace and the Nature of History

1. Myrdal, 1944.
2. For lengthy discussions and examples of this model see Shermer 1993, 1995, 1996, 1997.
3. RES (Royal Entomological Society, 396–97. No referencing designation given. Number designates page in the article).
4. Ibid. The enumeration is mine.
5. Ibid., 398–400, enumeration added.
6. Ibid., 399.
7. Ibid., 400.
8. Shermer, 1993.
9. Shermer, 2001c.
10. Pantin, 1959a, 77.
11. Ibid., 76.
12. Ibid., 83.
13. WB17, I, 106.
14. Ibid., 106–7.
15. WB9, 544.
16. WB16, 103.
17. In Marchant, 1916, 63.

Epilogue. Psychobiography and the Science of History

1. In Weaver, 1987, 122.
2. Kerlinger and Pedhazur, 1973, 4.
3. Thurston, 1947, 53.
4. See Erikson, 1942 and 1958; Freud, 1910. For a critique of Freudian psychohistory, see Crews, 1966.
5. Stone, 1981, 220. For a more positive view of history as a science, see Fogel and Elton, 1983; Kousser, 1980; Tilly, 1978.
6. Runyan, 1988, 40; this edited volume provides an excellent wide perspective on psychohistory, both its benefits and failings.
7. Ibid., 47.
8. Kuhn, 1962, 10.
9. See, for example, Lakatos and Musgrave, 1970.
10. Kuhn, 1977, 319.
11. In Shermer, 1997b.
12. Kloppenberg, 1989, 1030.
13. For example, Beard, 1935; Becker, 1955.
14. For example, Foucault, 1972; Harlan, 1989.
15. For example, Feyerabend, 1975; 1978.
16. Gould, 1989a, 282.
17. Kloppenberg, 1989, 1030.
18. Blalock, 1961, 7.
19. Ibid.

20. Frank, 1957, 8.
21. Blalock, 1961, 7.
22. Eysenck, 1957, 234.
23. Ibid.
24. Brown and Ghiselli, 1955, 48.
25. Whewell, 1840, 230.
26. Shermer, 1994, 1997b.
27. Gumerman and Gell-Mann, 1994, 17.
28. Ibid., 19.
29. Cowan, Pines, and Meltzer, 1994; Hawkins and Gell-Mann, 1994.
30. Mayr, 1954, 1970.
31. Eldredge and Gould, 1972.
32. Cavalli-Sforza and his colleagues, 1994.
33. Ibid., 19.
34. Diamond, 1997.
35. Ibid., 12.
36. Ibid., 364–65.
37. Horner, 1988, 168.
38. Ibid.
39. Ibid., 129.
40. Ibid., 129–133.
41. Sulloway, 1996.
42. Shermer, 1996b.
43. Bronowski, 1977, 81.

APPENDIX I

WALLACE ARCHIVAL SOURCES

There are, briefly, fifteen primary-source archives in England that contain varying amounts of Wallace material, ranging from a couple to a couple of thousand letters. A few comments here may help the reader understand the general historiography of this book, and direct future historians of Wallace to the specific source they may need. First, it should be noted that there is, as yet, no central clearinghouse for Wallace archival materials. Unlike the Darwin industry, where virtually all the correspondence (or copies thereof) are at the Cambridge University Library (and most of his personal artifacts and library are at his home in Down), Wallace's correspondences are scattered hither and yon. The British Library contains the most at approximately 1,400 letters, seven book manuscripts, and thirteen journal article manuscripts, while the Wellcome Institute, the Imperial College of Science and Technology, and University College, London, house the least, with just a handful of letters each.

Working approximately eight to ten hours a day, six days a week, it took my wife and me five and a half weeks to read thoroughly through all the primary source materials. It was a surprisingly easy task, enhanced by Wallace's legible handwriting, along with the majority of those who wrote to Wallace, with the exception of Darwin, Huxley, Spencer, and Lyell. All of the archivists were quite helpful, and some of them very knowledgeable on both Wallace and nineteenth-century evolutionary thought, particularly Gina Douglas at the Linnean Society, an excellent research facility and a warm introduction to the Wallace archives.

The Linnean Society is one of the gold mines of archival materials, housing Wallace's letters, superb nature sketches, and annotated books from his personal library (including a copy of Darwin's *Descent of Man*). More significant, the original notebooks in Wallace's hand from the Malay Archipelago are available, as well as his never-before-published American journal, containing fascinating descriptions of the American landscape, cityscape, universities, and his spiritualism experiences abroad. A small museum contains the skin of a giant python Wallace killed in the jungles of the Malay Archipelago (described in his book of the same name), as well as Darwin's satchel in which he kept his scientific instruments on the *Beagle* voyage. Although the room in which the Darwin–Wallace papers were read into the record on July 1, 1858, is now gone, another room is set up with most of the original furniture, such that one can touch history by sitting in the Linnean Society president's chair and, with imagination, hear the groundbreaking ideas presented publicly (and generally ignored) for the first time. Likewise, the Museum of Mankind contains artifacts Wallace brought back from Malaya, and the British Museum of Natural History has many of Wallace's butterfly and other entomological collections, as well as the magnificent color portrait painted in 1923 by J. W. Beaufort, a facsimile of which hangs in my office. The British Library contains a fascinating article written by Wallace in 1900, simply entitled "Evolution." Although it is reported by the archives catalogue as never published, it first appeared in the New York–based *The Sun* under the title "The Passing Century" on December 23, 1900, and was reprinted in *The Progress of the Century* in 1901. It is a splendid summary of the state of the science at the fin de siècle, by one of its major scientific players.

One of the most valuable sources of Wallace letters, books, and general memorabilia are his descendants. While there is no extant Wallace home, his grandsons Alfred John Russel Wallace and Richard Russel Wallace, from Bournemouth and Lymington, respectively, have carefully saved a number of important items of their grandfather. Among these is included a binder of early correspondence that contains the famed letter to Bates, apparently sent the same time as the 1858 letter to Darwin with the paper on natural selection, which Darwin claims arrived on June 18. One can clearly see on the Wallace–Bates letter the postmark of "London, June 3." The grandsons also have Wallace's sextant that he used on the Amazon excursion, as well as numerous family photographs, portraits, and even their grandfather's grandfather clock. Both grandsons were extremely helpful and cordial, and John's hospitality in extending a luncheon for us was above and beyond the call of historical duty. John, now retired, was a science teacher and fully understands the importance of his grandfather's work within the larger context of the history of science. He incurred a not-inconsiderable expense in photocopying a number of letters for this book, for which I am grateful. The warm reception dispelled a rumor that a prior Wallace biographer had apparently started about the unapproachability of the grandsons, done, it is speculated, to discourage future historians from getting to these archives. If so, the plan backfired, as John steered us to a set of letters henceforth lost to the historical community. As we were leaving his home I inquired whether there were any other archives that might contain Wallace documents that were not on my list. He replied that he thought he remembered his father, William, mentioning that he had turned over a couple of boxes of his father's letters to Oxford University.

A couple of phone calls and a train trip later my wife and I were at the *Hope Entomological Collections* in the University Museum at Oxford University. There were two boxes of letters, about 200 items in all, of exchanges between Wallace and primarily E. B. Poulton and Raphael Meldola (and a handful of others, including a letter from Darwin about the feeding habits of caterpillars and the coloration of insects, not mentioned in any Darwin correspondence source). It was immediately obvious that no one had looked at these before, or at least for a very long time. (They are mentioned in no primary- or secondary-source bibliography.) William Wallace had joined each letter together with its envelope and a thin piece of tissue paper with a straight pin to protect the letter from others piled on top. This produced two holes in each letter and envelope. Not only were the pins rusty and a little difficult to pull out, but to put them back together in such a way that the holes were all aligned was very difficult. It was my impression that no one had done so before and there was no indication at the museum that anyone had ever requested or seen these letters. (The content of these letters, which is important, is discussed in Chapter 7.)

The following appendix is the most complete archival source bibliography to date on Wallace. It was compiled during the research for the book because the previously published sources of primary documents were incomplete and impractical for research purposes. As much information as possible was recorded at each location so that historians will have only to turn to a single sourcebook to begin archival research. Due to space limitations, not every letter is listed (though they were all recorded at every archive except the British Library, whose over 2,400 letters made this task impossible). General summaries and categories of correspondence are provided, as well as details of particularly important or interesting archival material, and lists of letters and manuscripts where space or importance warrants.

Alfred John Russel Wallace/Richard Russel Wallace Archives: Alfred Russel Wallace and his wife Annie had three children: Herbert Spencer Wallace, William Randolf Wallace, and Violet Wallace. Only William had children: Alfred John Russel Wallace and Richard Russel Wallace. Richard has two children: Richard and William Wallace. John has one daughter, Susan, whose daughter Rosamund is the great-great-granddaughter of ARW. The grandsons live in Bournemouth and Lymington, respectively, and have carefully preserved much of their grandfather's archival material, including a binder of fifty-eight letters, dated July

Figure A-1 *Top:* The grandsons of Alfred Russel Wallace: Alfred John Russel Wallace, left, and Mr. and Mrs. Richard Russel Wallace, on the right, surround the author in front of the home of Richard. *Bottom:* The author with John Wallace, going through the notebook binder of Wallace letters, dating from 1835 to 1869. In the background is John's daughter Susan (the great-granddaughter of ARW) and her daughter Rosamund (the great-great-granddaughter of ARW), three generations of Wallace in one room reaching back over 150 years to the man who helped launch one of the greatest intellectual revolutions in history. (Photographs by Kim Ziel Shermer. Author collection)

1835–May 1869. These include correspondences to and from his father, mother, George Silk, his brothers John, H. E., and W. G. Wallace, his sister Frances Sims, her husband and Wallace's friend, Thomas Sims, R. Spruce, and Henry Walter and Frederick Bates, including the now famous letter allegedly sent the same time as the letter and essay to Darwin, and arriving in London on June 3. They also retained some of Wallace's personal library (109 books), including foreign translations of *The World of Life* (Spanish and Dutch), with some minor marginalia in mostly the later editions. Many of Wallace's books were sold off after his death. A hand catalogue of the library was produced by RRW. Also featured: a pencil sketch of Wallace; a watercolor of Wallace's birthplace (the home is still extant); a watercolor and painting of Wallace, his parents, and some children (in one frame); a sextant used on the Amazon trip; a butterfly collection; a grandfather clock; a three-dimensional likeness of the Westminster Abbey monument; an article on the BBC docudrama on Wallace; original photographs, many never before published. A handlist of the letters by AJRW, entitled *Old letters to and from A. R. Wallace and other members of his family,* reads as follows:

1.	Mrs. Wallace (ARW's mother) to Thos Wilson	July 1835
2.	Mrs. Wallace to Miss Draper	Aug. 1835
3.	ARW to G. Silk [childhood friend]	Jan. 1840
4.	ARW to John Wallace (brother)	Jan. 1840
5.	ARW to G. Silk	Jan. 1840
6.	T. V. Wallace (ARW's father) to Miss Wallace	June 1841
7.	ARW to H. E. Wallace (brother)	Mar. 1842
8.	W. G. Wallace (brother) to Fanny (sister)	Aug. 1844
9.	ARW to H. W. Bates [entomologist/Amazon companion]	Apr. 1845
10.	ARW to H. W. Bates	May 1845
11.	ARW to H. W. Bates	June 1845
12.	ARW to H. W. Bates	Oct. 1845
13.	ARW to H. W. Bates	Oct. 1845
14.	ARW to H. W. Bates	Nov. 1845
15.	ARW to H. W. Bates	Dec. 1845
16.	ARW to H. W. Bates	Aug. 1846
17.	ARW to H. W. Bates	Oct. 1847
18.	ARW to G. Silk	June 1848
19.	H. E. Wallace to R. Spruce	Dec. 1850
20.	H. E. Wallace to R. Spruce	Mar. 1851
21.	H. W. Bates to Mrs. Wallace	Oct. 1851
22.	A. R. Wallace to R. Spruce [botanist]	Sep. 1852
23.	R. Spruce to A. R. Wallace	Oct. 1852
24.	R. Spruce to A. R. Wallace	July 1853
25.	A. R. Wallace to G. Silk (copy)	Mar. 1854
26.	A. R. Wallace to Mrs. Wallace	Apr. 1854
27.	A. R. Wallace to Mrs. Wallace	May 1854
28.	A. R. Wallace to Mrs. Wallace	July 1854
29.	John Wallace to Mrs. Wallace	Aug. 1854
30.	A. R. Wallace to Mrs. Wallace	Sep. 1854
31.	A. R. Wallace to G. Silk	Oct. 1854
32.	A. R. Wallace to Fanny, Mrs. Sims	June 1855
33.	A. R. Wallace to Fanny, Mrs. Sims	Sept. 1855
34.	A. R. Wallace to Mrs. Wallace	Dec. 1855
35.	A. R. Wallace to Fanny, Mrs. Sims	Feb. 1856

36. A. R. Wallace to Fanny, Mrs. Sims — Apr. 1856
37. A. R. Wallace to H. W. Bates — Apr. 1856
38. A. R. Wallace to Fanny, Mrs. Sims — Dec. 1856
39. A. R. Wallace to H. W. Bates — Jan. 1858
40. A. R. Wallace to Frederick Bates — Mar. 1858

Used to date letter and essay sent to Darwin allegedly the same date, postmarked Singapore March 21, London, June 3, Leicester, June 3.

41. A. R. Wallace to Fanny, Mrs. Sims — Sep. 1858
42. A. R. Wallace to Mrs. Wallace Extract only — Oct. 1858
43. A. R. Wallace to G. Silk — Nov. 1858
44. A. R. Wallace to Thomas Sims — Apr. 1859
45. A. R. Wallace to H. W. Bates — Nov. 1859
46. A. R. Wallace to G. Silk — Sep. 1860
47. A. R. Wallace to H. W. Bates — Dec. 1860
48. A. R. Wallace to Mrs. Wallace — July 1861
49. A. R. Wallace to Fanny, Mrs. Sims — Oct. 1861
50. A. R. Wallace to H. W. Bates — Dec. 1861
51. A. R. Wallace to G. Silk — Dec. 1861
52. A. R. Wallace to John Wallace — Jan. 1863
53. R. Spruce to A. R. Wallace — Nov. 1863
54. R. Spruce to A. R. Wallace — Jan. 1866
55. R. Spruce to Mrs. Sims — Feb. 1867
56. R. Spruce to A. R. Wallace — Apr. 1867
57. A. R. Wallace to John Wallace — May 1869
58. A. R. Wallace to G. Silk — 1858

Note on the smoke nuisance [no explanation for out-of-order sequence]

British Library, Department of Manuscripts: A total of 3,071 folio pages of letters, dated 1848–1913, to and from Wallace, including:

349 folio page letters to and from Charles Darwin (including a few from George and Francis Darwin to Wallace)
14 letters to and from Herbert Spencer
1,395 letters "On Scientific Subjects" grouped by the following dates:
1848–1878: 432 letters (including 21 from C. Lyell on biogeography and never published)
1879–1894: 348 letters
1895–1908: 310 letters
1909–1914: 305 letters
437 letters "On Spiritualism" (1864–1913)
292 letters "On Socialism, Land Nationalization" (1867–1913)
60 letters "On Opposition to Vaccination" (1883–1912)
261 letters "General Correspondence" (1853–1904, including two from W. E. Gladstone, 1895 and 1898)
239 letters "General Correspondence" (1905–1913)

There are also seven book manuscripts, including:

Bad Times (1886, autograph draft with corrections)
Darwinism (1889, autograph draft with corrections)

The Wonderful Century (1898, autograph draft with corrections)
Man's Place in the Universe (1903, extracts from revised edition)
My Life: A Record of Events and Opinion (1905, original text same as published)
Is Mars Habitable? (1907, autograph draft with revisions)
The World of Life (1910, autograph draft with revisions)
Contributions to Periodicals 1890–1908 (some of which are republished in *Studies Scientific and Social* in 1900)

These *Contributions* include thirteen journal article manuscripts:

"Progress without Poverty" (*Fortnightly Review [FR],* 9/90)
"English and American Flowers" (*FR,* 10/91)
"Our Molten Globe" (*FR,* 11/92)
"Spiritualism" (*Chambers' Encyclopaedia, 1892*)
"The Ice-Age and its Work i; Erratic Blocks and Ice-Sheets" (*FR,* 11/93)
"The Ice-Age and its Work ii; Glacial Erosion of Lake Basins" (*FR,* 12/93)
"The Palaearctic and Nearctic Regions Compared as Regards the Families and Genera of their Mammalia and Birds" (*Natural Science,* 6/94)
"How to Preserve the House of Lords" (*Contemporary Review,* 1/94)
"Revd. George Henslow on Natural Selection" (*Natural Science,* 9/94)
"The Method of Organic Evolution" (*FR,* 2/95)
"The Expressivenes of Speech, or Mouth-Gesture as a Factor in the Evolution of Language" (*FR,* 10/95)
"Evolution" (possibly published in *The Sun,* 1900, and reprinted in *Progress of the Century,* 1900)
"The Legend of the Birds of Paradise in the Arabian Nights" (*Independent Review,* 3/04)
"The Nature Problem in South Africa and Elsewhere" (*Independent Review,* 11/06)
"The Remedy for Unemployment" (*Socialist Review,* 6/08)
"The Present Position of Darwinism" (*Contemporary Review,* 8/08)

Also stored in the archives is a picture postcard of Wallace, along with an intriguing printed press release and proposal to raise funds to finance a utopian society in Africa, based on Dr. Theodor Hertzka's novel *Freeland,* for which Wallace founded the *British Freeland Association,* with an office address of London. Most of this material was donated by Wallace's son William George Wallace.

British Museum (Natural History): Manuscript materials are under "Wallace Collection." Twenty letters. Two Species Registries: "Bird and Insect Register 1858–1862" and "Insect, Bird and Mammal Register 1855–1860." Drawings of fish of the Amazon. A large color portrait commissioned and painted ten years after Wallace's death (featured on the cover of this book).

Darwin Archive, Cambridge University Library: Original letters between Wallace and Darwin, all recorded in the *Handlist of the Darwin Papers,* Cambridge University Press, 1960. Included is the letter clipping from Wallace to Darwin, dated September 27, 1857, in which Wallace discusses his "plan" for a "detailed proof" of his theory first proposed in 1855. See also the *Calendar* of Darwin's correspondence for cross-referencing Wallace and tangential figures.

Hope Entomological Collections of Oxford University Museum: Approximately 200 letters in two boxes between Wallace and primarily E. B. Poulton, dated 1886–1913, and

Raphael Meldola, dated 1879–1910. There are a handful of others, including one from Darwin to Wallace (July 9–,) about caterpillar feeding habits and coloration of insects of Sydney, and from C. Lloyd Morgan (1891). The Poulton and Meldola letters are particularly important for understanding Wallace's *hyper-selectionism,* and his stance against use-inheritance, Mendelian genetics, and mutationism. Also, a manuscript by R. Meldola on Wallace and his theories; a copy of *The Scientific Aspects of the Supernatural,* with the account by Frances Sims of spirit writing on the frontispiece (see Chapter 8 for a photocopy and reprint); "A. Russel Wallace Spiritualist Library"; and a sizable insect collection from Malaya, the listing in detail of which fills an entire page in Audrey Smith's "Lists of Archives and Collections," an appendix in her *A History of the Hope Entomological Collections in the University Museum Oxford* (Oxford University Press, 1986). There are also photos of Wallace.

Imperial College of Science, Technology & Medicine, London: Letters are in the "Huxley Collection." Eight letters from Wallace to Huxley, dated 1863–1891, most dealing with either the evolution of man, or fossils and their use as evidence for evolution. There is one letter from Raphael Meldola and others related to the Alfred Russel Wallace Memorial Fund.

Kew Royal Botanic Gardens: A total of 139 letters of correspondence, dated 1848–1913 and listed in the "Director's Correspondence," mostly between Wallace and the directors of Kew, including D. Prain, W. T. Thiselton-Dyer, and, of course, J. D. Hooker. The subject of most of the letters is, of course, of a botanical nature, dealing with the identification of seeds, plants, or flowers, the investigation of specimens, plants for personal use (gifts), biogeography of plants, requests for seeds for gardening experiments, revisions on Wallace's *Island Life,* recollections for *My Life* from Hooker, information for book on Richard Spruce Wallace edited, orchids, extinct animals and missing links, request for letter of recommendation from Hooker, proofread *Darwinism,* and request for lectern slide of bird pollinating.

Linnean Society of London: Letters; four notebooks from the Malay Archipelago; the American journal; four other journals from both the Amazon and Malay Archipelago; manuscript of *Palm Trees of the Amazon;* "Some of My Original Sketches on the Amazon"; books from Wallace's library with marginalia (including Darwin's *Descent of Man*); Registry of Consignments (specimens sent to London agent Stevens). Also, a small museum contains the skin of a giant python Wallace killed in Malaya, as well as Darwin's satchel in which he kept his scientific instruments on board the *Beagle.* The room where the Darwin–Wallace joint papers were read on July 1, 1858, is no longer extant, but the original furniture and other room artifacts are set up exactly as before in a room at the current Linnean Society building. The Journals and Notebooks contain the following:

Wallace Journals (approximately 25,000 words per volume):
V. I: 13 June 1856–9 March 1857
V. II: 13 March 1857–1 March 1859
V. III: 25 March 1859–August 1859
V. IV: 29 October 1859–10 May, 1861

Wallace Notebooks:
(i) The Species Register
(ii) Financial Accounts
(iii) Notes on Habitats, etc.
(iv) Notes for Papers

These have all been transcribed, though it is advisable to check the transcripts with the original source because, not infrequently, there is obvious misreading of words, or blank spaces where words could not be read.

Museum of Mankind: Anthropological artifacts brought back from Malaya, including carvings, tools, and trinkets made by native peoples with whom Wallace interacted or lived. Many of these are mentioned or discussed in Wallace's *Malay Archipelago.*

Royal Entomological Society: Materials listed in "Wallace Collection." The original copy of the article Wallace wrote on "The Origin of the Theory of Natural Selection"; Obituary Notice (with portrait) by E. B. Poulton and published in the *Proceedings of the Royal Society* (1923, Vol. 95); a hard-to-come-by copy of Harry Clement's *Alfred Russel Wallace: Biologist and Social Reformer* (1983); and a paper by Gerald Henderson on "The Present Position of Darwinism," in which he discusses his 1959 dissertation on *Wallace and Darwin: Diverging Currents in 19th Century Evolutionary Thought.*

Royal Geographical Society: Materials listed in the "Wallace Collection." Twenty-three items total, including letters from Wallace, dated 1853–1878; a request for funding or assistance for the Malay Archipelago trip; and the correspondence from John Hampden and the Flat-Earth Society.

Royal Society of London: Wallace materials are scattered throughout the archives and listed in the card catalogue under "Wallace." Four letters, dated 1868–1908, regarding the Copley Medal, the £50 Copley gift, expenses in copying letters and journals, dust particles carried by the wind worldwide, and one to Professor A. Schuster on biogeography. Also, an article by C.F.A. Pantin on "ARW, FRS and his Essays of 1858 and 1855, RS, Notes and Records"; a good collection of books by and about Wallace and Darwin; as well as busts of Darwin, Lyell, and other luminaries in the history of science, including Newton.

University College London, Bloomsbury Science Library: Letters and assorted materials are in the Manuscript Room under "Galton Papers." Correspondence of Wallace and F. Galton, and of particular interest Galton's attempt to survey the eminent men and "noteworthy families" of his time, including Wallace, and Wallace's reticent (and modest) response declining participation in the study. Letters discuss an "Evolution Committee on Breeding" and a request for Wallace's fingerprints. There is also a "Scrapbook of Darwinia," including a letter from Darwin to Wallace.

University of London: Letters listed by MSS number in the "Spencer Collection." Limited correspondence between Wallace and Herbert Spencer, listed under Wallace letters in Spencer collection.

Wellcome Institute: Letters in the "Wallace Collection." Nine letters, dated 1863–1910, to T. C. Eyton, J. D. Hooker, D. Cook, Mrs. Dammeather, Mrs. Alice K. Wyme, S. Waddington, and three to P. L. Sclater. Mostly botanical and zoological observations and comments. No theoretical discussions.

Zoological Society of London: Letters in the "Sclater Letters." Thirty-three letters, dated 1850–1901, mostly to P. L. Sclater on zoological matters, zoogeography, and the sales of specimens in London, particularly birds, to finance research. Also, an interesting "Proposal as to a Joint Residential Estate," by Wallace and his partners in this proposal, for a planned community outside London in which the natural beauty of the landscape would remain preserved.

APPENDIX II

WALLACE'S PUBLISHED WORKS

The following bibliography includes all of Wallace's books, papers, essays, reviews, interviews, and letters to the editor, totaling 749 references. A bibliography of Wallace's published works was first compiled in 1916 by his friend James Marchant, as part of a collected volume of Wallace's letters entitled *Alfred Russel Wallace, Letters and Reminiscences.* H. L. McKinney's 1972 biography, *Wallace and Natural Selection,* added to it, but the definitive bibliography can be found in Charles H. Smith's 1991 *Alfred Russel Wallace: An Anthology of His Shorter Writings.* Smith tracked down every known published work of Wallace's and compiled a publishing history of each item, including reprints, second serials, anthologies, and the like, with pagination, illustrations, and other details provided. It is a remarkable achievement but one that need not be repeated here in its entirety for space considerations. I have, instead, provided shorter citations for each reference, following some of Smith's conventions (although abbreviated to provide only the essential information for readers), and refer scholars to Smith's volume for additional information, as well as to his Web page on Wallace at http://www.wku.edu/~smithch/index1.htm, where one will find a veritable treasure trove of Wallace materials. Smith's archive is an invaluable service to historians of science and it continues to grow as he posts additional materials on a regular basis, including complete book and article manuscripts, extended quotations, interviews, and answers to frequently asked questions about Wallace. No historian should attempt a Wallace biography without consulting Smith's Web page. For Wallace's unpublished writings see Appendix I: Wallace Archival Sources, the most complete to date. Both the archives and the bibliography attest to the depth and scope of Wallace's knowledge, his wide-ranging interests, and his fearless literary ventures into fringe and heretical science and the borderlands of knowledge.

Abbreviations

ASL: Anthropological Society of London
A-VL: Anti-Vaccination League
BAAS: British Association for the Advancement of Science
CTNS: Contributions to the Theory of Natural Selection
ESL: Entomological Society of London
LNS: Land Nationalisation Society
LTTE: Letter to the Editor
NSTN: Natural Selection and Tropical Nature
RGS: Royal Geographical Society
SPR: Society for Psychical Research
SSS: Studies Scientific and Social
TNOE: Tropical Nature and Other Essays
ZSL: Zoological Society of London

In addition:

n.s.: new series
o.s.: old series
2nd s.: second series

n.d.: no date
n.p.: no pagination
LTTE: letter to the editor

Obliques (/) signify publication in sequential journal issues. I have retained Smith's coding for abstract summaries prepared by Wallace (/Y), possibly prepared by Wallace (/P), unlikely prepared by Wallace (/U), or not prepared by Wallace (/N).

Samuel Stevens, referenced in some article citations, was Wallace's sales agent in England to whom he mailed his specimens from his various natural history expeditions around the world.

Books

1. *Palm Trees of the Amazon and their Uses.* John Van Voorst, London, Oct. 1853.
2. *A Narrative of Travels on the Amazon and Rio Negro, with an Account of the Native Tribes, and Observations on the Climate, Geology, and Natural History of the Amazon Valley.* Reeve & Co., London, Dec. 1853.
3. *The Malay Archipelago; The Land of the Orang-utan and the Bird of Paradise; A Narrative of Travel with Studies of Man and Nature.* 2 volumes. Macmillan & Co., London, Feb. 1869.
4. *Contributions to the Theory of Natural Selection. A Series of Essays.* Macmillan & Co., London & New York; April 1870.
5. *Miracles and Modern Spiritualism. Three Essays.* James Burns, London, March 1875. (Second Edition. Trubner & Co., London, Oct. 1881. Third Edition "with chapters on apparitions and phantasms" and new Preface. George Redway, London, Jan. 1896.)
6. *The Geographical Distribution of Animals; with A Study of the Relations of Living and Extinct Faunas as Elucidating the Past Changes of the Earth's Surface.* 2 volumes. Macmillan & Co., London, May 1876.
7. *Tropical Nature, and Other Essays.* Macmillan & Co., London & New York, April 1878.
8. *Australasia* ("edited and extended by Alfred R. Wallace, with Ethnological Appendix by A. H. Keane"). Stanford's Compendium of Geography and Travel. Edward Stanford, London, May 1879.
9. *Island Life: or, The Phenomena and Causes of Insular Faunas and Floras, Including a Revision and Attempted Solution of the Problem of Geological Climates.* Macmillan & Co., London, Oct. 1880.
10. *Land Nationalisation; Its Necessity and its Aims; Being a Comparison of the System of Landlord and Tenant with that of Occupying Ownership in their Influence on the Well-being of the People.* Trubner & Co., London, May 1882.
11. *Bad Times: An Essay on the Present Depression of Trade, Tracing It to its Sources in Enormous Foreign Loans, Excessive War Expenditure, the Increase of Speculation and of Millionaires, and the Depopulation of the Rural Districts; with Suggested Remedies.* Macmillan & Co., London & New York, Nov. 1885.
12. *Darwinism, An Exposition of the Theory of Natural Selection with Some of its Applications.* Macmillan & Co., London & New York, May 1889.
13. *Natural Selection and Tropical Nature; Essays on Descriptive and Theoretical Biology* (essay collection combining essays from *CTNS* and *TNOE,* plus newly reprinted material). Macmillan & Co., London & New York, May 1891.
14. *The Wonderful Century; Its Successes and its Failures.* Swan Sonnenschein & Co., London, June 1898.
15. *Studies Scientific and Social* (essay collection). 2 volumes. Macmillan & Co., Ltd., London, Nov. 1900/New York, Dec. 1900.

16. *Man's Place in the Universe; A Study of the Results of Scientific Research in Relation to the Unity or Plurality of Worlds.* Chapman & Hall, Ltd., London, Oct. 1903.
17. *My Life, A Record of Events* and *Opinions.* 2 volumes. Chapman & Hall, Ltd., London, Oct. 1905. Published in an abridged one-volume edition in 1908.
18. *Is Mars Habitable? A Critical Examination of Professor Percival Lowell's Book "Mars* and *its Canals," with an Alternative Explanation.* Macmillan & Co., Ltd., London, Dec. 1907.
19. *Notes of a Botanist on the Amazon and Andes* (by Richard Spruce, "edited and condensed by Alfred Russel Wallace"). 2 volumes. Macmillan & Co., Ltd., London, Dec. 1908.
20. *The World of Life; A Manifestation of Creative Power, Directive Mind and Ultimate Purpose.* Chapman & Hall, Ltd., London, Dec. 1910.
21. *Social Environment and Moral Progress.* Cassell & Co., Ltd., London, New York, Toronto & Melbourne, March 1913.
22. *The Revolt of Democracy.* Cassell & Co., Ltd., London, New York, Toronto & Melbourne, Oct. 1913.

Articles, Essays, Reviews, Letters, Interviews

1843

1. The Advantages of Varied Knowledge (excerpts from this essay, composed in late 1843, and possibly presented as a public lecture at that time), first published in *My Life,* Vol. 1: 201–4, Oct. 1905.

1845

2. An Essay, On the Best Method of Conducting the Kington Mechanic's Institution, in *The History of Kington*, ed. by Richard Parry (Kington), 1845: 66–70.

1847

3. Capture of *Trichius fasciatus* near Neath (excerpt from a letter sent from Neath, Wales). *Zoologist* 5: 1676, April? 1847.

1849

4. Journey to Explore the Province of Pará (extract from a letter dated 23 Oct., Pará, from Wallace and Henry W. Bates to Samuel Stevens). *Ann. & Mag. Nat. Hist.* 3: 74–75 Jan. 1849.

1850

5. Journey to Explore the Natural History of South America (extracts from a letter dated 12 Sept. 1849, Santarem, to Samuel Stevens). *Ann. Mag. Nat. Hist.* 5 (2nd s.): 156–157, Feb. 1850: no. 26, 2nd s.
6. On the Umbrella Bird (*Cephalopterus ornatus*), "Ueramimbé," L. G. (from a letter dated 10 March 1850, Barra do Rio Negro, Manaus; communicated to the ZSL meeting of 23 July 1850). *Proc. ZSL* 18: 206–7, 1850.
7. Journey to Explore the Natural History of the Amazon River (extracts from letters dated 15 Nov. 1849, Santarem, and 20 March 1850, Barra do Rio Negro, to Samuel Stevens). *Ann. & Mag. Nat. Hist.* 6: 494–96, Dec. 1850.

1852

8. Letter (about the fire on the ship during Wallace's return voyage from the Amazon, dated 19 Oct. 1852). *Zoologist* 10: 3641–43, Nov.? 1852.

9. On the Monkeys of the Amazon (paper read at the ZSL meeting of 14 Dec. 1852). *Proc. ZSL* 20: 107–10, 1852.

1853

10. Some Remarks on the Habits of the Hesperidae (dated March 1853). *Zoologist* 11: 3884–85, May? 1853.
11. On the Rio Negro (paper read at the RGS meeting of 13 June 1853). *J. RGS* 23: 212–17, 1853.
12. On some Fishes Allied to *Gymnotus* (a paper read at the ZSL meeting of 12 July 1853). *Proc. ZSL* 21: 75–76, 1853.

1854

13. On the Insects Used for Food by the Indians of the Amazon (paper read at the ESL meeting of 6 June 1853). *Trans. ESL* 2 (n.s.), part VIII: 241–44, April 1854.
14. On the Habits of the Butterflies of the Amazon Valley (paper read at the ESL meetings of 7 Nov. and 5 Dec. 1853). *Trans. ESL* 2 (n.s.), part VIII: 253–64, April 1854.
15. Letter (concerning collecting dated 9 May 1854, Singapore). *Zoologist* 4395–97, Aug.? 1854.

1855

16. On the Ornithology of Malacca. *Ann. & Mag. Nat. Hist.* 15, 95–99, Feb. 1855: no. 86.
17. Description of a New Species of *Ornithoptera. Ornithoptera brookiana* Wallace (communicated to the ESL meeting of 2 April 1855). *Proc. ESL,* 1854–1855: 104–5.
18. The Entomology of Malacca (dated 25 Nov. 1854, Sarawak, Borneo). *Zoologist* 13: 4636–39, April? 1855.
19. Letter (concerning Wallace's collecting environs in Sarawak dated 1854, Sarawak, Borneo; anonymous, but easily associated with Wallace's activities at this time and place). *Literary Gazette* (London) no. 2003: 366b–c, 9 June 1855.
20. Extracts of a Letter from Mr. Wallace (from a letter concerning collecting dated 10 Oct. 1854, Singapore). *(Hooker's) J. Botany* 7(7): 200–209, July? 1855.
21. On the Law which has Regulated the Introduction of New Species (dated Feb. 1855, Sarawak, Borneo). *Ann. & Mag. Nat. Hist.* 16 (2nd s.): 184–96, Sept. 1855: no. 93, 2nd s.
22. Letter (concerning collecting dated 8 April 1855, Si Munjon Coal Works, Borneo). *Zoologist* 13: 4803–7, Sept.? 1855.
23. Borneo (letter concerning Wallace's collecting environs in Sarawak dated 25 May 1855, Si Munjon Coal Works, Borneo). *Literary Gazette* (London) no. 2023: 683b–84a, 27 Oct. 1855.

1856

24. Some Account of an Infant "Orang-utan." *Ann. & Mag. Nat. Hist.* 17 (2nd s.): 386–90, May 1856: no. 101, 2nd s.
25. On the Orang-utan or Mias of Borneo (dated Dec. 1855, Sarawak). *Ann. & Mag. Nat. Hist.* 17 (2nd s.): 471–76, June 1856: no. 102, 2nd s.
26. Observations on the Zoology of Borneo (dated 10 March 1856, Singapore). *Zoologist* 14: 5113–17, June? 1856.
27. On the Habits of the Orang-utan of Borneo. *Ann. & Mag. Nat. Hist.* 18 (2nd s.): 26–32, July 1856: no. 103, 2nd s.
28. On the Bamboo and Durian of Borneo (part of a letter to W. J. Hooker). *(Hooker's) J. Botany* 8(8): 225–30, Aug.? 1856.

29. Attempts at a Natural Arrangement of Birds. *Ann. & Mag. Nat. Hist.* 18 (2nd s.): 193–216, Sept. 1856: no. 105, 2nd s.
30. Notes of a Journey up the Sadong River, in North-west Borneo (communicated to the RGS meeting of 10 Nov. 1856). *Proc. RGS Lon.* 1 (1855–1856 & 1856–1857), no. 6 193–205, 1857.
31. A New Kind of Baby (concerning a baby orangutan; anonymous, but in Wallace's style and listed in the bibliography of Marchant, 1916). *Chambers's J.* 6 (3rd s.): 325–27, 22 Nov. 1856: no. 151, 3rd s.
32. Letter (concerning collecting dated 21 Aug. 1856, Ampanam, Lombock). *Zoologist* 15: 5414–16, Feb.? 1857.
33. Letter (concerning collecting dated 27 Sept. 1856, Macassar). *Zoologist* 15: 5559–60, June? 1857.
34. Letter (concerning collecting dated 1 Dec. 1856, Macassar). *Zoologist* 15: 5652–57, Aug.? 1857.

1857

35. The Dyaks (anonymous, but containing various details strongly suggesting it is by Wallace). *Chambers's J.* 8 (3rd s.): 201–4, 26 Sept. 1857: no. 195, 3rd s.
36. Letter and postscript (concerning collecting to Samuel Stevens; dated 10 March 1857 Dobbo, Aru Islands, and 15 May, Dobbo; communicated to the ESL meeting of 5 Oct. 1857). *Proc. ESL,* 1856–1857: 91–93.
37. On the Habits and Transformations of a Species of *Ornithoptera,* Allied to *O. priamus,* Inhabiting the Aru Islands, near New Guinea (communicated to the ESL meeting of 7 Dec. 1857). *Trans. ESL* 4 (n.s.), part VII: 272–73, April 1858.
38. On the Great Bird of Paradise, *Paradisea apoda,* Linn.; *"Burong mati" (Dead Bird)* of the Malays; *"Fanéhan"* of the Natives of Aru. *Ann. & Mag. Nat. Hist.* 20 (2nd s.): 411–16, Dec. 1857: no. 120, 2nd s.
39. On the Natural History of the Aru Islands. *Ann. & Mag. Nat. Hist.,* Supplement to vol. 20 (2nd s.): 473–85, Dec. 1857: 110–21, 2nd s.

1858

40. Note on the Theory of Permanent and Geographical Varieties. *Zoologist* 16: 5887–88, Jan.? 1858.
41. On the Entomology of the Aru Islands. *Zoologist* 16: 5889–94, Jan.? 1858.
42. On the Arru Islands (communicated to the RGS meeting of 22 Feb. 1858). *Proc. RGS Lon.* 2 (1857–1858), no. 3: 163–70 (selection followed by an account of related discussion on pp. 170–71), 1858.
43. A Disputed Case of Priority in Nomenclature (dated 1 Jan. 1858, Amboyna; communicated to the ESL meeting of 3 May 1858). *Proc. ESL,* 1858–1859: 23–24.
44. On the Tendency of Varieties to Depart Indefinitely from the Original Type (dated Feb. 1858, Ternate; third part of 'On the Tendency of Species to Form Varieties; and On the Perpetuation of Varieties and Species by Natural Means of Selection' by Charles Darwin and Alfred Wallace; communicated by Sir Charles Lyell and Joseph D. Hooker to the LSL meeting of 1 July 1858). i. *Proc. Linn Soc. Zool.* 3(9): 53–62 (45–62), 20 Aug. 1858.
45. Letter (concerning collecting dated 20 Dec. 1857, Amboyna). *Zoologist* 16: 6120–24, July? 1858.

1859

46. Letters (extracts from letters concerning collecting dated 2 Sept. 1858, Ternate). *Ibis* 1: 111–13, Jan. 1859: no. 1.

47. Correction of an Important Error Affecting the Classification of the Psittacidae. *Ann. & Mag. Nat. Hist.* 3 (3rd s.): 147–48, Feb. 1859: no. 14, 3rds.
48. Letter (extract from letter concerning collecting dated 29 Oct. 1858, Batchian, Moluccas; communicated to the ESL meeting of 7 March 1859). *Proc. ESL,* 1858–1859: 61.
49. Letter (extract from letter concerning collecting dated 29 Oct. 1858, Batchian, Moluccas; communicated to the ZSI meeting of 22 March 1859). *Proc. ZSL 27:* 129, 1859.
50. Remarks on Enlarged Coloured Figures of Insects (dated Nov. 1858, Batchian, Moluccas; communicated to the ESL meeting of 2 May 1859). *Proc. ESL,* 1858–1859: 66–67.
51. Letter (extracts from letter concerning collecting dated 28 Jan. 1859, Batchian; communicated to the ESL meeting of 6 June 1859). *Proc. ESL,* 1858–1859: 70.
52. Notes of a Voyage to New Guinea (communicated to the RGS meeting of 27 June 1859). *J. RGS 30:* 172–77, 1860.
53. Letter from Mr. Wallace concerning the Geographical Distribution of Birds (dated March 1859, Batchian). *Ibis* 1: 449–54, Oct. 1859: no. 4.

1860

54. On the Zoological Geography of the Malay Archipelago (communicated by Charles Darwin to the LSL meeting of 3 Nov. 1859). *J. Proc. Linn. Soc. Zool.* 4: 172–84, 1860.
55. Note on the Habits of Scolytidae and Bostrichidae (communicated to the ESL meeting of 5 Dec.1859). *Trans. ESL* 5 (n.s.), part VI: 218–20, July 1860.
56. Notes on *Semioptera wallacii,* Gray (extract from a letter dated 30 Sept. 1859, Amboyna; communicated to the ZSL meeting of 24 Jan. 1860). *Proc. ZSL* 28:61, 1860.
57. Note on the Sexual Differences in the Genus *Lomaptera* (communicated to the ESL meeting of ó Feb. 1860). *Proc. ESL* 1858–1859: 107.
58. The Ornithology of Northern Celebes (part of a LTTE dated Oct. 1859, Amboyna). *Ibis* 2: 140–47, April 1860: no. 6.
59. Letters (extracts from letters concerning collecting dated 22 Oct. 1859, Amboyna). *Ibis* 2: 197–99, April 1860: no. 6.
60. Letters (extracts from letters concerning collecting dated 26 Nov. 1859, Awaiya, Ceram, 31 Dec. 1859, Passo, Amboyna, and 14 Feb. 1860, Passo). *Ibis* 2: 305–6, July 1860: no. 7.

1861

61. Letters (note/N describing letters concerning collecting dated June 1860, Ceram). *Ibis* 3: 118, Jan. 1861: no. 9.
62. Letter (extracts from letter concerning collecting dated 7 Dec. 1860, Ternate). *Ibis* 3: 211–12, April 1861: no. 10.
63. On the Ornithology of Ceram and Waigiou (dated 20 Dec.1860), Ternate). *Ibis* 3: 283–91, July 1861: no. 11.
64. Letters (note/N containing extracts from letter dated 10 Dec. 1860, Ternate, and describing letter dated 6 Feb. 1861, Delli, Timor, both concerning collecting). *Ibis* 3: 310–11, July 1861: no. 11.
65. Notes on the Ornithology of Timor (dated 20 April 1861, Delli, Timor). *Ibis* 3: 347–51, Oct. 1861: no. 12.

1862

66. On the Trade of the Eastern Archipelago with New Guinea and its Islands (communicated to the RGS meeting of 13 Jan. 1862). *J. RGS* 32: 127–37.

67. Letters (extracts from letters concerning collecting dated 20 Sept. 1861, Batavia). *Ibis* 4: 95–96, Jan. 1862: no. 13.
68. Narrative of Search after Birds of Paradise (paper read at the ZSL meeting of 27 May 1862). *Proc. ZSL, 1862:* 153–61, 1862.
69. On some New and Rare Birds from New Guinea (paper read at the ZSL meeting of 10 June 1862). *Proc. ZSL, 1862:* 164–66, 1862.
70. Descriptions of Three New Species of *Pitta* from the Moluccas (paper read at the ZSL meeting of 24 June 1862). *Proc. ZSL, 1862:* 187–88, 1862.
71. On some New Birds from the Northern Moluccas. *Ibis* 4: 348–51, Oct. 1862: no. 16.
72. List of Birds from the Sula Islands (East of Celebes), with Descriptions of the New Species (paper read at the ZSL meeting of 9 Dec. 1862). *Proc. ZSL, 1862:* 333–46, 1862.

1863

73. List of Birds Collected in the Island of Bourn (One of the Moluccas), with Descriptions of the New Species (paper read at the ZSL meeting of 13 Jan. 1863). *Proc. ZSL, 1863:* 18–36, 1863.
74. Note on *Conus senex,* Garn. & Less., and *Corvas fuscicapillus,* G. R. Gray. *Ibis* 5: 100–102, Jan. 1863: no. 17.
75. Notes on the Genus *Iphias*; with Descriptions of Two New Species from the Moluccas. *J. Entomology* 2: 1–5, Jan. 1863: no. 7.
76. On the Proposed Change in Name of *Gracula pectoralis. Ann. & Mag. Nat. Hist.* 11 (3rd s.): 15–17, Jan. 1863: no. 61, 3rd s.
77. Discussion/N (of three papers on Australia read by J. M. Stuart, "Mr. Lands borough" and J. M'Kinlay at the RGS meeting of 9 March 1863). *Proc. RGS 7* (1862–1863), no. 3: 88–89 (82–90), 1863.
78. Who are the Humming Bird's Relations? *Zoologist* 21: 8486–91, April? 1863.
79. On the Physical Geography of the Malay Archipelago (paper read at the RGS meeting of 8 June 1863). *J. RGS 33:* 217–34, 1863.
80. The Bucerotidae, or Hornbills. *Intellectual Observer* 3(5): 309–17, June 1863.
81. Notes on the Fruit-pigeons of the Genus *Treron. Ibis* 5: 318–20, July 1863: no. 19.
82. On the Geographical Distribution of Animal Life (paper read in Newcastle upon-Tyne at the 31 Aug. 1863 meeting of Section D, Zoology and Botany of the BAAS), abstract/P printed in "Notes and Abstracts of Miscellaneous Communications to the Sections" portion of the *Rept. BAAS* 33, 1863, John Murray, London, 1864: 108–109.
83. On the Varieties of Men in the Malay Archipelago (paper read in Newcastle upon-Tyne at the I Sept. 1863 meeting of Section E, Geography and Ethnology, of the BAAS), abstract/P printed in "Notices and Abstracts of Miscellaneous Communications to the Sections" portion of the *Rept. BAAS* 33, 1863, John Murray, London, 1864: 147–48.
84. Remarks on the Rev. S. Haughton's Paper on the Bee's Cell, and on the Origin of Species. *Ann. & Mag. Nat. Hist.* 12 (3rd s.): 303–9, Oct. 1863: no. 70, 3rd s.
85. On the Identification of the *Hirundo esculenta* of Linnaeus, with a Synopsis of the Described Species of *Collocalia* (paper read at the ZSL meeting of 10 Nov. 1863). *Proc. ZSL, 1863:* 382–85, 1863.
86. A List of the Birds Inhabiting the Islands of Timor, Flores, and Lombock, with Descriptions of the New Species (paper read at the ZSL meeting of 24 Nov. 1863). *Proc. ZSL, 1863:* 480–97, 1863.

1864

87. Discussion/N (of "On the Vitality of the Black Race," a paper by Count Oscar Reichenbach read at the ASL meeting of 15 Dec. 1863). *J. ASL* 2: Ixxiii (lxv–lxxiii), 1864.

88. Discussion/NT of papers on the extinction of races by Richard Lee and T. Bendyshe read at the ASL meeting of 19 Jan. 1864. *J. ASL* 2: cx–cxi (xcv–cxiii), 1864.
89. Remarks on the Value of Osteological Characters in the Classification of Birds. *Ibis* 6: 36–41, Jan. 1864: no. 21.
90. Remarks on the Habits, Distribution, and Affinities of the Genus *Pitta. Ibis* 6: 100–14, Jan. 1864: no. 21.
91. Discussion/N (of "On Anthropological Desiderata," a paper by James Reddie read at the ASL meeting of 2 Feb. 1864). *J. ASL* 2: cxxix–cxxx (cxv–cxxxv), 1864.
92. Discussion/N (of paper on North Peru by Don Antonio Raimondy read at the RGS meeting of 8 Feb. 1864). *Proc. RGS* 8 (1863–1864), no. 3: 61–62 (58–62), 28 April 1864.
93. Discussion/N (of paper on the Neanderthal Man skull by C. Carter Blake read at the ASL meeting of 16 Feb. 1864). *J. ASL* 2: clvi (cxxxix–clvii), 1864.
94. The Origin of Human Races and the Antiquity of Man Deduced from the Theory of "Natural Selection" (paper read at the ASL meeting of I March 1864). *J. ASL* 2: clviii–clxx (followed by an account of related discussion on pp. clxx–clxxxvii), 1864.
95. Discussion/N (of "On the Place of the Sciences of Mind and Language in the Science of Man," a paper by Luke Owen Pike read at the ASL meeting of 15 March 1864). *J. ASL* 2: ccviii (cxcii–ccviii), 1864.
96. Discussion/N (of "Notes on the Capabilities of the Negro for Civilisation," a paper by Henry F. J. Guppy read at the ASL meeting of 15 March 1864). *J. ASL* 2: ccxiii–ccxiv (ccix–ccxvi), 1864.
97. On the Phenomena of Variation and Geographical Distribution as Illustrated by the Papilionidae of the Malayan Region (paper read at the LSL meeting of 17 March 1864). *Trans. LSL* 25, part I: 1–71, 1865.
98. Bone-caves in Borneo (LTTE). *Reader* 3: 367 (19 March 1864: no. 64), also printed in *Natural History Rev.* 4: 308–11, April 1864: no. 14.
99. Discussion/N (regarding mimicry in insects, at the ESL meeting of 4 April 1864). *J. Proc. ESL, 1864–1865:* 14–15.
100. Discussion/N (of "On the Universality of Belief in God, and in a Future State," a paper by Rev. F. W. Farrar read at the ASL meeting of 5 April 1864). *J. ASL* 2: ccxx (ccxvii–ccxxii), 1864.
101. Discussion/N (of "On Hybridity," a paper by Rev. F. W. Farrar read at the ASL meeting of 5 April 1864). *J. ASL* 2: ccxxvi–ccxxvii (ccxxii–ccxxix), 1864.
102. Note on *Astur griseiceps,* Schlegel. *Ibis* 6: 184, April 1864: no. 22.
103. On the Parrots of the Malayan Region, with Remarks on their Habits, Distribution, and Affinities, and the Descriptions of Two New Species (paper read at the ZSL meeting of 28 June 1864). *Proc. ZSL,* 1864: 272–95 (includes map of the Malay Archipelago), 1864.
104. Discussion/N (of paper on the delta of the Amazons by Henry W. Bates read in Bath at the 17 Sept. 1864 meeting of Section E, Geography and Ethnology, of the *BAAS*). *Athenaeum* no. 1928: 470, 8 Oct. 1864.
105. On the Progress of Civilization in Northern Celebes (paper read in Bath at the 19 Sept. 1864 meeting of Section E, Geography and Ethnology, of the BAAS), abstract/P printed in "Notices and Abstracts of Miscellaneous Communications to the Sections" portion of the *Rept. BAAS* 34 (1864), John Murray, London, 1865: 149–50.
106. Discussion/N (of paper on islands north of Flores by John Cameron read at the RGS meeting of 12 Dec. 1864). *Proc. RGS 9* (1864–1865), no. 2: 31–32 (30–32), 1865.

1865

107. Discussion/N (of "Linga Puja, or Phallic Worship in India," a paper by E. Sellon read at the ASL meeting of 17 Jan. 1865. *J. ASL* 3: cxviii–cxix, 1865.

108. The *British Quarterly* and Darwin (LTTE). *Reader* 5: 77c–78a, 21 Jan. 1865: no. 108.
109. The *British Quarterly* Reviewer and Darwin (LTTE). *Reader* 5: 173a–b, 11 Feb. 1865: no. 111.
110. List of the Land Shells Collected by Mr. Wallace in the Malay Archipelago with Descriptions of the New Species by Mr. Henry Adams (paper read at the ZSL meeting of 25 April 1865). *Proc. ZSL, 1865:* 405–16, 1865.
111. Public Responsibility and the Ballot (LTTE responding to comments by John Stuart Mill). *Reader* 5: 517a–b, 6 May 1865: no. 123.
112. Discussion/N (of "On the Efforts of Missionaries among Savages," a paper by Rev. J. W. Colenso read at the ASL meeting of 16 May 1865). *J. ASL* 3: cclxxxviii (ccxlviii–cclxxxix), 1865.
113. Descriptions of New Birds from the Malay Archipelago (paper read at the ZSL meeting of 13 June 1865). *Proc. ZSL, 1865:* 474–81 (& 2 colour plates), 1865.
114. How to Civilize Savages (signed "W." but referred to in *My Life* (Vol. 2, 24, 52). *Reader* 5: 671a–72a, 17 June 1865: no. 129.
115. On the Pigeons of the Malay Archipelago. *Ibis* I (n.s.): 365–400 (& I color plate), Oct. 1865: no. 4, n.s.

1866

116. Is the Earth an Oblate or a Prolate Spheroid? (LTTE). *Reader* 7: 497b–c, 19 May 1866: no. 177.
117. LTTE (one of two concerning the nature of the sphericity of the earth printed as "The Figure of the Earth" (not Archdeacon Pratt's). *Reader* 7: 546b (546a–b), 2 June 1866: no. 179.
118. List of Lepidopterous Insects Collected at Takow, Formosa, by Mr. Robert Swinhoe (joint paper by Wallace & Frederic Moore read at the ZSL meeting of 12 June 1866). *Proc. ZSL, 1866:* 355–65, 1866.
119. The Scientific Aspect of the Supernatural. *The English Leader* 2(4): 59–60 / 2(5): 75–76 / 2(6): 91–93 / 2(7): 107–108 / 2(8): 123–25 / 2(9): 139–40 / 2(10): 156–57 / 2(11): 171–73, 11 Aug. 1866 to 29 Sept. 1866: nos. 5259.
120. Address (given in Nottingham on 23 Aug. 1866 as President of the Dept. of Anthropology, Section D, Biology, of the BAAS), in "Notices and Abstracts of Miscellaneous Communications to the Sections" portion of the *Rept. BAAS 36 1866,* John Murray, London, 1867: 93–94.
121. Discussion/N (of various remarks made by others while serving as president of the Aug. 1866 meeting of the Dept. of Anthropology, Section D, Biology, of the BAAS; part of an article/N entitled "Anthropology at the British Association"). *Anthro. Rev.* 4: 391–408 (386–408), Oct. 1866: no. 15.
122. On Reversed Sexual Characters in a Butterfly, and their Interpretation on the Theory of Modifications and Adaptive Mimicry (paper read in Nottingham at the 27 Aug. 1866 meeting of Section D, Biology, of the BAAS), abstract/P printed in "Notices and Abstracts of Miscellaneous Communications to the Sections" portion of the *Rept.* BAAS 36 (1866), John Murray, London, 1867: 79.
123. Discussion/N (of "On Flint Implements Recently Discovered in North Devon," a paper by H. S. Ellis read in Nottingham at the Aug. 1866 meeting of the Dept. of Anthropology, Section D, Biology, of the BAAS). *Reader 7:* c, 13 Oct. 1866: no. 198.
124. Natural Selection (LTTE concerning mimicry). *Athenaeum* no. 2040: 716–17, 1 Dec. 1866.

1867

125. Ice Marks in North Wales (With a Sketch of Glacial Theories and Controversies). *Q. J. Science* 4: 33–51 Jan. 1867: no. 13.
126. Mr. Wallace on Natural Selection Applied to Anthropology (LTTE). *Anthro. Rev.* 5: 103–5 Jan. 1867: no. 16.
127. Postscript by Alfred R. Wallace (to an account of a séance written by Francis Sims, Wallace's brother-in-law, entitled "A New Medium"). *The Spiritual Mag* 2 (n.s.): 51–52 (49–52), 1 Feb. 1867.
128. On the Pieridae of the Indian and Australian Regions (paper read at the ESL meeting of 18 Feb. 1867). *Trans. ESL* 4 (3rd s.), part 111: 301–416 (& 4 color plates), Nov. 1867.
129. Discussion/N (of paper concerning a tributary of the Rio Purus in Peru by W. Chandless read at the RGS meeting of 25 Feb. 1867). *Proc. RGS* 11 (1866–1867), no. 3: 108–9 (100–110), 14 June 1867.
130. Discussion/N (Wallace's explanation of brilliant colors in caterpillar larvae, and others' comments thereon, presented at the ESL meeting of 4 March 1867). *J. Proc. ESL, 1866–1867:* lxxx–1xxxi.
131. Caterpillars and Birds (LTTE). *The Field, The Country Gentleman's Newspaper* 29: 206a–b, 23 March 1867: no. 743.
132. The Polynesians and their Migrations (review of *Les Polynesiens et leurs Migrations* by Armand de Quatrefages de Breau, 1864). *Q. J. Science* 4: 161–66, April 1867: no. 14.
133. Annotations to List of Birds Collected by Mr. Wallace on the Lower Amazons and Rio Negro by P. L. Sclater and Osbert Salvin (read at the ZSL meeting of 23 May 1867). *Proceedings of the Scientific Meetings of the Zoological Society of London* 1867: 566–96, 1867.
134. Notes of a Seance with Miss Nicholl at the House of Mr. A. S.—*The Spiritual Mag* 2 (n.s.): 254–55, 1 June 1867.
135. Discussion/N (of "On Physio-anthropology, its Aim and Method," a paper by James Hunt read at the 4 June 1867 meeting of the ASL; Wallace's discussion presented at the 18 June 1867 meeting). *J. ASL* 5: cclvi–cclvii (ccix–cclvii), 1867.
136. Mimicry, and other Protective Resemblances among Animals (notice of works by Henry W. Bates, Alfred R. Wallace, Andrew Murray, and Charles Darwin; anonymous, but referred to in a letter from Darwin to Wallace dated 6 July 1867 (Merchant, 1916, 154); written in 1865–1866, according to *My Life* (Vol. 1, 407). *Westminster Rev.* 32 (n.s.), no. 1: 1–43 (London ed.), 1 July 1867: no. 173.
137. A Catalogue of the Cetoniidae of the Malayan Archipelago, with Descriptions of the New Species (paper read at the ESL meeting of I July 1867). *Trans. ESL,* 4 (3rd s.), part *V:* 519–601, May 1868.
138. The Philosophy of Birds' Nests. *Intellectual Observer* 11(6): 413–20, July 1867.
139. Letter (concerning séance experiences to Benjamin Coleman; described and quoted from in Coleman's article "Passing Events—The Spread of Spiritualism"). *The Spiritual Mag.* 2 (n.s.) 349–50 (342–54), 1 Aug. 1867.
140. The Disguises of Insects. *Hardwicke's Science-Gossip* 3: 193–98, 1 Sept. 1867.
141. On Birds' Nests and their Plumage; or the Relation between Sexual Differences of Colour and the Mode of Nidification in Birds (paper read in Dundee at the 9 Sept. 1867 meeting of Section D, Biology, of the BAAS), abstract/U printed in "Notices and Abstracts of Miscellaneous Communications to the Sections" portion of the *Rept. BAAS 37* (1867), John Murray, London, 1868: 97.
142. Creation by Law (in effect a review of *The Reign of Law* by the Duke of Argyll, 1867). *Q. J. Science* 4: 471–88 (& 1 plate), Oct. 1867: no. 16.

1868

143. On the Raptorial Birds of the Malay Archipelago. *Ibis* 4 (n.s.): 1–27, Jan. 1868: no. 13, n.s.
144. Corrections of, and Additions to, the Catalogue of the Raptorial Birds of the Malay Archipelago. *Ibis* 4 (n.s.): 215–16, April 1868: no. 14, n.s.

1869

145. Museums for the People. *MacMillan's Mag.* 19: 244–50, Jan. 1869: no. 111.
146. LTTE (one of two printed as "A Scientific Club"). *Scientific Opinion* 1: 378b (378b–c), 17 March 1869: no. 20.
147. Notes on Eastern Butterflies (paper read at the ESL meeting of 5 April 1869). *Trans. ESL 1869,* part I: 77–81, April 1869.
148. Sir Charles Lyell on Geological Climates and the Origin of Species (running title for review of *Principles of Geology* (10th ed.), 1867–1868, and *Elements of Geology* (6th ed.), 1865, both by Sir Charles Lyell; anonymous, but referred to in *My Life* (Vol. 1, 406). *Quarterly Rev.* 126: 359–94, April 1869: no. 252.
149. Notes on Eastern Butterflies (continued) (paper read at the ESL meeting of 3 May 1869). *Trans. ESL 1869,* part IV: 277–88, Aug. 1869.
150. Notes on Eastern Butterflies (continued) (paper read at the ESL meetings of 7 June and 5 July 1869). *Trans. ESL 1869,* part IV: 321–49, Aug. 1869.
151. Questions (posed as member of the Committee on Spiritualism of the London Dialectical Society; concerning evidence presented by Manual Eyre and "Miss Douglass" at the Committee meetings of 8 June 1869 & 6 July 1869, respectively), in *Report on Spiritualism, of the Committee of the London Dialectical Society, together with the Evidence, Oral and Written, and a Selection from the Correspondence* (Longmans, Green, Reader & Dyer, London, Nov. 1871): 183 (178–84), 210, 209–11.
152. Discussion/N (of "On the Primitive Condition of Man," a paper by Sir John Lubbock read in Exeter at the 23 Aug. 1869 meeting of Section D, Biology, of. the BAAS; part of an article/N entitled "Anthropology at the British Association," 1869). *Anthro. Rev. 7:* 420–21 (414–32), Oct. 1869: no. 27.
153. Discussion/N (of "The Occasional Definition of the Convolutions of the Brain on the Exterior of the Head," a paper by T. S. Prideaux read in Exeter at the late August meeting of the BAAS; part of an article/N entitled "Anthropology at the British Association, 1869"). *Anthro. Rev. 7:* 430–31 (414–32), Oct. 1869: no. 27.
154. The Origin of Civilization (LTTE). *Spectator* 42: 1072–73, 11 Sept. 1869: no. 215.
155. The Origin of Moral Intuitions (LTTE). *Scientific Opinion* 2: 336b–37a, 15 Sept. 1869: no. 46.
156. Notes on the Localities Given in *Longicornia Malayana,* with an Estimate of the Comparative Value of the Collections Made at Each of Them. *Trans. ESL* 3 (3rd s.), part VII: 691–96, Oct. 1869.
157. The Origin of Species Controversy (review of *Habit and Intelligence in their Connexion with the Laws of Matter and Force* by Joseph John Murphy, 1869). I. *Nature* 1: 105–7, 25 Nov. 1869 / II.1: 132–33, 2 Dec. 1859.
158. Introductory Remarks to *A Catalogue of the Aculeate Hymenoptera and Ichneumonidae of India and the Eastern Archipelago* by Frederick Smith (paper read at the LSL meeting of 16 Dec. 1869). *J. Linn. Soc. Zool.* 11: 285–302, 1873.

1870

159. Government Aid to Science (LTTE). *Nature* 1: 288–89, 13 Jan. 1870.
160. Government Aid to Science (LTTE). *Nature* 1: 315, 20 Jan. 1870.
161. The Measurement of Geological Time (according to *Island Life* (3rd ed., 223), first

presented in 1869 as a paper to Section C, Geology, at the annual BAAS meetings). I. *Nature* 1: 399–401, 17 Feb. 1870 / II. 1: 452–55, 3 March 1870.

162. Review (of *A Geographical Handbook of all the Known Ferns with Tables to Show their Distribution* by Katharine M. Lyell, 1870). *Nature* 1: 428, 24 Feb. 1870.
163. Review (of *Hereditary Genius: An Inquiry into its Laws and Consequences* by Francis Galton, 1869). *Nature* 1: 501–3, 17 March 1870.
164. LTTE (one of several printed as "Experiments on the Convexity of Water"). *The Field, The Country Gentleman's Newspaper* 35: 305a–b (305a–c), 2 April 1870: no. 901.
165. LTTE (one of two printed as "The Convexity of Water"). *The Field, The Country Gentleman's Newspaper* 35: 317b–c (317b–c), 16 April 1870: no. 903.
166. On Instinct in Man and Animals, in *CTNS:* 201–10, April 1870.
167. The Limits of Natural Selection as Applied to Man, in *CTNS:* 332–71, April 1870.
168. Review (of *The Handy Book of Bees; Being a Practical Treatise on their Profitable Management* by A. Pettigrew, 1870). *Nature* 2: 82–83, 2 June 1870.
169. Discussion/N (of "On the Geographical Distribution of the Chief Modifications of Mankind," a paper by Thomas Huxley read at the Ethnological Society of London meeting of 7 June 1870). *J. Ethnol. Soc. Lon.* 2 (n.s.): 411–12 (404–12), 1870.
170. Twelve-wired Bird of Paradise (LTTE). *Nature* 2: 234, 21 July 1870.
171. Review (of *Researches into the Early History of Mankind and the Development of Civilization,* 2nd ed.) by Edward B. Tylor, 1870). *Nature* 2: 350–51, 1 Sept. 1870.
172. Discussion/N (of "Certain Principles to be Observed in the Establishment of a National Museum of Natural History," a paper by P. L. Sclater read in Liverpool at the 16 Sept. 1870 meeting of Section D, Biology, of the BAAS). *Nature* 2: 465 (465–66), 6 Oct. 1870.
173. On a Diagram of the Earth's Eccentricity and the Precession of the Equinoxes, Illustrating their Relation to Geological Climate and the Rate of Organic Change (paper read in Liverpool at the 20 Sept. 1870 meeting of Section C, Geology, of the BAAS), abstract/P printed in the "Notices and Abstracts of Miscellaneous Communications to the Sections" portion of the *Rept. BAAS* 40 (1870), John Murray, London, 1871: 89.
174. The Glaciation of Brazil (review of *Thayer Expedition: Scientific Results of a Journey in Brazil. By Louis Agassiz and his Travelling Companions. Geology and Physical Geography of Brazil* by Charles Frederick Hartt, 1870). *Nature* 2: 510–12, 27 Oct. 1870.
175. Man and Natural Selection. *Nature* 3: 8–9, 3 Nov. 1870.
176. An Answer to the Arguments of Hume, Lecky, and Others, against Miracles (paper read in London as the first of a series of weekly "winter soirées" on 14 Nov. 1870). *The Spiritualist* (London) 1(15): 113c–16b, 15 Nov. 1870.
177. Natural Selection—Mr. Wallace's Reply to Mr. Bennett. *Nature* 3: 49–50, 17 Nov. 1870.
178. LTTE (one of several printed as "The Difficulties of Natural Selection"). *Nature* 3: 85–86, 1 Dec. 1870.
179. The Difficulties of Natural Selection (LTTE). *Nature* 3: 107, 8 Dec. 1870.
180. Review (of *Observations on the Geology and Zoology of Abyssinia* by W. T. Blanford, *1870). Academy* 2: 66–67 (15 Dec. 1870: no. 15).
181. LTTE (one of two printed as "Mimicry versus Hybridity"). *Nature* 3: 165–66, 29 Dec. 1870.

1871

182. Review (of *The Intelligence and Perfectibility of Animals from a Philosophic Point of View. With a Few Letters on Man* by Charles Georges Leroy, 1870). *Nature* 3: 182–83, 5 Jan. 1871.

183. The President's Address (read at the ESL annual meeting of 23 Jan. 1871). *Proc. ESL,* 1870: xliv–lxix.
184. Review (of *A Voyage Round the World* by the Marquis de Beauvoir, 1870). *Nature* 3: 244, 26 Jan. 1871.
185. Review (of *A Ride through the Disturbed Districts of New Zealand; Together with some Account of the South Sea Islands. Being Selections from the Journals and Letters of Lieut. the Hon. Herbert Meade, R. N. Edited by his Brother,* 1870). *Nature* 3: 264, 2 Feb. 1871.
186. The Theory of Glacial Motion (comments on two papers by James Croll: "On Ocean Currents" and "On the Cause of the Motion of Glaciers"). *Nature* 3: 309–10, 16 Feb. 1871.
187. The Metamorphoses of Insects (review of *The Transformations (or Metamorphoses) of Insects (Insecta, Myriapoda, Arachnida, and Crustacea)* by M. Emile Blanchard, adapted to English by P. Martin Duncan, 1870). *Nature* 3: 329–31, 23 Feb. 1871.
188. Review (of *The Descent of Man and Selection in Relation to Sex* by Charles Darwin, 1871). *Academy* 2: 177–83, 15 March 1871: no. 20.
189. Review (of *The Honey Bee: Its Nature History, Physiology, and Management,* rev. ed.) by Edward Bevan, 1870). *Nature* 3: 385, 16 March 1871.
190. Discussion/N (of "On Additions to the Atlantic Coleoptera," a paper by T. Vernon Wollaston read at the ESL meeting of 20 March 1871). *Proc. ESL,* 1871: xiii (xi–xiii), 1871.
191. Review (of *A Monograph of the Alcedinidae: or, Family of Kingfishers* by Richard B. Sharpe, 1868–1871). *Nature* 3: 466–67, 13 April 1871.
192. Review (of *A History of the Birds of Europe, Including all the Species Inhabiting the Western Palaearctic Region, Part I* by Richard B. Sharpe & Henry E. Dresser, 1871). *Nature* 3: 505, 27 April 1871.
193. On the Attitude of Men of Science towards the Investigators of Spiritualism, in *The Year-book of Spiritualism for 1871* ed. by Hudson Tuttle & J. M. Peebles (William White & Co., Boston), April 1871: 28–31.
194. Review (of *British Insects. A Familiar Description of the Form, Structure, Habits and Transformations of Insects* by E. F. Staveley, 1871). *Nature* 4: 22–24, 11 May 1871.
195. Review (of *Natural History of the Azores, or Western Islands* by Frederick Du Cane Godman, 1870). *Academy* 2: 266–67, 15 May 1871: no. 24.
196. Review (of *Notes of a Naturalist in the Nile Valley and Malta* by Andrew Leith Adams, 1870). *Academy* 2: 336–37, July 1871: no. 27.
197. Review (of *Journeys in North China, Manchuria, and Eastern Mongolia; with some Account of Corea* by Rev. Alexander Williamson, 1870). *Academy* 2: 337, 1 July 1871: 110. 27.
198. Bastian on the Origin of Life (review of *The Modes of Origin of Lowest Organisms: Including a Discussion of the Experiments of M. Pasteur, and a Reply to some Statements by Professors Huxley and Tyndall* by H. Charlton Bastian, 1871). *Nature* 4: 178–79, 6 July 1871.
199. LTTE (one of two printed as "A New View of Darwinism"). *Nature* 4: 181 (180–81), 6 July 1871.
200. LTTE (one of several printed as "Mr. Howorth on Darwinism"). *Nature* 4: 221 (221–22), 20 July 1871.
201. LTTE (one of several printed as "Recent Neologisms"). *Nature* 4: 222, 20 July 1871.
202. Review (of *At Last: 4 Christmas in the West Indies* by Charles Kingsley, 1871). *Nature* 4: 282–84, 10 Aug. 1871.
203. Review (of *Insects at Home: Being a Popular Account of British Insects, their Structures, Habits and Transformations* by Rev. John G. Wood, 1871). *Nature* 5: 65–68, 23 Nov. 1871.

204. *Reply to Mr. Hampden's Charges against Mr. Wallace* (privately printed pamphlet responding to John Hampden's slanderous remarks): see *My Life* (Vol. 2, 364–76), J. J. Tiver, London, Nov. 1871: 1–8.
205. Review (of *A Synonymic Catalogue of Diurnal Lepidoptera* by William F. Kirby, 1871). *Academy* 2: 538, 1 Dec. 1871: no. 37.
206. Letter (to James Edmunds, included as part of a letter by Edmunds entitled "The Paper by Dr. Edmunds"). *The Spiritualist* (London) 1(28): 221b (221b–c), 15 Dec. 1871.

1872

207. The President's Address (read at the ESL's annual meeting of 22 Jan. *1872*). *Proc. ESL,* 1871: li–lxxv.
208. Dr. Carpenter and Psychic Force (LTTE originally submitted to, but not printed by, *The Daily Telegraph). The Spiritualist* (London) 2(2): 13a, 15 Feb. 1872: no. 30.
209. Review (of review of *Primitive Culture: Researches into the Development of Mythology, Philosophy, Religion, Art, and Custom* by Edward B. Tylor, 1871). *Academy* 3: 69–71, 15 Feb. 1872: no. 42.
210. Ethnology and Spiritualism (LTTE responding to Edward Tylor's comments in *Nature* 5: 343). *Nature* 5: 363–64, 7 March 1872.
211. Review (of *The Debatable Land Between this World and the Next* by Robert Dale Owen, 1871). *Q. J. Science* 2 (n.s.; 9, o.s.): 237–47, April 1872: no. 34, o.s.
212. The Last Attack on Darwinism (review of *An Exposition of Fallacies in the Hypothesis of Mr. Darwin* by Charles R. Bree, 1872). *Nature* 6: 237–39, 25 July 1872.
213. Review (of *The Beginnings of Life: Being some Account of the Nature, Modes of Origin, and Transformations of Lower Organisms* by H. Charlton Bastian, 1872).1. *Nature* 6: 284–87 (8 Aug.1872) / II. 6: 299–303, 15 Aug. 1872.
214. Discussion/N (of "On the Diversity of Evolution under Uniform External Conditions," a paper by Rev. John T. Gulick read in Brighton at the 19 Aug. 1872 meeting of the Dept. of Zoology and Botany, Section D, Biology, of the BAAS). *Nature* 6: 407 (406–7), 12 Sept. 1872.
215. Discussion/N (of "On the Scientific Value of Beauty," a paper by F. T. Mott read in Brighton at the 20 Aug. 1872 meeting of the Dept. of Zoology and Botany, Section D, Biology of the BAAS). *Athenaeum* no. 2340: 275, 31 Aug. 1872.
216. LTTE (one of two printed as "Ocean Circulation"). *Nature* 6: 328–29, 22 Aug. 1872.
217. Review (of *The Birds of Europe, Parts 11 & 12* by Richard B. Sharpe & Henry E. Dresser). *Nature* 6: 390–91, 12 Sept. 1872.
218. Instinct (LTTE). *The Spiritualist* (London) 2(9): 70c, 15 Sept. 1872: no. 37.
219. Houzeau on the Faculties of Man and Animals (review of *Etudes sur les Facultes Mentales des Animaux Compare'es a celles de l'Homme, par un Voyageur Naturaliste* by Jean Charles Houzeau, 1872). *Nature* 6: 469–71, 10 Oct. 1872.
220. Misleading Cyclopaedias (LTTE). *Nature* 7: 68, 28 Nov. 1872: no. 161.
221. Spiritualism and Science (LTTE). *The Times* (London) no. 27578:10e, 4 Jan. 1873.
222. Review (of *The Expression of the Emotions in Man and Animals* by Charles Darwin, 1872). *Q. J. Science* 3 (n.s.; 10, o.s.): 113–18, Jan. 1873: no. 37.

1873

223. Modern Applications of the Doctrine of Natural Selection (review of *Physics and Politics; or, Thoughts on the Application of the Principles of "Natural Selection" and "Inheritance" to Political Society* by Walter Bagehot, 1872; and *Histoire des sciences et des savants depuis deux siècles suivie d'autres études sur les sujets scientifiques en particulier sur la selection dans l'espèce humaine* by Alphonse de Candolle, 1873). *Nature* 7: 277–79, 13 Feb. 1873: no. 172.

224. LTTE (one of several printed as "Inherited Feeling"). *Nature 7:* 303, 20 Feb. 1873: no. 173.
225. Review (of *Harvesting Ants and Trap-door Spiders. Notes and Observations on their Habits and Dwellings* by J. Traherne Moggridge, 1873). *Nature* 7: 337, 6 March 1873: no. 175.
226. Cave-deposits of Borneo (LTTE introducing remarks by A. Everett). *Nature* 7: 461–62, 17 April 1873: no. 181.
227. Disestablishment and Disendowment: with a Proposal for a really National Church of England. *Macmillan's Mag* 27: 498–507, April 1873: no. 162.
228. LTTE (one of two printed as "East India Museum"). *Nature* 8: 5 (5–6), 1 May 1873: no. 183.
229. Perception and Instinct in the Lower Animals (LTTE). *Nature* 8: 65–66, 22 May 1873: no. 186.
230. Perception and Instinct in the Lower Animals (LTTE). *Nature* 8: 302, 14 Aug. 1873: no. 198.
231. Review (of *Advanced Text-book of Physical Geography* (2nd ed.) by David Page, 1873). *Nature* 8: 358–61, 4 Sept. 1873: no. 201.
232. Review (of *A History of the Birds of Europe, Parts 18, 19* & 20 by Henry E. Dresser). *Nature* 8: 380–81, 11 Sept. 1873: no. 202.
233. Free-trade Principles and the Coal Question (LTTE). *The Daily News* (London) no. 8546: 6a–e, 16 Sept. 1873.
234. African Travel (review of *The Lands of Cazembe. Lacerda's Journey to Cazembe in 1798, etc.* translated and annotated by Capt. R. F. Burton and published by the Royal Geographical Society, 1873; and *The African Sketch Book* by Winwood Reade, 1873). *Nature* 8: 429–31, 25 Sept. 1873: no. 204.
235. Lyell's *Antiquity of Man* (review of *The Geological Evidences of the Antiquity of Man,* 4th ed., by Charles Lyell, 1873). *Nature* 8: 462–64, 2 Oct. 1873: no. 205.
236. A Primeval Race (review of *A Phrenologist amongst the Todas; or, The Study of a Primitive Tribe in South India: History, Character, Customs, Religion, Infanticide, Polyandry, Language* by William E. Marshall, 1873; anonymous, but in Wallace's writing style and listed in the bibliography of Marchant, 1916). *Athenaeum* no. 2403: 624–25, 15 Nov. 1873.
237. Meyer's Exploration of New Guinea (LTTE). *Nature* 9: 102, 11 Dec. 1873: no. 215.
238. Limitation of State Functions in the Administration of Justice. *Contemporary Rev.* 23: 43–52, Dec. 1873.

1874

239. The Origin of Man and of Civilisation (review of *Man and Apes: An Exposition of Structural Resemblances and Differences Bearing upon Questions of Affinity and Origin* by St. George Mivart, 1873; and "On the Origin of Savage Life: Opening Address Read before the Literary and Philosophical Society of Liverpool, Oct. 6th, 1873" by Albert J. Mott, President, *Academy* 5: 66–67, 17 Jan. 1874: no. 89.
240. Review (of *The Naturalist in Nicaragua by* Thomas Belt, 1873). *Nature* 9: 218–21, 22 Jan. 1874: no. 221.
241. Review (of *The Object and Method of Zoological Nomenclature* by David Sharp, 1873). *Nature* 9: 258–60, 5 Feb. 1874: no. 223.
242. LTTE (one of two printed as "Animal Locomotion"). *Nature* 9: 301, 19 Feb. 1874: no. 225.
243. Animal Locomotion (LTTE). *Nature* 9: 403, 26 March 1874: no. 230.
244. Review (of *Darwinism and Design; or Creation by Evolution* by George St. Clair, 1873; anonymous, but in Wallace's writing style and appears at time listed in the bibliography of Marchant, 1916). *Spectator* 47: 535–36, 25 April 1874: no. 2391.
245. A Defence of Modern Spiritualism. *Fortnightly Rev.* 15 (n.s.; 21, o.s.): 630–57, 1

May 1874: no. 89, n.s. / A Defence of Modern Spiritualism. Part II. Spirit-photographs.15 (n.s.; 21, o.s.): 785–807, 1 June 1874: no. 90, n.s.

246. Migration of Birds (LTTE). *Nature* 10: 459, 8 Oct. 1874: no. 258.
247. Automatism of Animals (LTTE). *Nature* 10: 502–3, 22 Oct. 1874: no. 260.
248. On the Arrangement of the Families Constituting the Order Passeres. *Ibis* 4 (3rd s.): 406–16, Oct. 1874: no. 16, 3rd s.

1875

249. Review (of *Supplement to Harvesting Ants and Trap-door Spiders* by J. Traherne Moggridge, 1874). *Nature* 11: 245–46, 28 Jan. 1875: no. 274.
250. Acclimatisation, in *Encyclopaedia Britannica* (9th ed., 25 vols., Adam & Charles Black, Edinburgh, 1875–1889), Vol. 1: 84–90, Jan. 1875.
251. Review (of *A History of the Birds of Europe, Parts 35 & 36* by Henry E. Dresser). *Nature* 11: 485–86, 22 April 1875: no. 286.
252. Review (of *Wanderings in the Interior of New Guinea* by John A. Lawson, 1875). *Nature* 12: 83–84, 3 June 1875: no. 292.
253. Wallace and Spiritualism (note/N including extract from letter to Henry S. Olcott concerning Wallace's continuing interest in spiritualism). *Banner of Light* (Boston) 37(10): 2e, 5 June 1875.
254. Review (of *The Handy-book of Bees; Being a Practical Treatise on their Profitable Management,* 2nd ed., by A. Pettigrew, 1875; and A *Manual of Bee-keeping* by John Hunter, 1874). *Nature* 12: 395, 9 Sept.1875: no. 306.

1876

255. Review (of *Lessons from Nature, as Manifested in Mind and Matter* by St. George Mivart, 1876). *Academy* 9: 562–63, 10 June 1876: no. 214, n.s. / 9: 587–88, 17 June 1876: no. 215, n.s.
256. Review (of *Notes on Collecting and Preserving Natural-history Objects* ed. by John E. Taylor, 1876). *Nature* 14: 168, 22 June 1876: no. 347.
257. A Sitting with Or. Slade. *The Spiritualist* (London) 9(4): 42, 25 Aug. 1876.
258. *Geographical Distribution of Animals* (LTTE correcting error made by a reviewer). *Athenaeum* no. 2549: 311, 2 Sept. 1876.
259. Address (given in Glasgow on 6 Sept. 1876 as President of Section D, Biology, of the BAAS; divided into parts entitled "On some Relations of Living Things to their Environment" and "Rise and Progress of Modern Views as to the Antiquity and Origin of Man"), in "Notices and Abstracts of Miscellaneous Communications to the Sections" portion of the *Rept. BAAS* 46 (1876)(John Murray, London), 1877: 100–119.
260. The British Association at Glasgow (text of the discussion of Prof. William F. Barrett's paper on spiritualism read at the 11 Sept. 1876 meeting of the Dept. of Anthropology, Section D, Biology, of the BAAS, with the remarks of Wallace as chair of the session). *The Spiritualist* (London) 9(8): 88–94, 22 Sept. 1876.
261. LTTE (dated 18 Sept., Glasgow; one of several printed as "A Spirit Medium"). *The Times* (London) no. 28738: 4f (4f), 19 Sept. 1876.
262. Erratum in Mr. Wallace's Address (LTTE). *Nature* 14: 473, 28 Sept. 1876: no. 361.
263. Evidence of Mr. A. R. Wallace, President of the Biological Section of the British Association for the Advancement of Science (transcript of Wallace's testimony in the Henry Slade fraud trial; part of an article/N entitled "Evidence in Defence of Dr. Slade"). *The Spiritualist* (London) 9(14): 161, 164 (160–61, 164–65), 3 Nov. 1876.
264. Mr. Wallace and his Reviewers (LTTE). *Nature* 15: 24, 9 Nov. 1876: no. 367.
265. Dr. Carpenter on Mesmerism, etc. (LTTE). *The Daily News* (London) no. 9559: 2c, 11 Dec. 1876.

266. Dr. Carpenter on Spiritualism (LTTE). *The Daily News* (London) no. 9566: 3f, 19 Dec. 1876.
267. Mr. G. H. Lewes's Exposure of Mrs. Hayden (LTTE). *Spectator* 49: 1608, 23 Dec. 1876: no. 2530.
268. Review (of *The Races of Man and their Geographical Distribution,* 2nd ed., from the German of Oscar Peschel, 1876). *Nature* 15: 174–76, 28 Dec. 1876: no. 374.
269. Comments (brief remarks specially solicited by Mr. Harris for inclusion in his work) in *A Philosophical Treatise on the Nature and Constitution of Man* by George Harris (George Bell & Sons, London, 1876): 372, 373 & 374 (Volume I), and 254 (Volume II).

1877

270. Glacial Drift in California (LTTE). *Nature* 15: 274–75, 25 Jan. 1877: no. 378.
271. The "Hog-wallows" of California (LTTE). *Nature* 15: 431–32, 15 March 1877: no 385.
272. The Comparative Antiquity of Continents, as Indicated by the Distribution of Living and Extinct Animals (paper read at the RGS meeting of 25 June 1877). *Proc. RGS* 21 (1876–1877), no. 6: 505–34 (followed by an account of related discussion on pp. 534–35), 19 Sept. 1877.
273. Review (of *Mesmerism, Spiritualism, etc., Historically & Scientifically Considered* by William B. Carpenter, 1877). *Q. J. Science* 7 (n.s.; 14, o.s.): 391–416, July 1877.
274. Spiritualism and Conjurors (LTTE). *The Spiritualist* (London) 11(7): 78, 17 August 1877: no. 260.
275. The Colours of Animals and Plants. I.—The Colours of Animals. *Macmillan's Mag* 36: 384–408, Sept. 1877: no. 215 / The Colours of Animals and Plants. II.—The Colours of Plants. 36: 464–71, Oct. 1877: no. 216.
276. Slate-writing Extraordinary (LTTE). *Spectator* 50: 1239–40, 6 Oct. 1877: no. 2571.
277. Hartlaub's Birds of Madagascar (review of *Die Vogel Madagascars und der Benachbarten Inselgruppen. Ein Beitrag zur Zoologie der Athiopischen Region.* by Gustav Hartlaub, 1877; anonymous, but obliquely referred to by Alfred Newton (*Nature* 17: 9–10). *Nature* 16: 498–99, 11 Oct. 1877: no. 415.
278. The Zoological Relations of Madagascar and Africa (LTTE). *Nature* 16: 548, 25 Oct. 1877: no. 417.
279. LTTE (concerning remarks by William B. Carpenter; one of several printed as "Mr. Wallace and Reichenbach's Odyle"). *Nature* 17: 8 (8–9), 1 Nov. 1877: no. 418.
280. LTTE (concerning remarks by William B. Carpenter; one of several printed as "The Radiometer and its Lessons"). *Nature* 17: 44 (43–44), 15 Nov. 1877: no. 420.
281. Bees Killed by Tritoma (LTTE). *Nature* 17: 45, 15 Nov. 1877: no. 420.
282. Humming-birds. *Fortnightly Rev.* 22 (n.s.; 28, o.s.): 773–91, 1 Dec. 1877: no. 132, n.s.
283. The Comparative Richness of Faunas and Floras Tested Numerically (LTTE). *Nature* 17: 100–101, 6 Dec. 1877: no. 423.
284. Mr. Crookes and Eva Fay (LTTE concerning William B. Carpenter's comments on the medium Eva Fay). *Nature* 17: 101, 6 Dec. 1877: no. 423.
285. Test Materialisation Seance with Mr. W. Eglinton (signed "Alfred R. Wallace, William Tebb, William Williams Clark"). *The Spiritualist* (London) 11(23): 272, 7 Dec. 1877.
286. Psychological Curiosities of Scepticism. A Reply to Dr. Carpenter. *Fraser's Mag* 16 (n.s.): 694–706, Dec. 1877: no. 96.

1878

287. Northern Affinities of Chilian Insects (LTTE). *Nature* 17: 182–83, 3 Jan. 1878: no. 427.

288. The Curiosities of Credulity. *Athenaeum* no. 2620: 54–55, 12 Jan. 1878.
289. Distribution (by Alfred Russel Wallace and William T. Thiselton Dyer), in *Encyclopaedia Britannica* (9th ed., 25 vols., Adam & Charles Black, Edinburgh, 1875–1889), Vol. 7: 267–90 (pp. 267–86 by Wallace; pp. 286–90 by Thiselton-Dyer), Jan. 1878.
290. Psychological Curiosities of Credulity (LTTE concerning remarks by William B. Carpenter). Athenaeum no. 2623: 157, 2 Feb. 1878.
291. The Climate and Physical Aspects of the Equatorial Zone. Essay I of *TNOE:* 1–26, April 1878.
292. Equatorial Vegetation. Essay II of *TNOE:* 27–68, April 1878.
293. Animal Life in the Tropical Forests. Essay III of *TNOE:* 69–123, April 1878.
294. A Twenty Years' Error in the Geography of Australia (LTTE). *Nature* 18: 193–94, 20 June 1878: no. 451.
295. Epping Forest and How to Deal with it. *Fortnightly Rev.* 24 (n.s.; 30, o.s.): 628–45, 1 Nov. 1878: no. 143, n.s.
296. Remarkable Local Colour-variation in Lizards (LTTE introducing remarks by the Baron de Basterot on color-variation in Capri lizards). *Nature* 19: 4, 7 Nov. 1878: no. 472.
297. *Scilla autumnalis* in Essex (note on distribution). *J. of Botany* 7 (n.s.): 346, Nov. 1878.
298. LTTE (one of two printed as "The Formation of Mountains"). *Nature* 19: 121, 12 Dec. 1878: no. 476.

1879

299. Discussion/N (of paper on the Australasian-Pacific races of man by S. J. Whitmee read at the Anthropological Institute meeting of 7 Jan. 1879). *J. Anthro. Instit. Great Britain & Ireland* 367–68 (360–69), 1879.
300. Discussion/N (of paper on New Guinea tribes by W. G. Lawes read at the Anthropological Institute meeting of 7 Jan. *1879). J. Anthro. Instit. Great Britain & Ireland* 8: 377 (369–77), 1879.
301. The Formation of Mountains (LTTE). *Nature* 19: 244, 16 Jan. 1879: no. 481.
302. Review (of *Leisure-time Studies, Chiefly Biological: A Series of Essays and Lectures* by Andrew Wilson, 1878). *Nature* 19: 286–87, 30 Jan. 1879: no. 483.
303. LTTE (one of two printed as "The Formation of Mountains"). *Nature* 19: 289, 30 Jan. 1879: no. 483.
304. New Guinea and its Inhabitants. *Contemporary Rev.* 34: 421–41, Feb. 1879.
305. Animals and their Native Countries. *Nineteenth Century* 5: 247–59, Feb. 1879: no. 24.
306. Organisation and Intelligence (review of *Habit and Intelligence: Essays on the Laws of Life and Mind,* 2nd ed., by Joseph John Murphy, 1879; and *Life and Habit* by Samuel Butler, 1877). *Nature* 19: 477–80, 27 March 1879: no. 491.
307. Colour in Nature (review of *The Colour Sense, its Origin and Development; An Essay in Comparative Psychology* by Grant Allen, 1879). *Nature* 9: 501–505, 3 April 1879: no. 492.
308. Review (of *The Evolution of Man: A Popular Exposition of the Principal Points of Human Ontogeny and Phylogeny* by Ernst Haeckel, 1879). *Academy* 15: 326–27, 12 April 1879: no. 362, n.s./15: 351–52, 19 April 1879: no. 363, n.s.
309. Reciprocity the True Free Trade. *Nineteenth Century* 5: 638–49, April 1879: no. 26.
310. Waterton's Life and Travels (review of *Wanderings in South America, the Northwest of the United States, and the Antilles, in the Years 1812, 1816, 1820, and 1824* (new ed., ed. by Rev. John G. Wood) by Charles Waterton, 1879). *Nature* 19: 576–78, 24 April 1879: no. 495.
311. Discussion (of comments by Grant Allen referencing Allen's letter as "Colour in Nature"). *Nature* 19: 581 (580–81), 24 April 1879: no. 495.

312. Did Flowers Exist during the Carboniferous Epoch? (LTTE). *Nature* 19: 582, 24 April 1879: no. 495.
313. Bounties and Countervailing Duties (LTTE followed by reply by the Editor). *Spectator* 52: 531, 26 April 1879: no. 2652.
314. Review (of *Evolution, Old and New or, The Theories of Buffon, Dr. Erasmus Darwin, and Lamarck, as Compared with that of Mr. Charles Darwin* by Samuel Butler, 1879). *Nature* 20: 141–44, 12 June 1879: no. 502.
315. A Few Words in Reply to Mr. Lowe. *Nineteenth Century* 6: 179–81, July 1879: no. 29.
316. Glacial Epochs and Warm Polar Climates (running title for notice of books on glacial theory by James Croll, James Geikie, George S. Nares, Oswald Heer & Julius Payer; anonymous, but referred to in a letter from Wallace to Darwin dated 9 Jan. 1880) (Merchant, 1916, 250). *Quarterly Rev.* 148: 119–35, July 1879: no. 295.
317. Review (of *Scientific Lectures* by Sir John Lubbock, 1879). *Nature* 20: 335–36, 7 Aug. 1879: no. 510.
318. Review (of *The Natural History of the Agricultural Ant of Texas* by Henry Christopher McCook, 1879). *Nature* 20: 501, 25 Sept. 1879: no. 517.
319. Wallace's *Australasia* (LTTE). *Nature* 20: 625–26, 30 Oct. 1879: no. 522.
320. The Protective Colours of Animals, in *Science for All* ed. by Robert Brown (5 vol., Cassell, Petter, Galpin & Co., London, Paris & New York, 1877–1882), Vol. 2: 128–37, Oct. 1879.
321. Protective Mimicry in Animals, in *Science for All* ed. by Robert Brown (5 vol., Cassell, Petter, Galpin & Co., London, Paris & New York, 1877–1882), Vol. 2: 284–96, Oct. 1879.
322. Discussion/N (of paper on a Dutch expedition to Sumatra by P. J. Veth read at the RGS meeting of 10 Nov. 1879). *Proc. RGS & Monthly Record of Geography* 1(12) (new monthly s.): 775–77 (759–77), Dec. 1879.

1880

323. Discussion/N (of "The Future of Epping Forest," a paper by William Paul read at the Society of Arts meeting of 28 Jan. 1880). *Journal of the Society of Arts* (London) 28: 184 (177–85), 30 Jan. 1880.
324. Islands, as Illustrating the Laws of the Geographical Distribution of Animals (from a lecture delivered in the Hulme Town Hall, Manchester, 15 Oct. 1879), in *Science Lectures for the People.* Eleventh Series (Manchester & London, April 1880), no. 1: 1–18, Jan. 1880.
325. Popular Natural History (review of *Animal Life; Being a Series of Descriptions of the Various Sub-kingdoms of the Animal Kingdom* by E. Perceval Wright, 1879). *Nature* 21: 232–35, 8 Jan. 1880: no. 532.
326. The Origin of Species and Genera. *Nineteenth Century* 7: 93–106, Jan. 1880: no. 35.
327. Wallace's *Australasia* (LTTE responding to published criticism). *Nature* 21: 562, 15 April 1880: no. 546.
328. Two Darwinian Essays (review of *Studies in the Theory of Descent Pt. I* by August Weismann, 1880; and *Degeneration: A Chapter in Darwinism* by E. Ray Lankester, 1880). *Nature* 22: 141–42, 17 June 1880: no. 555.
329. Dr. Croll's Excentricity Theory (LTTE). *Geological Mag* 7 (Decade II, n.s.), no. 6: 284–85, June 1880.
330. European Caddis-flies (review of *A Monographic Revision and Synopsis of the Trichoptera of the European Fauna* by Robert McLachlan, 1874–1880). *Nature* 22: 314–15, 5 Aug. 1880: no. 562.
331. English and American Bee-keeping (review of books on bee-keeping by Henry Taylor, 1880, James Robinson, 1880, and Albert J. Cook, 1879). *Nature* 22: 433–34, 9 Sept. 1880: no. 567.

332. The Emotions and Senses of Insects (review of *Insect Variety, its Propagation and Distribution* by Archibald H. Swinton, 1880). *Academy* 18: 294–96, 23 Oct. 1880: no. 442, n.s.
333. How to Nationalize the Land: A Radical Solution of the Irish Land Problem. *Contemporary Rev.* 38: 716–36, Nov. 1880.
334. Review (of *Siberia in Europe: A Visit to the Valley of the Petchora, in North-East Russia. With Descriptions of the Natural History, Migrations of Birds, etc.* by Henry Seebohm, 1880). *Academy* 18: 408–9, 4 Dec. 1880: no. 448, n.s.
335. Geological Climates (LTTE). *Nature* 23: 124, 9 Dec. 1880: no. 580.
336. Review (of *New Guinea: What I Did and What I Saw* by Luigi M. D'Albertis, 1880). *Nature* 23: 152–55, 16 Dec. 1880: no. 581/23: 175–78 / 23 Dec. 1880: no. 582.
337. Climates of Vancouver Island and Bournemouth (LTTE). *Nature* 23: 169, 23 Dec. 1880: no. 582.
338. Correction of an Error in *Island Life* (LTTE). *Nature* 23: 195, 30 Dec. 1880: no. 583.

1881

339. Discussion/N (of various papers and motions presented by others at meetings of the Club, 1879–1880). *Proceedings and Transactions of the Croydon Microscopical and Natural History Club* from February 20th, 1878, to January 19th, 1881 (1881): vii, xx, xxv, xxxiv, xxxvi, xliii, xliv, xlv, lv–lvii, lviii.
340. On the Peculiar Species of the British Fauna and Flora (lengthy abstract/U of a paper read at the Club meeting of 17 March 1880). *Proceedings and Transactions of the Croydon Microscopical and Natural History Club* from February 20th, 1878, to January 19th, 1881 (1881): 58–60.
341. LTTE (one of two printed as "Geological Climates"). *Nature* 23: 217, 6 Jan. 1881: no. 584.
342. LTTE (one of two printed as "Geological Climates"). *Nature* 23: 266–67, 20 Jan. 1881: no. 586.
343. Review (of *Anthropology; An Introduction to the Study of Man and Civilization* by Edward B. Tylor, 1881). *Nature* 24: 242–45, 14 July 1881: no. 611.
344. Review (of *Studies in the Theory of Descent Pt. II* by August Weismann, 1881). *Nature* 24: 457–58, 15 Sept.1881: no. 620.
345. LTTE (written at the request of the Editor; one of two letters printed as "Nationalisation of the Land"). *Mark Lane Express* 51: 1351 a–b (1351 a–c), 3 Oct. 1881.
346. Abstract of Four Lectures on the Natural History of Islands (from a series of lectures given at Rugby School on 6, 13, 20, & 27(?) Oct. 1881; consists of four summaries/U composed of excerpts). *Report Rugby School Nat. Hist. Soc., 1881:* 1–17 ("Lecture I. Sea and Land," 1–4; "Lecture II. Recent Continental Islands," 4–9; "Lecture III. Ancient Continental Islands," 9–12; "Lecture IV. Oceanic Islands," 12–17), 1882.
347. Nationalisation of the Land (LTTE). *Mark Lane Express* 51: 1383a–c, 10 Oct. 1881.
348. Review (of *The Head-hunters of Borneo: A Narrative of Travel up the Mahakkam and down the Barito; also Journeyings in Sumatra* by Carl Bock, 1881). *Nature* 25: 3–4, 3 Nov. 1881: no. 627.
349. LTTE (one of several printed as "Nationalisation of the Land"). *Mark Lane Express* 51: 1544a–b, 14 Nov. 1881.
350. The Land Question (LTTE). *The Times* (London) no. 30362: 11f, 26 Nov. 1881.
351. The Land Question (LTTE). *The Times* (London) no. 30371: 10d, 7 Dec. 1881.

1882

352. The Land Question (LTTE). *The Times* (London) no. 30394: 3c, 3 Jan. 1882.
353. Land Nationalisation Society. Conference this Day. (note/N summarizing the meeting and Wallace's address to it). *The Echo* (London) no. 4075: 3c, 16 Jan. 1882.

354. Review (of *Vignettes from Nature* by Grant Allen, 1881). *Nature* 25: 381–82, 23 Feb. 1882: no. 643.
355. Review (of *The Story of our Museum; Showing How We Formed It, and What It Taught Us* by Rev. Henry Housman, 1881). *Nature* 25: 407–8, 2 March 1882: no. 644.
356. Monkeys. *Contemporary Rev.* 41: 417–30, March 1882.
357. Review (of *Rhopalocera Malayana: A Description of the Butterflies of the Malay Peninsula* by William L. Distant, 1882). *Nature* 26: 6–7, 4 May 1882: no. 653.
358. Review (of *Studies in the Theory of Descent Pt. 111* by August Weismann, 1882). *Nature* 26: 52–53, 18 May 1882: no. 655.
359. Dr. Fritz Muller on some Difficult Cases of Mimicry. *Nature* 26: 86–87, 25 May 1882: no. 656.
360. Review (of *Worked Examination Questions in Plane Geometrical Drawing* by Frederick E. Hulme, 1882). *Nature* 26: 103, 1 June 1882: no. 657.
361. Review (of *Wanderings South and East* by Walter Coote, 1882; and *Pioneering in the Far East, and Journeyings to California in 1849, and the White Sea in 1878* by Ludwig Verner Helms, 1882; anonymous, but attributed to Wallace in the Index for Vol. 26). *Nature* 26: 476, 14 Sept. 1882: no. 672.
362. LTTE (one of several printed as "Materialisation and Exposures"). *Light* (London) 2: 447–48, 7 Oct. 1882: no. 92.
363. Review (of *The Sportman's Handbook to Practical Collecting, Preserving, and Artistic Setting up of Trophies and Specimens,* 2nd ed., by Rowland Ward, 1882). *Nature* 27: 146, 14 Dec. 1882: no. 685.

1883

364. The Debt of Science to Darwin. *Century Mag.* 25(3): 420–32, Jan. 1883.
365. LTTE (one of two printed as "Difficult Cases of Mimicry"). *Nature* 27: 481–82, 22 March 1883: no. 699.
366. Letter (sent to a LNS meeting in Bristol, responding to Alfred Marshall's lectures on Henry George's *Progress and Poverty). Western Daily Press* (Bristol), 17 March 1883: 3.
367. LTTE (responding to a letter by Alfred Marshall, printed 17 March 1883, responding to earlier reference). *Western Daily Press* (Bristol), 23 March 1883: 3.
368. On the Value of the "Neoarctic" as One of the Primary Zoological Regions (LTTE). *Nature* 27: 482–83, 22 March 1883: no. 699.
369. President's Address, Summarised (summary/U of Wallace's address to the second annual meeting of the LNS, 27 June 1883), in *Report of the LNS. 1881–3.* (LNS, London), 1883: 5–6.
370. Review (of *Ants and their Ways* by Rev. W. Farren White, 1883). *Nature* 28: 293–94, 26 July 1883: no. 717.
371. Review (of *Manual of Taxidermy; A Complete Guide in Collecting and Preserving Birds and Mammals* by Charles J. Maynard, 1883). *Nature* 28: 317–18, 2 Aug. 1883: no. 718.
372. *How Land Nationalisation will Benefit Householders, Labourers, and Mechanics.* LNS Tract no. 3: LNS, London, n.d. (probably published in 1882 or 1883): 1–7.
373. The "Why" and the "How" of Land Nationalisation. I. *Macmillan's Mag* 48: 357–68, Sept. 1883: no. 287 / II. 48: 485–93, Oct. 1883: no. 288.
374. Lord Jersey on Land Nationalization (LTTE). *The Times* (London) no. 30964: 12a, 30 Oct. 1883.
375. Mr. Wallace's Reply to Mr. T. Mellard Reade on the Age of the Earth (LTTE). *Geological Mag.* 10 (Decade 11, n.s.), no. 10: 478–80, Oct. 1883.
376. Letter (of support to the Berne International Anti-Vaccination Congress held on 28

Sept. 1883 "and following days"). *The Vaccination Inquirer and Health Review* 5: 160, Nov. 1883: no. 56.

377. LTTE. Mr. S. Smith, M. P., on Land Nationalisation. Liverpool Daily Post no. 8866: 7h–i, 4 Dec. 1883. Reprinted in *The Nationalisation of the Land* by Samuel Smith (Kegan Paul, Trench & Co.), 1884: 39–43.

1884

378. LTTE (one of several printed as "Mr. [Henry] George's Theories"). *The Times* (London) no. 31042: 3a–b, 29 Jan. 1884.
379. The Morality of Interest—The Tyranny of Capital. *The Christian Socialist* no. 10: 150–51, March 1884.
380. *How to Experiment in Land Nationalisation.* LNS Tract no. 8: LNS, London, n.d. (probably published in late 1883 or 1884); 1–3.
381. Two Bee Books (review of *A Collection of Papers on Bee-keeping in India* published by the Dept. of Revenue and Agriculture, Government of India, 1883; and *The Honey-bee; Its Nature Homes, and Products* by William H. Harris, 1884). *Nature* 31: 1–2, 6 Nov. 1884: no. 784.
382. Review (of *Elements de Mecanique, avec de Nombreux Exercises* by "F.I.C.") *Nature* 31: 78, 27 Nov. 1884: no. 787.
383. *(To Members of Parliament and Others.) Forty-five Years of Registration Statistics, Proving Vaccination to be both Useless and Dangerous.* E. W. Allen, London, Dec. 1884 or Jan. 1885; 1–38.

1885

384. How to Cause Wealth to be more Equally Distributed (from a paper read in Prince's Hall, Piccadilly, on 30 Jan. 1885), in *Industrial Remuneration Conference; The Report of the Proceedings and Papers* (Cassell & Co., Ltd., London, Paris, New York, & Melbourne), June 1885: 368–92.
385. The Industrial Remuneration Conference (LTTE "correcting misapprehension" concerning the aims and methods of the LNS). *The Times* (London) no. 31360: 6d, 3 Feb. 1885.
386. Review (of *Differential Calculus for Beginners, with a Selection of Easy Examples* by Alexander Knox, 1885). *Nature* 31: 527, 9 April 1885: no. 806.
387. The Colours of Arctic Animals (LTTE). *Nature* 31: 552, 16 April 1885: no. 807.
388. Modern Spiritualism. Are its Phenomena in Harmony with Science? *The Sunday Herald* (Boston) 26 April 1885: 9c–d.
389. President's Address (to the fourth annual meeting of the LNS, 13 May 1885), in *Report of the LNS. 1884–85.* (LNS, London), 1885: 5–15.
390. Review (of *A Naturalist's Wanderings in the Eastern Archipelago, a Narrative of Travel and Exploration from 1878 to 1883* by Henry O. Forbes, 1885). *Nature* 32: 218–20, 9 July 1885: no. 819.
391. The *Journal of Science* on Spiritualism (reply to criticism by "R. M. N."). *Light* (London) 5: 327–28, 11 July 1885: no. 236.
392. Harmony of Spiritualism and Science (reply to criticism by Frederick F. Cook). *Light* (London) 5: 352, 25 July 1885: no. 238.
393. Review (of *The Wanderings of Plants and Animals from their First Home* by Victor Hehn, 1885). *Nature* 33: 170–71, 24 Dec. 1885: no. 843.
394. *State-tenants versus Freeholders.* LNS Tract no. 15: The LNS, London, n.d. (probably published in the second half of 1885 or the first half of 1886).

1886

395. President's Address (summary/U of Wallace's address to the fifth annual meeting of the LNS, 29 June 1886). In *Report of the LNS. 1885–86.* (LNS, London), 1886: 3–7.

396. The Depression of Trade, its Causes and its Remedies, in *The Claims of Labour, A Course of Lectures Delivered in Scotland in the Summer of 1886, on Various Aspects of the Labour Problem* by John Burnett and others (Cooperative Printing Co., Ltd., Edinburgh), Dec. 1886: 112–54.
397. Central American Entomology (review *of Biologia Centrali-Americana. Insecta: Coleoptera Vol. 111, Part 2, Malacodermata* by Rev. Henry Stephen Gorham, 1880–86, and *Vol. V, Longicornia* by Henry W. Bates and *Bruchides* by David Sharp, 1879–86). *Nature* 34: 333–34, 12 Aug. 1886: no. 876.
398. Romanes versus Darwin. An Episode in the History of the Evolution Theory. *Fortnightly Rev.* 40 (n.s.; 46, o.s.): 300–316, 1 Sept. 1886: no. 237, n.s.
399. LTTE (one of several printed as "Physiological Selection and the Origin of Species"). *Nature* 34: 467–68, 16 Sept. 1886: no. 881.
400. *Note on Compensation to Landlords.* LNS Tract no. 16: LNS, London, n.d. (probably published in late summer or early autumn 1886, per information provided in *Report of the LNS. 1886–7*): 1–4.
401. The English Naturalist (anonymous interview). *The Sunday Herald* (Boston), 31 Oct. 1886: 13a–c.
402. Review (of *Lessons in Elementary Dynamics* arranged by Henry G. Madan, 1886). *Nature* 35: 51, 18 Nov. 1886: no. 890.

1887

403. Oceanic Islands: Their Physical and Biological Relations (synopsis/P of a lecture delivered at the American Geographical Society meeting of 11 Jan. 1887). *Bull. Amer. Geogr. Soc.* 19(1): 1–17 (followed on pp. 17–21 by related material by D. Morris originally printed in *Nature*), 1887.
404. The American School of Evolutionists (review of *The Origin of the Fittest: Essays on Evolution* by Edward D. Cope, 1887). *The Nation* (New York) 44: 121–23, 10 Feb. 1887: no. 1, 128.
405. Mr. Romanes on Physiological Selection (letter dated 30 Jan., Washington, U.S.A.). *Nature* 35: 366, 17 Feb. 1887: no. 903.
406. Letter from Dr. Alfred R. Wallace, in re Mrs. Ross (dated 23 Feb., Washington, D.C.; concerning the medium Mrs. Ross). *Banner of Light* (Boston) 60(25): 4a–e, 5 March 1887.
407. Review (of *The Origin of the Fittest: Essays on Evolution* by Edward D. Cope, 1887), signed "A. H." but referred to in *My Life* (Vol. 2, 133). *The Independent* (New York) 39: 335–36, 17 March 1887: no. 1998.
408. If a Man Die, Shall He Live Again? (lecture given at Metropolitan Temple, San Francisco, on 5 June 1887). *Golden Gate* (San Francisco) (exact date of publication unknown, but undoubtedly in early June 1887).
409. Letter (extract from letter on American land taxes read at the sixth annual meeting of the LNS, 8 June 1887; dated 8 Feb., Washington, D.C.), in *Report of the LNS. 1886–87,* LNS, London, 1887: 7.
410. Account of slate-portrait drawing (referred to in *My Life,* Vol. 2, 349, and *The Medium and Daybreak,* 24 June 1887). *Golden Gate* (San Francisco) (exact date of publication unknown, but probably in mid-June 1887).
411. American Museums. The Museum of Comparative Zoology, Harvard University. *Fortnightly Rev.* 42 (n.s.; 48, o.s.): 347–59, 1 Sept. 1887: no. 249, n.s.
412. The British Museum and American Museums (LTTE). *Nature* 36: 530–31, 6 Oct. 1887: no. 936.
413. *Land Lessons from America* (from an address delivered at Essex Hall, Essex Street, Strand, on 1 Nov. 1887). Land Nationalisation Tract no. 18: LNS, London, n.d.; 1–15.

414. American Museums. Museums of American Pre-historic Archaeology. *Fortnightly Rev.* 42 (n.s.; 48, o.s.): 665–75, 1 Nov. 1887: no. 251, n.s.
415. Note on American Museums and the British Museum (LTTE). *Fortnightly Rev.* 42 (n.s., 48, o.s.): 740, 1 Nov. 1887: no. 251, n.s.
416. The Antiquity of Man in North America. *Nineteenth Century* 22: 667–79, Nov. 1887. no. 129.

1888

417. Address (Presidential Address to the seventh annual meeting of the LNS, 8 May 1888), in *Report of the LNS, 1887–88* (LNS, London), 1888: 10–16.
418. Replies (to questions regarding land purchase and rental posed during the seventh annual meeting of the LNS, 8 May 1888), in *Report of the LNS, 1887–88* (LNS, London, 1888): 23–24.
419. Account of a Spiritualistic Test. Nellie Morris. (letter by Wallace introducing a series of letters by others concerning this subject) *J. SPR* 3 (1887–88): 273–74 (273–88), June 1888: no. 51.
420. Mr. Gulick on Divergent Evolution (LTTE). *Nature* 38: 490–91, 20 Sept. 1888: no. 986.
421. Nellie Morris (edited discussion/N concerning the "spirit" Nellie Morris consisting of three letters by Wallace and several by others). *J. SPR* 3 (1887–1888): 313–14, 315, 316–17 (312–18), Oct. 1888: no. 55.
422. Introductory Note by Alfred R. Wallace to *A Colonist's Plea for Land Nationalisation* by Arthur J. Ogilvy, Land Nationalisation Tract no. 23: LNS, London, n.d. (probably published in 1888): 2.

1889

423. Which are the Highest Butterflies? (LTTE introducing remarks by W. H. Edwards on Colorado butterflies). *Nature* 39: 611–12, 25 April 1889: no. 1017.
424. Address (Presidential Address to the eighth annual meeting of the LNS, 20 June 1889; includes notice of *Poverty and the State* by Herbert V. Mills, 1889), in *Report of the LNS, 1888–89* (LNS, London), 1889: 15–23.
425. Lamarck versus Weismann (LTTE). *Nature* 40: 619–20, 24 Oct. 1889: no. 1043.
426. LTTE (one of two printed as "Mr. Angelo Lewis and Dr. Monck" concerning the medium Francis W. Monck, *J. SPR* 4 (1889–90): 143–44 (143–46), Oct. 1889: no. 63.
427. Protective Coloration of Eggs (LTTE introducing remarks by Rev. Fred. F. Grensted). *Nature* 41: 53, 21 Nov. 1889: no. 1047.
428. Letter (responding to an inquiry regarding *"The Star's"* [London?] comments on Wallace's support of land nationalisation and socialism). *Land and Labour* no. 1: 7–8, Nov. 1889.
429. The Action of Natural Selection in Producing Old Age, Decay, and Death (added as footnote to essay/N entitled "The Duration of Life"; written "between 1865 and 1870," according to source), in *Essays Upon Heredity and Kindred Biological Problems* by August Weismann, authorised translation ed. by Edward B. Poulton, Selmar Schonland, and Arthur E. Shipley (The Clarendon Press, Oxford), Dec. 1889: 23 (5–35).

1890

430. Testimony (the text of testimony Wallace presented before a Royal Commission on vaccination on 26 Feb., 5 March, 12 March, and 21 May 1890) in *Third Report of the Royal Commission Appointed to Inquire into the Subject of Vaccination; with*

Minutes of Evidence and Appendices (printed for Her Majesty's Stationery Office by Eyre & Spottiswoode, London, 24 Dec. 1890): 6–35, 121–31.

431. Review (of A *Naturalist among the Head-hunters. Being an Account of Three Visits to the Solomon Islands in the Years 1886, 1887, and 1888* by Charles Morris Woodford, 1890). *Nature* 41: 582–83 (24 April 1890: no. 1069).
432. The Instability of Peasant-proprietorship—The Necessity of Rent (first part of an article/N entitled "The New Round Table: Land Nationalisation" with contributions from 10 writers). *Westminster Rez.* 133(5): 541–43 (541–59), 1890.
433. *Presidential Address Delivered by Dr. Alfred R. Wallace, at the Ninth Annual Meeting, June 19th, 1890* (LNS Tract no. 33: LNS, London), June 1890; 1–12.
434. Review (of *The Colours of Animals: Their Meaning and Use especially Considered in the Case of Insects* by Edward B. Poulton, 1890). *Nature* 42: 289–91, 24 July 1890: no. 1082.
435. Birds and Flowers (LTTE). *Nature* 42: 295, 24 July 1890: no. 1082.
436. Taxation or Compensation (LTTE sent in response to queries regarding Wallace's LNS annual meeting address). *The Democrat* (London) 6(8): 192, 1 Aug. 1890.
437. Human Selection. *Fortnightly Rev.* 48 (n.s.; 54, o.s.): 325–37, 1 Sept. 1890: no. 285, n.s.
438. Dr. Romanes on Physiological Selection (LTTE). *Nature* 43: 79, 27 Nov. 1890: no. 1100.
439. Dr. Romanes on Physiological Selection (LTTE). *Nature* 43: 150, 18 Dec. 1890: no. 1103.

1891

440. Are there Objective Apparitions? *Arena* 3: 129–46, Jan. 1891: no. 14.
441. An English Nationalist. Alfred Russell [*sic*] Wallace Converted by Bellamy's Book. (note/N containing extracts from two Wallace letters concerning Richard T. Ely's book *An Introduction to Political Economy,* 1889). *New York Times* no. 12303: 9d, 1 Feb. 1891.
442. Modern Biology and Psychology (review of *Animal Life and Intelligence* by C. Lloyd Morgan, 1890). *Nature* 43: 337–41, 12 Feb. 1891: no. 1111.
443. Remarkable Ancient Sculptures from North-west America (notice of *Sculptured Anthropoid Ape Heads Found in or near the Valley of the John Day River, a Tributary of the Columbia River, Oregon* by James Terry, 1891). *Nature* 43: 396, 26 Feb. 1891: no. 1113.
444. What Are Phantasms, and Why Do They Appear? *Arena* 3: 257–74, Feb. 1891: no. 15.
445. LTTE. *Light* (London) 1: 133–34, 21 March 1891: no. 533.
446. Mr. S. J. Davey's Experiments (LTTE concerning the medium S. J. Davey followed by comment by the Editor). *J. SPR* 5 (1891–92): 43 (43–45), March 1891: no. 78.
447. Another Darwinian Critic (review of *On the Modification of Organisms* by David Syme, 1890). *Nature* 43: 529–30, 9 April 1891: no. 1119.
448. Presidential Address of Dr. Alfred Russel Wallace (to the tenth annual meeting of the LNS, 18 June 1891), in *Report of the LNS.* 1890–91. (LNS, London 1891): 16–23; also printed in *Land and Labour* no. 21: 1–4, July 1891.
449. Introductory Note to *The History of Human Marriage* by Edward Westermarck (Macmillan & Co., London & New York), July 1891: v–vi.
450. Variation and Natural Selection (LTTE commenting on an address given by C. Lloyd Morgan to the Bristol Naturalists' Society). *Nature* 44: 518–19, 1 Oct. 1891: no. 1144.
451. English and American Flowers. 1. *Fortnightly Rev.* 50 (n.s.; 56, o.s.): 525–34 (I Oct.

1891: no. 298, n.s.) / English and American Flowers. 11. Flowers and Forests of the Far West.50 (n.s.; 56, o.s.): 796–810, 1 Dec. 1891: no. 300, n.s.

452. Discussion (of comments by David Syme; printed with Syme's letter as "Topical Selection and Mimicry"). *Nature* 45: 31 (30–31), 12 Nov. 1891: no. 1150.
453. Commons (letter). *Land and Labour* no. 25: 2, Nov. 1891.

1892

454. Popular Zoology (review of *Animal Sketches* by C. Lloyd Morgan, 1891). *Nature* 45: 291–92, 28 Jan. 1892: no. 1161.
455. Human Progress: Past and Future. *Arena* 5: 145–59, Jan. 1892: no. 26.
456. H. W. Bates, the Naturalist of the Amazons (obituary). *Nature* 45: 398–99, 25 Feb. 1892: no. 1165.
457. A Remarkable Book on the Habits of Animals (review of *The Naturalist in LaPlata* by William H. Hudson, 1892). *Nature* 45: 553–56, 14 April 1892: no. 1172.
458. Correction in *Island Life* (LTTE). *Nature* 46: 56, 19 May 1892: no. 1177.
459. Remarks (on receiving the Founder's Medal during the RGS meeting of 23 May 1892). *Proc. RGS & Montly Record of Geography* 14(7) (new monthly s.): 485–86, July 1892.
460. Presidential Address (to the eleventh annual meeting of the LNS, 23 June 1892; consists in large part of a discussion of *Justice* by Herbert Spencer, 1891, and *Freeland; A Social Anticipation* by Theodor Hertzka, 1891), in *Report of the LNS.* 1891–92. (LNS Tract no. 48: LNS, London), July 1892: 15–26.
461. Spiritualism, in *Chambers' Encyclopaedia* (new ed., 10 vols., William & Robert Chambers, Ltd., London & Edinburgh, 1888–92), Vol. 9: 645–49, July 1892.
462. Psychography in the Presence of Mr. Keeler (description of experiences at three séances). *Psychical Rev.* 1(1): 16–18, July 1892.
463. The Permanence of the Great Oceanic Basins. *Natural Science 1:* 418–26, Aug. 1892: no. 6.
464. Why I Voted for Mr. Gladstone. IV (one of eight invited replies to an inquiry). *Nineteenth Century* 32: 182–85 (177–93), Aug. 1892: no. 186.
465. Our Molten Globe (an analysis of *Physics of the Earth's Crust,* 2nd ed., by Rev. Osmond Fisher, 1891). *Fortnightly Rev.* 52 (n.s.; 58, o.s.): 572–84, 1 Nov. 1892: no. 311, n.s.
466. An Ancient Glacial Epoch in Australia (notice of *Notes on the Glacial Conglomerate, Wild Duck Creek* by Edward J. Dunn, 1892). *Nature* 47: 55–56, 17 Nov. 1892: no. 1203.
467. The Permanence of Ocean Basins (LTTE). *Natural Science 1:* 717–18, Nov. 1892: no. 9.
468. LTTE (one of two printed as "The Earth's Age"). *Nature* 47: 175, 22 Dec. 1892: no. 1208.
469. Note on Sexual Selection. *Natural Science 1:* 749–50, Dec. 1892: no. 10.

1893

470. LTTE (one of two printed as "The Earth's Age"). *Nature* 47: 227 (226–27), 5 Jan. 1893: no. 1210.
471. Man and Evolution (review of *Evolution and Man's Place in Nature* by Henry Calderwood, 1893). *Nature* 47: 385–86, 23 Feb. 1893: no. 1217.
472. The Glacier Theory of Alpine Lakes (LTTE). *Nature* 47: 437–38, 9 March 1893: no. 1219.
473. Reveries of a Naturalist (review of *Idle Days in Patagonia* by William H. Hudson, 1893). *Nature* 47: 483–84, 23 March 1893: no. 1221.

474. Inaccessible Valleys; A Study in Physical Geography. *Nineteenth Century* 33: 391–404, March 1893: no. 193.
475. Note on Mr. Jukes-Browne's Paper (concerning the origin and classification of islands). *Natural Science* 2: 193–94, March 1893: no. 13.
476. The Social Quagmire and the Way out of It. I. The *Farmers Arena 7:* 395–410, March 1893: no. 40 / The Social Quagmire and the Way out of It. II. Wage-workers. 7: 525–42, April 1893: no. 41.
477. The Late Mr. S. J. Davey's Experiments (LTTE followed by a lengthy reply from Richard Hodgson). *J. SPR* 6 (1893–94): 33–36 (33–47), March 1893: no. 98.
478. Are Individually Acquired Characters Inherited? I. *Fortnightly Rev.* 53 (n.s.; 59, o.s.): 490–98, 1 April 1893: no. 316, n.s. / II. 53 (n.s.; 59, o.s.): 655–68, 1 May 1893: no. 317, n.s.
479. Mr. H. O. Forbes's Discoveries in the Chatham Islands (LTTE). *Nature* 48: 27–28, 11 May 1893: no. 1228.
480. Reason versus Instinct (review of *The Intelligence of Animals* by Charles William Purnell, 1893). *Nature* 48: 73–74, 25 May 1893: no. 1230.
481. President's Address, 1893. The Conditions Essential to the Success of Small Holdings (read by William Volckman, Chairman, at the twelfth annual meeting of the LNS, 15 June 1893), in *Report of the LNS, 1892–93* (LNS Tract no. 50: LNS, London), June 1893: 15–23.
482. Discussion (of letter by Graham Officer; printed with Officer's letter as "The Glacier Theory of Alpine Lakes"). *Nature* 48: 198, 29 June 1893: no. 1235.
483. The Non-inheritance of Acquired Characters (LTTE). *Nature* 48: 267, 20 July 1893: no. 1238.
484. The Response to the Appeal. From Prelates, Pundits and Persons of Distinction (article/N posting results of a survey of opinion concerning the stated intended purpose of the journal; Wallace's one of many responses printed). *Borderland* 1: 17 (10–23), July 1893: no. 1.
485. The Bacon-Shakespeare Case. Verdict no. I. (one of many solicited responses concerning the authorship of Shakespeare's works). *Arena* 8: 222–25, July 1893: no. 44.
486. Prenatal Influences on Character (LTTE). *Nature* 48: 389–90, 24 Aug. 1893: no. 1243.
487. Habits of South African Animals (LTTE introducing comments by R. R. Mortimer on some South African birds). *Nature* 48: 390–91, 24 Aug. 1893: no. 1243.
488. Notes on the Growth of Opinion as to Obscure Psychical Phenomena during the Last Fifty Years (paper communicated to the Psychical Congress in Chicago, held 21–25 Aug. 1893; sent from Parkstone, Dorset, England). *Religio-Philosophical.* (Chicago) 4 (n.s.), no. 15: 229a–30a, 2 Sept. 1893.
489. On Malformation from Pre-natal Influence on the Mother (letter communicated to the 15 Sept. 1893 Nottingham meeting of Section D, Biology, of the BAAS, in *Rept BAAS* 63, John Murray, London), 1894: 798–99.
490. The Supposed Glaciation of Brazil (LTTE). *Nature* 48: 589–90, 19 Oct. 1893: no. 1251.
491. The Ice Age and its Work. I. Erratic Blocks and Ice-sheets. *Fortnightly Rev.* 54 (n.s.; 60, o.s.): 616–33, 1 Nov. 1893: no. 323, n.s. / The Ice Age and its Work. II. Erosion of Lake Basins. 54 (n.s.; 60, o.s.): 750–74, 1 Dec. 1893: no. 324, n.s.
492. The Recent Glaciation of Tasmania (LTTE). *Nature* 49: 3–4, 2 Nov. 1893: no. 1253.
493. The Programme of Land Nationalisers (paper read at a general meeting of the LNS held 15 Nov. 1893). *Land and Labour* no. 50: 1–2, Dec. 1893.
494. Sir Henry H. Howorth on "Geology in Nubibus" (LTTE concerning glacial movement). *Nature* 49: 52, 16 Nov. 1893: no. 1255.

495. Recognition Marks (letter responding to inquiry regarding rabbits' use of their tails as danger signals). *Nature* 49: 53, 16 Nov. 1893: no. 1255.
496. Geology in Nubibus (letter responding to remarks by Henry H. Howorth concerning glacial movement). *Nature* 49: 101, 30 Nov. 1893: no. 1257.
497. LTTE (one of two printed as "The Origin of Lake Basins"). *Nature* 49: 197, 28 Dec. 1893: no. 1261.
498. Preface to *The Dispersal of Shells; An Inquiry into the Means of Dispersal Possessed by Fresh-water and Land Mollusca* by Harry Wallis Kew (International Scientific Series, Vol. LXXV: Kegan Paul, Trench, Trubner & Co., Ltd. London), Dec. 1893: v–viii.
499. The Problem of the Unemployed (relating the ideas in Herbert V. Mills's book *Poverty and the State). The Daily Chronicle* (London) no. 9921: 7a–b, 28 Dec. 1893.

1894

500. LTTE (one of two printed as "The Origin of Lake Basins"). *Nature* 49: 220–21 (220–21), 4 Jan. 1894: no. 1262.
501. *Why Does Man Exist?* (letter to Arthur J. Bell, author of *Why Does Man Exist?* 1890); *Borderland* 1: 272, Jan. 1894: no. 3.
502. How to Preserve the House of Lords. *Contemporary Rev.* 65: 114–22, Jan. 1894.
503. Richard Spruce, Ph.D., F.R.G.S. (obituary). *Nature* 49: 317–19, 1 Feb. 1894: no. 1266.
504. A Critic Criticised (review of *Darwinianism: Workmen and Work* by James Hutchison Stirling, 1893). *Nature* 49: 333–36, 8 Feb. 1894: no. 1267.
505. Interview. *The Daily Chronicle* (London), referred to in *My Life,* Vol. 2, 210.
506. Heredity and Pre-natal Influences. An Interview with Dr. Alfred Russel Wallace (by Sarah A. Tooley). *Humanitarian* 4 (n.s.), no. 2: 80–88, Feb. 1894.
507. What Are Zoological Regions? (paper read at the 500th meeting of the Cambridge Natural Science Club, 12 March 1894). *Nature* 49: 610–13, 26 April 1894: no. 1278.
508. President's Address (presented at the thirteenth annual meeting of the LNS 9 April 1894; includes a notice of *Social Evolution* by Benjamin Kidd, 1894), in *13th Report of the LNS, 1893–94* (LNS Tract no. 56: LNS, London), April 1894: 15–24.
509. The Future of Civilisation (review of *Social Evolution* by Benjamin Kidd, 1894). *Nature* 49: 549–51, 12 April 1894: no. 1276.
510. LTTE (one of two printed as "Woman and Natural Selection"). *Humanitarian* (n.s.), no. 4: 315 (315–19), April 1894.
511. Economic and Social Justice, in Vox Clamantium; *The Gospel of the People* by "writers, preachers & workers brought together by Andrew Reid" (A. D. Innes & Co., London), April 1894: 166–97.
512. Panmixia and Natural Selection (LTTE). *Nature* 50: 196–97, 28 June 1894: no. 87.
513. The Palaearctic and Nearctic Regions Compared as regards the Families and Genera of their Mammalia and Birds. *Natural Science* 4: 433–44, June 1894: no. 28.
514. A New Book on Socialism (review of *The Great Revolution of 1905—or The Story of the Phalanx* by Frederick W. Hayes, 1893). *Land and Labour* no. 57: 52–54, July 1894.
515. The Influence of Previous Fertilisation of the Female on her Subsequent Offspring, and the Effect of Maternal Impressions during Pregnancy on the Offspring, in *Report of the Sixty-fourth Meeting of the BAAS* (1894), John Murray, London, 1894: 346.
516. Rev. George Henslow on Natural Selection. *Natural Science* 5: 177–83, Sept. 1894: no. 31.
517. Another Substitute for Darwinism (review of *Nature's Method in the Evolution of Life* by "Anonymous" James W. Barclay), *Nature* 50: 541–42, 4 Oct. 1894: no. 1301.
518. A Suggestion to Sabbath-keepers. *Nineteenth Century* 36: 604–11, Oct. 1894: no. 212.

519. Why Live a Moral Life? The Answer of Rationalism, in *The Agnostic Annual* 1895 ed. by Charles A. Watts (London, W. Stewart & Co.), Dec.? 1894: 6–12.
520. The Social Economy of the Future, in *The New Party Described by Some of its Members* (new ed.), ed. by Andrew Reid (Hodder Brothers, London), Dec. 1894: 177–211.

1895

521. Note on Compensation to Landlords. *Land and Labour* no. 63:5 Jan. 1895.
522. Forty-five Years' Registration Statistics. A Correction. (LTTE). *The Vaccination Inquirer and Health Review* 16: 159–60, 1 Feb. 1895: no. 191.
523. The Method of Organic Evolution. I. *Fortnightly Rev.* 57 (n.s.; 63, o.s.): 211–24, 1 Feb. 1895: no. 338, n.s. / II. 57 (n.s.; 63, o.s.): 435–45, 1 March 1895: no. 339, n.s.
524. Tan-spots over Dogs' Eyes (LTTE responding to an inquiry). *Nature* 51: 533, 4 April 1895: no. 1327.
525. *Suggestions for Solving the Problem of the Unemployed, etc., etc.* (Presidential Address read at the fourteenth annual meeting of the LNS, 8 April 1895). Land Nationalisation Tract no. 64 (LNS, London), April? 1895; 1–20.
526. LTTE (one of two printed as "The Age of the Earth"). *Nature* 51: 607 (607–608), 25 April 1895: no. 1330.
527. Uniformitarianism in Geology (LTTE). *Nature* 52: 4, 2 May 1895: no. 1331.
528. Another Book on Social Evolution (review of *The Evolution of Industry* by Henry Dyer, 1895). *Nature* 52: 386–87, 22 Aug. 1895: no. 1347.
529. Discussion (concerning a letter Wallace sent to Adolf B. Meyer in 1869; Meyer's letter (containing an excerpt from the 1869 letter) and Wallace's comments thereon (printed as "How Was Wallace Led to the Discovery of Natural Selection?"). *Nature* 52: 415, 29 Aug. 1895: no. 1348.
530. Our Native Birds (review of *British Birds* by William H. Hudson, 1895). *Saturday Rev.* 80: 342–43, 14 Sept. 1895: no. 2081.
531. The Expressiveness of Speech, or Mouth-gesture as a Factor in the Origin of Language. *Fortnightly Rev.* 58 (n.s.; 64, o.s.): 528–43, 1 Oct. 1895: no. 346, n.s.
532. Introductory Note to *Psychic Philosophy as the Foundation of a Religion of Natural Law* by V. C. Desertis (pseudonym of Stanley DeBrath) (George Redway, London, Dec. 1895 / Bellairs & Co., London), Nov. 1895: v–vi.

1896

533. Eusapia Palladino (LTTE). *The Daily Chronicle* (London) no. 10572: 3g, 24 Jan. 1896.
534. LTTE (one of two printed as "The Cause of an Ice Age"). *Nature* 53: 220–21, 9 Jan. 1896: no. 1367.
535. The Astronomical Theory of a Glacial Period (LTTE). *Nature* 53: 317, 6 Feb. 1896: no. 1371.
536. The Theory of the Double. *Light* (London) 16: 87–88, 22 Feb. 1896: no. 789.
537. Philosophy and Evolution (review of *Evolution and Man's Place in Nature,* 2nd ed., by Henry Calderwood, 1896). *Nature* 53: 435, 12 March 1896: no. 1376.
538. Old and New Theories of Evolution (review of *The Primary Factors of Organic Evolution* by Edward D. Cope, 1896; and *The Present Evolution of Man* by G. Archdall Reid, 1896). *Nature* 53: 553–55, 16 April 1896: no. 1381.
539. Letter from the President (to Joseph Hyder, read at the fifteenth annual meeting of the LNS, 20 April 1896). *Land and Labour* no. 79: 33–34, May 1896.
540. The Proposed Gigantic Model of the Earth. *Contemporary Rev.* 69: 730–40, May 1896.
541. The Problem of Utility: Are Specific Characters Always or Generally Useful? (paper read at the LSL meeting of 18 June 1896). *J. Linn. Soc. Zool.* 25: 481–96, 1896.

542. *Miracles and Modern Spiritualism* (LTTE concerning the new edition of this book). *Light* (London) no. 806: 298, 20 June 1896.
543. LTTE (concerning labour and militarism to Keir Hardie, on the occasion of the International Labour Congress). *Labour Leader* 8: 251, 25 July 1896: no. 121, n.s.
544. The Gorge of the Aar and its Teachings. *Fortnightly Rev.* 60 (n.s.; 66, o.s.,): 175–82, 1 Aug. 1896: no. 356, n.s.
545. Spiritualism (LTTE). *The Echo* (London) no. 8637: 4a, 12 Sept. 1896.
546. Methods of Land Nationalisation. *Land and Labour* no. 85: 82–83, Nov. 1896.

1897

547. Darwin and Darwinism (review of *Charles Darwin and the Theory of Natural Selection* by Edward B. Poulton, 1896). *Nature* 55: 289–90, 28 Jan. 1897: no. 1422.
548. Lord Penrhyn and the Quarrymen (full version of a LTTE first printed in edited form in *The Daily Chronicle,* London, n.d.). *Land and Labour* no. 88: 9–10, Feb. 1897.
549. The Problem of *Instinct* (review of *Habit and Instinct* by C. Lloyd Morgan, 1896). *Natural Science* 10: 161–68, March 1897: no. 61.
550. On the Colour and Colour-patterns of Moths and Butterflies (comments on a paper on this subject by Alfred Goldsborough Mayer). *Nature* 55: 618–19, 29 April 1897: no. 1435.

1898

551. *Vaccination a Delusion; Its Penal Enforcement a Crime: Proved by the Official Evidence in the Reports of the Royal Commission.* Swan Sonnenschein & Co., Ltd., London, March 1898.
552. *The Eagle and the Serpent* (letter commending the aims of *The Eagle and the Serpent*). *The Clarion* (London) no. 328: 95b, 19 March 1898.
553. Mr. A. R. Wallace and Vaccination (LTTE). *Lancet,* Vol. I for 1898: 894, 26 March 1898: no. 3891.
554. LTTE (commending the aims of the journal). *The Eagle and the Serpent* 1(2): 21, 15 April 1898.
555. Nietzsche as a Social Reformer, or, The Joys of Fleecing and being Fleeced (article/N posting results of a general inquiry; Wallace's one of several responses printed). *The Eagle and the Serpent* 1(2): 26–27 (25–28), 15 April 1898.
556. Letter (to Joseph Hyder, read at the seventeenth annual meeting of the LNS, 22? April 1898). *Land and Labour* no. 103: 35, May 1898.
557. Vaccination and Its Enforcement (LTTE). Dr. Bond and Mr. A. R. Wallace. *The Daily Chronicle* (London) no. 11291: 3b, 12 May 1898.
558. Dr. Bond v. Mr. A. R. Wallace (LTTE concerning vaccination). *Shrewsbury Chronicle* 20 May 1898: 3b.
559. Letter (included in note/N entitled "Dr. Alfred R. Wallace on Altruism: Mr. Platt's Repudiation"). *The Eagle and the Serpent* 1(3): 36, 15 June 1898.
560. Dr. Bond and Mr. A. R. Wallace. Mr. Wallace Replies (LTTE concerning vaccination). *The Echo* (London) no. 9183: Ic, 16 June 1898.
561. Spiritualism and Social Duty (from an address delivered at the International Congress of Spiritualists on 23 June 1898, at St. James Hall, London). *Light* (London) 18: 334–36 (selection followed by an account of related discussion on pp. 336–37), 9 July 1898: no. 913.
562. Spiritualism—and Things (interview by columnist "The Whatnot"). *The Clarion* (London) no. 343: 213b–e, 2 July 1898.
563. Discussion/N (of paper on Brazilian spiritism by A. Alexander read at the 23 June 1898 session of the International Congress of Spiritualists). *Light* (London) 18: 337, 9 July 1898: no. 913.

564. The Importance of Dust: A Source of Beauty and Essential to Life. Chapter 9 of *The Wonderful Century:* 69–85, June 1898.
565. Dr. Bond and Mr. A. R. Wallace. Mr. Wallace's Final Reply. (LTTE). *The Echo* (London) no. 9234: Ic, 15 Aug. 1898.
566. Darwinism in Sociology: Dr. Alfred Russell [*sic*] Wallace Replies to Mr. Thomas Common. *The Eagle and the Serpent* 1(4): 57–59, 1 Sept. 1898.
567. Letter (referring to the intent of the journal and to a book by Ragnar Redbeard). *The Eagle and the Serpent* 1 (4): 62, 1 Sept. 1898.
568. The Vaccination Question (LTTE). *The Times* (London) no. 35610: 10e, 1 Sept. 1898.
569. Letter (to columnist "Dangle" concerning socialist reforms and the monopoly of money). *The Clarion* (London) no. 357: 325c–d, 8 Oct. 1898.
570. Is there Scarcity or Monopoly of Money? (letter to columnist "Dangler"). *The Clarion* (London) no. 360: 348c–e, 29 Oct. 1898.
571. *The Wonderful Century.*—A Correction. *Land and Labour* no. 108: 82, Oct. 1898.
572. A Visit to Dr. Alfred Russel Wallace, F.R.S. (interview by "A. D."). *The Bookman* (London) 13: 121–24, Jan. 1898: no. 76.
573. The Problem of the Tropics (LTTE concerning the matter of whether white men can work effectively in tropical climates). *The Daily Chronicle* (London) no. 11440: 3g, 2 Nov. 1898.
574. Introductory Note to *The Third Factor of Production and other Essays* by Arthur J. Ogilvy (Swan Sonnenschein & Co., Ltd., London), Nov. 1898: vii–viii.
575. A Complete System of Paper Money (letter to columnist "Dangler"). *The Clarion* (London) no. 365: 389e–f, 3 Dec. 1898.
576. Paper Money as a Standard of *Value Academy* 55: 549–50, 31 Dec. 1898: no. 1391, n.s.

1899

577. The Utility of Specific Characters (LTTE concerning recognition marks). *Nature* 59: 246, 12 Jan. 1899: no. 1524.
578. America, Cuba, and the Philippines (LTTE). *The Daily Chronicle* (London) 19 Jan. 1899: 3g.
579. The Inefficiency of Strikes: Is There Not a Better Way? in *The Labour Annual:* 1899 ed. and published by Joseph Edwards (Wallasey, Cheshire), Jan.? 1899: 105.
580. Mr. Podmore on Clairvoyance and Poltergeists (LTTE) *J. SPR* 9 (1899–1900): 22–30, Feb. 1899: no. 156.
581. White Men in the Tropics. *The Independent* (New York) 51: 667–70, 9 March 1899: no. 2623.
582. The Storage of Gunpowder (LTTE referring to Wallace's c1882 idea that explosives might safely be stored under water). *The Daily Chronicle* (London) no. 11562: 11c, 24 March 1899.
583. Letter from the President (to Joseph Hyder, read at the eighteenth annual meeting of the LNS), 26 April 1899. *Land and Labour* no. 115: 46, May 1899.
584. Letter (to columnist Julia Dawson, accompanying a donation to the *Clarion* Van). *The Clarion* (London) no. 386: 130e, 29 April 1899.
585. Clairvoyance and Poltergeists (LTTE). *J. SPR* 9 (1899–1900): 56–57, April 1899: no. 158.
586. Garden City (letter to Ebenezer Howard concerning planned communities). *Land and Labour* no. 114:38, April 1899.
587. Les Causes de la guerre—Comment y remedier (abridged translation of one of many essays included in a special supplementary publication entitled "Enquête sur la guerre et le militarisme"). *L 'Humanite' Nouvelle* Mai 1899: 245–50.
588. Introductory Note to "Extract from Js-E de Mirville's *Des Esprits et de leurs man-*

ifestations fluidiques" (read as "Clairvoyance of Alexis Didier" at the SPR meeting of 23 June 1899). *Proc. SPR* 14 (1898–99): 373–74 (373–81), Supplement 5 to Part XXXV, July? 1899.

589. LTTE (one of many printed as "Protests against War"). *Manchester Guardian* no. 16557: 7f (7f–h), 2 Sept. 1899.
590. Letter (reply to remarks by Thomas Common concerning the "might vs. right" issue). *The Eagle and the Serpent* 1(9): 136, 15 Oct. 1899.
591. The Transvaal War. Wanted Facts. (LTTE). *The Clarion* (London) no. 415: 365a–b, 18 Nov. 1899.
592. Facts from the Transvaal (LTTE introducing remarks by an observer back from the Boer War). *The Clarion* (London) no. 417: 380f, 2 Dec. 1899.
593. Mottoes for the New Year. Wise Words from Famous People (article/N posting results of a general inquiry; Wallace's one of many responses printed). *The Daily News Weekly* (London) no. 27: 11b–c (11a–d), 30 Dec. 1899.
594. Introductory Note to *The Ascent of Man* by Mathilde Blind (T. Fisher Unwin, London), Dec. 1899: v–xii.

1900

595. Is New Zealand a Zoological Region? (LTTE). *Nature* 61: 273, 18 Jan. 1900: no. 1577.
596. Labour and the next General Election (article/N posting results of a general inquiry; Wallace's one of several responses printed), in *The Labour Annual: The Reformers' Year-book for 1900* ed. and published by Joseph Edwards, Wallasey, Cheshire, Jan. 1900: 29 (26–30).
597. Letter (to columnist "Nunquam" concerning the death of *Clarion* columnist "The Bounder"). *The Clarion* (London) no. 447: 205b, 30 June 1900.
598. LTTE (concerning Edward Jenner's observations on the cuckoo; part of a discussion including another letter and comments by the Editor entitled "Jenner and the Cuckoo"). *The Vaccination Inquirer and Health Rev.* 22: 58 (58–61), 2 July 1900: no. 256.
599. Imperial Might and Human Right (comments on an article by George Bernard Shaw concerning imperialism). *The Clarion* (London) no. 450: 230b, 21 July 1900.
600. Letter (reply to remarks by Ragnar Redbeard concerning the "might vs. right" issue). *The Eagle and the Serpent* 1(11): 164, July 1900.
601. Letter from Dr. A. R. Wallace (concerning social problems), in *The Anatomy of Misery: Plain Lectures on Economics* (2nd ed.) by John Coleman Kenworthy (Simpkin, Marshall, Hamilton, Kent, & Co., Ltd., London, Sept. 1900 / Small, Maynard & Co., Boston, 6 April 1901): 98–100 (as part of the Appendix).
602. Letter extracts (in Part XI and Appendix G of "On the so-called Divining Rod. A Psycho-Physical Research on a Peculiar Faculty Alleged to Exist in Certain Persons Locally Known as Dowsers. Book II" by William F. Barrett). *Proc. SPR* 15 (1900–1901): 277, 374–75, Part XXXVIII, Oct.? 1900.
603. Affinities and Origin of the Australian and Polynesian Races (the first part of this article, pp. 461–73, forms the concluding section of Vol. 1, Chapter 5 of *Australasia,* ed. of Nov. 1893).
604. Interest-bearing Funds Injurious and Unjust, in *SSS,* Vol. 2: 254–64, Nov. 1900.
605. Some Objections to Land Nationalization Answered (derived "from Tracts issued by the LNS"), in *SSS,* Vol. 2: 345–63, Nov. 1900.
606. Ralahine and its Teachings (comments on *The Irish Land & Labour Question, Illustrated in the History of Ralahine and Co-operative Farming* by Edward T. Craig, 1882), in *SSS,* Vol. 2: 455–77, Nov. 1900.
607. True Individualism—The Essential Preliminary of a Real Social Advance, in *SSS,* Vol. 2: 510–20, Nov. 1900.

1901

608. Professor Alfred Russel Wallace (letter of support sent to the Eastbourne Conference of the National A-VL, held 5 Dec. 1900). *The Vaccination Inquirer and Health Rev.* 22: 156, 1 Jan. 1901: no. 262.
609. Evolution (the first in a series of articles published in *The Sun* under the overall title "The Passing Century"). *The Sun* (New York) 68(114): 4a–8,5a, 23 Dec. 1900. Reprint: nearly verbatim in *The Progress of the Century* (Harper & Brothers, New York & London), 20 April 1901: 3–29.
610. Article concerning social evolution in the twentieth century (not seen; part of a special feature entitled "Dawn of the Century"). *The (Sunday) Journal* (New York) no. 6611: ?23 Dec. 1900.
611. A Message to my Fellow Spiritualists for the New Century. *The Two Worlds* (Manchester) 13: 867, 28 Dec. 1900: no. 685.
612. Letter (dated Nov. 1900; written in response to an inquiry). *"The Sermon"* (title of serial not traced) Dec. 1900.
613. Words of Counsel (article/N posting results of a general inquiry; Wallace's one of several responses printed). *The Morning Leader* (London) no. 2689: 3a (3a), 2 Jan. 1901.
614. Letter (to columnist "Nunquam," concerning his novel *Julie*). *The Clarion* (London) no. 485: 92e, 23 March 1901.
615. Letter (to columnist Julia Dawson, concerning the subjects of peace and the *Clarion* Van). *The Clarion* (London) no. 493: 160d, 18 May 1901.

1902

616. Are Plant Diseases Hereditary? *The Garden* 61: 317, 17 May 1902: no. 1591.
617. *Eucalyptus gunnii. The Garden* 62: 47, 19 July 1902: no. 1600.
618. Dr. Russel Wallace's Advice (letter read at an A-VL meeting held 22 Oct. 1902). *Vaccination Inquirer* 24: 158–59, 1 Nov. 1902: no. 284.

1903

619. The Dawn of a Great Discovery (My Relations with Darwin in Reference to the Theory of Natural Selection). *Black and White* 25: 78–79, 17 Jan. 1903: no. 64.
620. Genius and the Struggle for Existence (LTTE). *Nature* 67: 296, 29 Jan. 1903: no. 1735.
621. Our Sphinx's Fatal Question: Why Do the Ungodly Prosper? Is Might Right? Can the Poor be Saved through the Pity of the Rich? (article/N posting results of a general inquiry; Wallace's one of several responses printed). *The Eagle and the Serpent no.* 18:69 (68–73), Feb.? 1903.
622. Man's Place in the Universe. *The Independent* (New York) 55: 473–83, 26 Feb. 1903: no. 2830; also printed as "Man's Place in the Universe: As Indicated by the New Astronomy" in *Fortnightly Rev.* 73 (n.s.; 79, o.s.): 395 411, 1 March 1903: no. 435, n.s.
623. Dr. Russel Wallace on the Use of the Vote (letter read at the annual meeting of the A-VL held 18 March 1903). *Vaccination Inquirer* 25: 14, April 1903: no. 289.
624. Letter (to T.D.A. Cockerell, concerning Wallace's early influences; printed in Cockerell's article "The Making of Biologists"). *Popular Science Monthly* 62(6): 517 (512–20), April 1903.
625. LTTE (one of several printed as "Man's Place in the Universe"). *Knowledge* 26: 107–108 (107–10), May 1903: no. 211.
626. Dr. A. R. Wallace on the Strenuous Policy (LTTE). *Vaccination Inquirer* 25: 49 (comment by the Editor follows on pp. 49–50), 1 June 1903: no. 291.
627. Man's Place in the Universe: A Reply to Criticisms. *The Independent* (New York)

55: 2024–31, 27 Aug. 1903: no. 2856; also printed in *Fortnightly Rev.* 74 (n.s.; 80, o.s.): 380–90, 1 Sept. 1903: no. 441, n.s.

628. Letter (of support read at a meeting of the Gloucester League held 22 Oct. 1903). *Vaccination Inquirer* 25: 153, 2 Nov. 1903: no. 296.
629. *The Wonderful Century* (letter responding to comments by a reviewer on the new edition of *The Wonderful Century,* Sept. 1903). *Academy* 65: 453, 24 Oct. 1903: no. 1642, n.s.
630. Does Man Exist in Other Worlds? "A Reply to my Critics." *The Daily Mail* (London) no. 2362: 4d, 12 Nov. 1903.
631. Interview. *The Christian Commonwealth* 10 Dec. 1903, 176a, & 12 Nov. 1913, 112c.
632. A Visit to Dr. Alfred Russel Wallace (by Albert Dawson). *The Christian Commonwealth* 23: 176a–77d, 10 Dec. 1903: no. 1156.

1904

633. Anticipations and Hopes for the Immediate Future (invited for publication by the *Berliner-Lokal-Anzeiger* but rejected by it). *The Clarion* (London) no. 630: I c–d, 1 Jan. 1904.
634. A Letter from Wallace (excerpt from letter concerning Wallace's continued support of spiritualism). *The Sentinel* (Wood Green, etc., England) no. 451: 3d, 15 Jan. 1904.
635. An Unpublished Poem by Edgar Allan Poe. Leonaine [*sic*]. *Fortnightly Rev.* 75 (n.s.; 81, o.s.): 329–32, 1 Feb. 1904: no. 446, n.s.
636. From Dr. Russel Wallace (letter of support sent to the annual meeting of the A-VL held 15–16 March 1904). *Vaccination Inquirer* 26: 18, 1 April 1904: no. 301.
637. The "Leonainie" Problem (LTTE). *Fortnightly Rev.* 75 (n.s.; 81, o.s.): 706–11, 1 April 1904: no. 448, n.s.
638. The Birds of Paradise in the Arabian Nights. I. *Independent Rev.* 2(7): 379–91, April 1904 / II. 2(8): 561–71, May 1904.
639. *A Summary of the Proofs that Vaccination Does Not Prevent Small-pox but Really Increases It.* National A-VL, London, 1904 (probably April, May or June); 1–24.
640. Master Workers. XVII.—Dr. Alfred Russel Wallace (by Harold Begbie). *The Pall Mall Mag* 34: 73–79, Sept. 1904: no. 137.
641. Practical Politics (reply to Editor Robert Blatchford's comments on military spending). *The Clarion* (London) no. 669: lb–c, 30 Sept. 1904.
642. Letter (excerpt from letter concerning John F. Burton's *Story of the Vaccination Crusade in Hackney & Stoke Newington 1902–1904, and What Came of It*). *Vaccination Inquirer 26:* 162, 1 Nov. 1904: no. 308.
643. Have We Lived on Earth Before? Shall We Live on Earth Again? (Article/N posting results of a survey; Wallace's one of several responses printed). *The London: A Magazine of Human Interest* 13: 401–403 (401–408), Nov. 1904: no. 76.

1905

644. Letter (of support read at the annual meeting of the A-VL held 15–16 March 1905). *Vaccination Inquirer* 27: 6, 1 April 1905: no. 313.
645. In Memoriam A. C. Swinton (obituary). *Land and Labour* 16(4): 33–34, April 1905.
646. A Man of the Time; Dr. Alfred Russel Wallace and his Coming Autobiography (interview by James Marchant). *The Book Monthly* 2(8): 545–49, May 1905.
647. A Message from Dr. A. R. Wallace (concerning the inconsistency of the government's position on vaccination). *Vaccination Inquirer* 27: 49, 1 June 1905: no. 315.
648. If there were a Socialist Government—How should It Begin? *The Clarion* (London) no. 715: 5a–f, 18 Aug. 1905.
649. The South-Wales Farmer (written for publication about 1843 but not appearing at that time), in *My Life,* Vol. 1: 206–22, Oct. 1905.

1906

650. Alfred R. Wallace on *Isis Unveiled* (letter to Helena P. Blavatsky dated 1 Jan. 1878, commending her book). *The Theosophist* (Adyar, India) 27(7): 559, April 1906.
651. Letter (of support to Joseph Hyder, read at the 25th annual meeting of the LNS, 9 May 1906). *Land and Labour* 1 7(6): 61–62, June 1906.
652. *A Statement of the Reasons for Opposing the Death Penalty.* Leaflet no. 2: Society for the Abolition of Capital Punishment, London, n.d.; 2.
653. The Nativity of Dr. Alfred Russel Wallace (article by the Editor containing short excerpt from Wallace letter commenting on the results of an astrological charting). *Modern Astrology* 3 (n.s.; o.s., 17), no. 5: 206–7, May 1906.
654. How to Nationalise Railroads. *The Daily News* (London) no. 18883: 6e, 24 Sept. 1906.
655. How to Buy the Railways. Dr. A. R. Wallace's Reply to Critics (LTTE). *The Daily News* (London) no. 18888: 4f, 29 Sept. 1906.
656. Should Women Have Votes: A Symposium (article/N posting the results of a general inquiry; Wallace's one of many responses printed). *The Daily* News (London) no. 18925: 8b (8), 12 Nov. 1906.
657. Our Black Brother (LTTE). *The Clarion* (London) no. 781: 5c, 23 Nov. 1906.
658. The Native Problem in South Africa and Elsewhere. *Independent Rev.* 11: 174–82, Nov. 1906: no. 38.
659. Britain's Greatest Benefactor (article/N posting results of an opinion survey concerning the most important figures in British history; Wallace's one of many responses printed; his choice: Shakespeare). *The Clarion* (London) no. 786: 3a (3a–c), 28 Dec. 1906.

1907

660. Personal Suffrage. A Rational System of Representation and Election. *Fortnightly Rev.* 81 (n.s.; 87, o.s.): 3–9, 1 Jan. 1907: no. 481, n.s.
661. Fertilisation of Flowers by Insects (LTTE sent in response to an inquiry). *Nature* 75: 320, 31 Jan. 1907: no. 1944.
662. The Railways for the Nation. *Arena* 37: 1–6, Jan. 1907: no. 206.
663. A New House of Lords: Representative of the Best Intellect and Character of the Nation. *Fortnightly Rev.* 81 (n.s.; 87, o.s.): 205–14, 1 Feb. 1907: no. 482, n.s.
664. Dr. Alfred Russel Wallace (excerpt from letter of support read at the annual meeting of the A-VL). *Vaccination Inquirer* 29: 4, 1 April 1907: no. 337.
665. Archdeacon Colley and Mr. Maskelyne (article/N including part of the transcript of Wallace's testimony at the libel trial of J. N. Maskelyne in late April 1907). *Light (London)* 27: 208–209 (207–209, 213) (4 May 1907: no. 1373).
666. Dr. Wallace's Letter (of support to Joseph Hyder, read at the 26th annual meeting of the LNS, 14 May 1907, *Land and Labour* 18(6): 65, June 1907.
667. "Economic Chivalry" (article/N posting survey of opinions regarding Bishop Gore's report "The Moral Witness of the Church on Economic Subjects"; Wallace's one of several criticisms printed). *Public Opinion* (London) 91: (639–40), 24 May 1907: no. 2383.
668. The New Vaccination Bill (LTTE). *The Clarion* (London) no. 808: 2e–f (31 May 1907); highly edited version printed as "A Simple Declaration" in *The Daily News* (London) no. 19093: 4e, 27 May 1907.
669. *Britain's Hope:* An Appreciation (notice of *Britain's Hope; An Open Letter concerning the Pressing Social Problems to the Rt. Hon. John Burns, M. P.* by Julie Sutter, 1907). *Land and Labour* 18(6): 73, June 1907.
670. LTTE (one of two printed as "The 'Double Drift' Theory of Star Motions"). *Nature* 76: 293 (293–94), 25 July 1907: no. 1969.

671. Letter (to Editor Robert Blatchford; included in a Blatchford article entitled "The Socialist Ideal"). *The Clarion* (London) no. 819: 1f (1f), 16 Aug. 1907.
672. Dr. Alfred Russel Wallace on Socialist Poets (letter to columnist A. E. Fletcher). *The Clarion* (London) no. 823: 2a, 13 Sept. 1907.
673. Dr. A. R. Wallace & Sir W. M. Ramsay's Theory; Did Man Reach his Highest testimony summarized as part of note/N entitled "Colley v. Development in the Past?" (letter). *Public Opinion* (London) 92: 336, 13 Sept. 1907: no. 2399.
674. How Life became Possible on the Earth, in *Harmsworth History of the World* ed. by Arthur Mee, J. A. Hammerton, & A. D. Innes (8 vol., Carmelite House, London, 1907–1909), Vol. 1: 91–98, 11 Oct. 1907.
675. Conversations (portions of conversations with William Allingham on the dates 2 August 1884 and 6 Nov. 1884, noted down in Allingham's diary, the second conversation including Lord Tennyson), in *William Allingham; A Diary* ed. by Helen P. Allingham & Dollie Radford (Macmillan & Co., Ltd., London), Oct. 1907: 329–30, 332–35.
676. Letter (to columnist Julia Dawson, giving advice to Socialist women of Great Britain). *The Clarion* (London) no. 838: 7c, 27 Dec. 1907.
677. Dr. Wallace Pleased with Advance in Natural History Exploration (remarks contributed to a feature article/N entitled "Leaders of Thought Tell *The World* of Achievements of 1907"; Wallace's one of many invited responses printed). *The World* (New York) no. 16931: 2a–b ec. 1907.

1908

678. Evolution and Character. *Fortnightly Rev.,* 83 (n.s.; 89, o.s.): 1–24, 1 Jan. 1908: no. 493, n.s.
679. LTTE (one of several printed as "Dr. Russel Wallace and Woman"). *The Outlook* (London) 21:89 (89–90), 18 Jan. 1908: no. 520.
680. What to Eat, Drink, and Avoid (article/N posting results of a general inquiry; Wallace's one of many responses printed). *The Review of Reviews* (London) 37: 137–38 (136–46), Feb. 1908: no. 218.
681. Dr. A. R. Wallace's Message (letter of support read at the annual meeting of the A-VL held 11 March 1908). *Vaccination Inquirer* 30: 8, 1 April 1908: no. 349.
682. Letter from Dr. A. R. Wallace (to Joseph Hyder, read at the 27th annual meeting of the LNS, 29 April 1908). *Land and Labour* 19(6): 65, June 1908.
683. Letter (to columnist Julie Dawson concerning the book that converted Wallace to Socialism, *Looking Backward*). *The Clarion* (London) no. 861: 9c–d, 5 June 1908.
684. The Remedy for Unemployment (in large part an account of *Poverty and the State* by Herbert V. Mills, 1889). I. *Socialist Rev.* 1: 310–20), June 1908 / II. 1: 390–400, July 1908.
685. Address (acceptance speech on receiving the Darwin–Wallace Medal on 1 July 1908), in *The Darwin–Wallace Celebration Held on Thursday, 1st July 1908, by the Linnean Society of London* (printed for the Linnean Society by Burlington House, Longmans, Green & Co., London), Feb. 1909: 5–11.
686. The First Paper on Natural Selection (LTTE). *The Times* (London) no. 38689: 12b, 3 July 1908.
687. Dr. A. R. Wallace and Honours (LTTE concerning an erroneous report that Wallace was soon to be knighted). *Public Opinion* (London) 94: 78a, 17 July 1908: no. 2443.
688. Is It Peace or War?; A Reply by Dr. Alfred R. Wallace (letter responding to article printed the week before). *Public Opinion* (London) 94: 202–3, 14 Aug. 1908: no. 2447.
689. The Present Position of Darwinism. *Contemporary Rev.* 94: 129–41, Aug. 1908.
690. Nationalisation, not Purchase, of Railways (LTTE). *The New Age* 3: 417–18, 19 Sept. 1908: no. 21, n.s.; 732, o.s.

691. The Facts Beat Me. *Delineator* 72: 542, Oct. 1908 / The Dead have never really Died. 72: 852–53, Nov. 1908.
692. Letter (concerning Wallace's notification of winning the Copley Medal; read at the Royal Society's annual meeting of 30 Nov. 1908). *The Times* (London) no. 38818: 9e, 1 Dec. 1908.
693. A Veteran Scientist's Testimony (note responding to an inquiry regarding Wallace's thoughts on Jesus Christ and his teachings). *The Christian Commonwealth* 29: 166d, 9 Dec. 1908: no. 1417.
694. 1909. For What Should We Strive? (article/N posting results of an opinion survey; Wallace's one of many responses printed). *The Christian Commonwealth* 29: 231a (231a–b), 30 Dec. 1908: no. 1420.
695. Darwinism versus Wallaceism. *Contemporary Rev.* 94: 716–17, Dec. 1908.
696. Preface to *Notes of a Botanist on the Amazon and Andes* by Richard Spruce (ed. and condensed by Alfred Russel Wallace), Dec. 1908, Vol. 1: v–ix.
697. Biography, in *Notes of a Botanist on the Amazon and Andes* by Richard Spruce (ed. and condensed by Alfred Russel Wallace), Dec. 1908, Vol. 1: xxi–xlvii.

1909

698. Dr. Alfred Russel Wallace at Home (interview by Ernest H. Rann). *The Pall Mall Mag.* 43: 274–84, March 1909: no. 191.
699. The World of Life: As Visualised and Interpreted by Darwinism (revision of a lecture a portion of which was delivered at the Royal Institution on 22 Jan. 1909). *Fortnightly Rev.* 85 (n.s.; 91, o.s.): 411–34, 1 March 1909: no. 507, n.s.
700. Flying Machines in War. Dr. A. R. Wallace Calls to Action (LTTE). *The Daily News* (London) no. 19626: 4f, 6 Feb. 1909.
701. Dr. A. R. Wallace and Woman Suffrage (note/N including letter from Wallace read at a meeting to support woman suffrage held 10 Feb. 1909 at Godalming). *The Times* (London) no. 38880: 10d, 11 Feb. 1909.
702. To-day's Centenaries. Charles Darwin. *The Daily Mail* (London) no. 4007: 4d, 12 Feb. 1909.
703. The Centenary of Darwin. *The Clarion* (London) no. 897: 5c–d, 12 Feb. 1909.
704. Dr. A. R. Wallace (letter of support read at the annual meeting of the A-VL held 24 Feb. 1909). *Vaccination Inquirer* 30: 202, 1 March 1909: no. 360.
705. Note to the Present Edition to *Psychic Philosophy as the Foundation of a Religion of Natural Law* (new ed.) by V. C. Desertis (pseudonym of Stanley DeBrath) (William Rider & Son, Ltd., London), Feb. 1909: vi.
706. Note on the Passages of Malthus's *Principles of Population* which Suggested the Idea of Natural Selection to Darwin and Myself, in *The Darwin–Wallace Celebration Held on Thursday, 1st July 1908, by the Linnean Society of London* (printed for the Linnean Society by Burlington House, Longmans, Green & Co., London), Feb. 1909: 111–18.
707. Aerial Fleets. Dr. Russel Wallace on Idle Panics (LTTE). *The Daily News* (London) no. 19678: 4e, 8 April 1909.
708. Letter (of support to Joseph Hyder, read at the 28th annual meeting of the LNS, 13 May 1909). *Land and Labour* 20(6): 64, June 1909.
709. The Development Fund. Letter from Dr. A. R. Wallace. *Land and Labour* 20(10): 112–13, Oct. 1909.

1910

710. Letter (of support as Honorary Vice-president read at the second annual conference of the A-VL of America, held 29 Dec. 1909 in Philadelphia). *Vaccination Inquirer* 31: 253, 1 March 1910: no. 372.
711. Dr. Wallace's High Praise (note/N including letter to Ernest McCormick regarding

latter's article "Is Vaccination a Disastrous Delusion?"). *Vaccination Inquirer 31:* 228, 1 Feb. 1910: no. 371.

712. Alfred Russel Wallace, LL.D., O.M., etc. (letter of support read at the annual meeting of the A-VL held 15 March 1910). *Vaccination Inquirer* 32: 5–6, 1 April 1910: no. 373.
713. Telegram (of support read at the 29th annual meeting of the LNS, 18 April 1910). *Land and Labour* 21(5): 51, May 1910.
714. A New Era in Public Opinion; Some Remarkable Changes in the Last Half Century. *Public Opinion* (London) 98: 377, 14 Oct. 1910: no. 2555.
715. New Thoughts on Evolution (interview by Harold Begbie). *The Daily Chronicle* (London) no. 15197: 4d, 3 Nov. 1910 / no. 15198: 4a–e, 4 Nov. 1910.
716. A Scientist on Politics (interview by Harold Begbie). *The Daily Chronicle* (London) no. 15206: 5a–c, 14 Nov. 1910.

1911

717. Scientist's 88th Birthday. Interview with Dr. Russel Wallace. Social Problems (anonymous interview). *The Daily News* (London) no. 20227: 1e–f, 9 Jan. 1911.
718. Life after Life. "Star-shine and Immortal Tears" (article/N including a letter by Wallace concerning spiritualism). *The Clarion* (London) no. 998: 5c (5ce), 20 Jan. 1911.
719. Letter (concerning the still-holding arguments posed in *Vaccination a Delusion;* presented at interview of Deputation of the A-VL with the Edinburgh Public Health Committee on 24 Jan. 1911). *Vaccination Inquirer* 32: 293, 1 March 1911: no. 384.
720. Alfred Russel Wallace, LL.D, O.M., etc. (letter of support read at the annual meeting of the A-VL held 28 March 1911). *Vaccination Inquirer* 33: 42, 1 May 1911: no. 386.
721. Letter (of support to Joseph Hyder, read at the 30th annual meeting of the LNS, 15 May 1911). *Land and Labour* 22(6): 61, June 1911.
722. Letter extracts (passages from a letter to Clement Reid concerning the relation of the present flora of the British Isles to the Glacial Period; communicated to the 4 Sept. 1911 meeting of Section K, Botany, of the BAAS), in *Report of the Eightieth Meeting of the BAAS* John Murray, London, 1912: 577–78.
723. Mr. A. R. Wallace and the Insurance Act (LTTE). *The Times* (London) no. 39781: 6c, 29 Dec. 1911.

1912

724. Dr. Russel Wallace on Insurance Act (letter). *The Daily Chronicle* (London) no. 15581: 4d, 25 Jan. 1912.
725. Naturalist Answers Birthday Greeting. A. R. Wallace Sends Letter. *Silver and Gold* (student newspaper at the University of Colorado, Boulder) 20(52): 4c, 2 Feb. 1912.
726. Dinner to Mr. Hyndman. Mr. Bernard Shaw on the Strike (note/N containing summary of letter Wallace sent on the occasion of Henry H. Hyndman's 70th birthday). *The Times* (London) no. 39841: 7c, 8 March 1912.
727. The Great Strike—and After. Hopes of a National Peace (interview by Harold Begbie). *The Daily Chronicle* (London) no. 15622: 4a–e, 13 March 1912.
728. Alfred Russel Wallace, LL.D. O.M., etc. (letter of support read at the annual meeting of the A-VL held 19 March 1912). *Vaccination Inquirer* 34: 11, 1 April 1912 no. 397.
729. Letter (to Isaac Bickerstaffe, printed in Part I of his article "Some Principles of Growth and Beauty"). *The Field, The Country Gentleman's Newspaper* 119: 946b (946–48), 11 May 1912: no. 3098.
730. Letter (of support to Joseph Hyder, read at the 31st annual meeting of the LNS, 16 May 1912). *Land and Labour* 23 (6): 63–64, June 1912.

731. Dr. Alfred Russel Wallace's Letter (letter to Bombay medical authorities concerning the accuracy of statistics Wallace presented before the Royal Commission on Vaccination). *Vaccination Inquirer* 34: 121, 1 Aug. 1912: no. 401.
732. A Policy of Defence (LTTE). *The Daily News & Leader* (London & Manchester) no. 20723: 4f, 9 Aug. 1912.
733. The Last of the Great Victorians. Special Interview with Dr. Alfred Russel Wallace (by Frederick Rockell). *The Millgate Monthly 7,* part 2: 657–63, Aug. 1912: no. 83.
734. The Problem of Life (anonymous interview regarding a paper on the origin of life given by Prof. E. A. Schafer in Dundee at the 82nd annual BAAS meeting). *The Daily News & Leader* (London & Manchester) no. 20748: 1a–b, 7 Sept. 1912.
735. Mr. Blatchford's Dogmatism (letter commenting on discussion on spiritualism between *The Clarion* editor Robert Blatchford and *The Christian Commonwealth*). *The Christian Commonwealth* 32: 815b, 11 Sept. 1912: no. 1613.
736. The Origin of Life. A Reply to Dr. Schafer. *Everyman* 1(1): 5–6, 18 Oct. 1912.

1913

737. The Spectre of Poverty (anonymous interview). *The Daily News & Leader* (London & Manchester) no. 20850: 1a–b (6 Jan. 1913).
738. Dr. A. R. Wallace's Birthday (note/N including one-line response by Wallace to letter sent by the London Spiritualist Alliance on his 90th birthday). *Light* (London) 33: 28, 18 Jan. 1913: no. 1671.
739. Letter (of support read at the annual meeting of the A-VL held 13 March 1913). *Vaccination Inquirer* 35: 9, 1 April 1913: no. 409.
740. Letter (of support to Joseph Hyder, read at the 32nd annual meeting of the LNS, 28 May 1913). *The Land Nationaliser* 24 (6): 88, June 1913.
741. A Scientist's Sleepless Hours (note/N printing a letter to columnist Solomon Eagle concerning the authorship of two poems; first referred to in Eagle's column in *New Statesman* [London] 1: 790, 27 Sept. 1913: no. 25). *Public Opinion* (London) 104: 568, 21 Nov. 1913: no. 2717.
742. Dr. A. R. Wallace's Faith (note/N containing one-sentence reply to an inquiry). *The Christian Commonwealth* 34: 120b, 12 Nov. 1913: no. 1674.
743. Dr. Wallace on the Genesis of the Soul (letter dated 5 April 1903 to an unnamed correspondent; submitted by S. H. Leonard). *Spectator* 111: 863, 22 Nov. 1913: no. 4456.
744. Preface to *The Case for Land Nationalisation* by Joseph Hyder (Simpkin, Marshall, Hamilton, Kent & Co., Ltd., London), Nov. 1913: v–viii.
745. Alfred Russel Wallace (interview by W. B. Northrop). *The Outlook* (New York) 105: 618–22, 22 Nov. 1913.
746. From Alfred Russel Wallace (single-sentence tribute to Eugene V. Debs) in *Debs and the Poets*, ed. by Ruth Le Prade (Upton Sinclair, Pasadena, California), 1920: 40.
747. *Edgar Allan Poe; A Series of Seventeen Letters concerning Poe's Scientific Erudition in Eureka and his Authorship of "Leonainie."* New York, n.d. (possibly, and no later than, 1930, its cataloguing date).

BIBLIOGRAPHY

Adams, R. L., and B. N. Phillips. 1972. "Motivation and Achievement Differences Among Children of Various Ordinal Birth Positions." *Child Development* 43: 155–64.

Axelrod, R. 1984. *The Evolution of Cooperation.* New York: Basic Books.

Axelrod, R., and W. D. Hamilton, 1981. "The Evolution of Cooperation." *Science* 211: 1390–96.

Aquinas, T. 1952. *Summa Theologica.* Great Books of the Western World. R. M. Hutchins (ed. in chief). Chicago: Encyclopaedia Britannica.

Bank, S. P., and M. D. Kahn. 1982. *The Sibling Bond.* New York: Basic Books.

Barber, B. 1961. "Resistance of Scientists to Scientific Discovery." *Science* 134: 596–602.

Barrett, P. H. (ed.) 1977. *The Collected Papers of Charles Darwin.* 2 vols. Chicago: University of Chicago Press.

Barrow, J., and F. Tipler. 1986. *The Anthropic Cosmological Principle.* Oxford: Oxford University Press.

Barzun, J. 1958. *Darwin, Marx, Wagner.* New York: Doubleday.

Bates, H. W. 1851. "Letter to Stevens." December 23 and 31, 1850, from Ega (Brazil). *Zoologist:* 3230–32.

———. 1863. *The Naturalist on the River Amazons.* London: Murray.

Beard, C. A. 1935. "That Noble Dream." In F. Stern (ed.)., *The Varieties of History,* 1973. New York: Vintage.

Becker, C. L. 1955. "What Are Historical Facts?" *The Western Political Quarterly* 3: 327–40.

Beddall, B. G. 1968. "Wallace, Darwin, and the Theory of Natural Selection: A Study in the Development of Ideas and Attitudes." *Journal of the History of Biology* 1: 261–323.

———. 1969. *Wallace and Bates in the Tropics.* London: Macmillan.

———. 1972. "Wallace, Darwin, and Edward Blyth: Further Notes on the Development of Evolutionary Theory." *Journal of the History of Biology* 5: 153–58.

———. 1988. "Darwin and Divergence: The Wallace Connection." *Journal of the History of Biology* 21, no. 1: 2–68.

Bell, T. 1859. "Presidental Address." *J. Linn. Soc. London (Zool.),* 4.

Bernstein, R. 1982. "Wallace: The Man Who Almost Pipped Darwin." *New Scientist* 94, no. 1308: 625–55.

Blalock, H. M. 1961. *Causal Inferences in Nonexperimental Research.* New York: W. W. Norton.

Blunt, W. 1971. *The Compleat Naturalist: A Life of Linnaeus.* New York: The Viking Press.

Boring, E. G., and G. Lindzey. 1950. *A History of Psychology in Autobiography.* New York: Appleton.

Bowlby, J. 1990. *Charles Darwin: A New Life.* New York: W. W. Norton.

Bowler, P. J. 1976. "Alfred Russel Wallace's Concepts of Variation." *Journal of the History of Medicine* 31: 17–29.

———. 1983. *The Eclipse of Darwinism. Anti-Darwinian Evolution Theories in the Decades around 1900.* Baltimore: Johns Hopkins University Press.

———. 1988. *The Non-Darwinian Revolution: Reinterpreting a Historical Myth.* Baltimore: Johns Hopkins University Press.

———. 1989. *Evolution: The History of an Idea.* Rev. ed. (1983). Berkeley: University of California Press.

Brackman, A. 1980. *A Delicate Arrangement: The Strange Case of Charles Darwin and Alfred Russel Wallace.* New York: Times Books.

Bronowski, J. 1977. *A Sense of the Future.* Cambridge: MIT Press.

Brooks, J. L. 1984. *Just Before the Origin: Alfred Russel Wallace's Theory of Evolution.* New York: Columbia University Press.

Brown, C. W., and E. E. Ghiselli. 1955. *Scientific Method in Psychology.* New York: McGraw-Hill.

Browne, J. 1981. "Review of Arnold C. Brackman. *A Delicate Arrangement: The Strange Case of Charles Darwin and Alfred Russel Wallace.*" In *ISIS* 72, no. 2: 262.

———. 1983. *The Secular Ark: Studies in the History of Biogeography.* New Haven: Yale University Press.

Buffon, G. L. 1749–1804. *Historie naturelle, generale et particuliere.* 44 vols. Paris: Imprimerie Royale, puis Plassan, Vol. VI, 1756.

Burkhardt, F., and S. Smith (eds.) 1985. *A Calendar of the Correspondence of Charles Darwin, 1821–1882.* New York: Garland.

———. 1984–1991. *The Correspondence of Charles Darwin.* 7 vols. Edited by Frederick Burkhardt and Sydney Smith. Cambridge: Cambridge University Press.

Burnet, T. 1691 (1965). *Sacred Theory of the Earth.* London: R. Norton (Carbondale: Southern Illinois University Press).

Byrne, D. 1974. *An Introduction to Personality: Research, Theory and Applications.* Englewood Cliffs, NJ: Prentice-Hall.

Cannon, S. 1978. *Science in Culture: The Early Victorian Period.* New York: Science History Publications.

Caroll-Sforya, L. L., P. Menozzi, and A. Piazza. 1994. *The History and Geography of Human Genes.* Princeton: Princeton University Press.

Carroll, P. T. 1976. *An Annotated Calendar of the Letters of Charles Darwin in the Library of the American Philosophical Society.* Wilmington, DE: Scholarly Resources.

Chambers, Robert. 1844 (1944). *Vestiges of the Natural History of Creation.* London (Chicago: University of Chicago Press).

Clements, H. 1983. *Alfred Russel Wallace: Biologist and Social Reformer.* London: Hutchinson & Co. Publishers, Ltd.

Cohen, I. B. 1985. *Revolution in Science.* Cambridge: Harvard University Press.

Collingwood, R. G. 1956. *The Idea of History.* New York: Oxford University Press.

Combe, G. 1835. *Essay on the Constitution of Man and Its Relation to External Objects.* Boston: Phillips and Sampson.

Cooter, R. 1984. *The Cultural Meaning of Popular Science. Phrenology and the Organization of Consent in Nineteenth-Century Britain.* Cambridge: Cambridge University Press.

Costa, P. T., and R. McRae. 1992. "Four Ways Five Factors are Basic." *Personality and Individual Differences* 13, no. 6: 653–45.

Cowan, G. A., D. Pines, and D. Meltzer. 1994. *Complexity: Metaphors, Models, and Reality.* Reading, MA: Addison-Wesley.

Crews, F. 1966. *The Sins of the Fathers.* New York: Oxford University Press.

Crouzet, F. 1982. *The Victorian Economy.* Cambridge: Cambridge University Press.

Darwin, C. 1839. *Journal of Researches into the Geology and Natural History of the Various Countries Visited by H.M.S. Beagle.* London: Henry Colburn.

———. 1859. *On the Origin of Species by Means of Natural Selection: Or the Preservation of Favoured Races in the Struggle for Life.* London: John Murray.

———. 1862. *On the Various Contrivances by Which British and Foreign Orchids are Fertilized by Insects.* London: John Murray.

———. 1868. *The Variation of Animals and Plants Under Domestication.* Vols. I & 2. London: John Murray.

———. 1871. *The Descent of Man.* Vols. I & II. London: John Murray.

———. 1892 (1958). *The Autobiography of Charles Darwin and Selected Letters.* F. Darwin (ed.). New York: Dover Publications.

———. *The Correspondence of Charles Darwin.* See Burkhardt and Smith, 1985–1991.

Darwin, C., and A. R. Wallace. 1858. "On the Tendency of Species to Form Varieties and on the Perpetuation of Varieties and Species by Natural Means of Selection." *Journal of the Proceedings of the Linnean Society (Zoology)* 3: 53–62.

Darwin, F. 1887. *The Life and Letters of Charles Darwin.* Vols. 1–3. London: John Murray.

———. 1903. *More Letters of Charles Darwin.* London: John Murray.

Dawkins, R. 1976. *The Selfish Gene.* Oxford: Oxford University Press.

de Vries, H. 1906. *Species and Varieties: Their Origin by Mutation.* 2nd ed. Chicago: Open Court Publishing Co.

Debus, A. 1978. *Man and Nature in the Renaissance.* Cambridge: Cambridge University Press.

Desmond, A. 1982. *Archetypes and Ancestors. Palaeontology in Victorian London 1850–1875.* Chicago: University of Chicago Press.

Desmond, A., and J. Moore. 1991. *Darwin. The Life of a Tormented Evolutionist.* New York: Warner Books.

Diamond, J. 1997. *Guns, Germs, and Steel: The Fates of Human Societies.* New York: W. W. Norton.

Digman, J. 1990. "Personality Structure: Emergence of the Five-Factor Model." *Annual Review of Psychology* 41: 417–40.

DiGregorio, M. A. 1984. *T. H. Huxley's Place in Natural Science.* Cambridge: Cambridge University Press.

Dobzhansky, T. 1937. *Genetics and the Origin of Species.* New York: Columbia University Press.

Dunn, J., and C. Kendrick. 1982. *Siblings: Love, Envy and Understanding.* Cambridge: Harvard University Press.

Durant, J. R. 1979. "Scientific Naturalism and Social Reform in the Thought of Alfred Russel Wallace." *The British Journal for the History of Science* 12, no. 40: 31–58.

Durkheim, E. 1960. *The Division of Labor in Society.* Glencoe, IL: Free Press.

Dyson, F. 1979. *Disturbing the Universe.* New York: Harper and Row.

Eaton, G. 1986. *Alfred Russel Wallace. 1823–1913. Biologist and Social Reformer. A Portrait of his Life and Work and a History of Neath Mechanics Institute and Museum.* Neath: W. Whittington, Ltd.

Edwards, W. H. 1847. *A Voyage Up the River Amazon, Including a Residence at Pará.* London: John Murray.

Eiseley, L. 1979. *Darwin and the Mysterious Mr. X.* London: J. M. Dent.

Eldredge, N. 1985. *Time Frames: The Rethinking of Darwinian Evolution and the Theory of Punctuated Equilibria.* New York: Simon and Schuster.

Eldredge, N., and S. J. Gould. 1972. "Punctuated Equilibria: An Alternative to Phyletic Gradualism." In T.J.M. Schopf (ed.), *Models in Paleobiology,* 82–115. San Francisco: Freeman.

Ellenberger, H. 1970. *The Discovery of the Unconscious: The History and Evolution of Dynamic Psychology.* New York: Basic Books.

Erikson, E. H. 1942. "Hitler's Imagery and German Youth." *Psychiatry* 5: 475–93.

———. 1958. *Young Man Luther: A Study in Psychoanalysis and History.* New York: Norton.

Eysenck, H. J. 1957. *Sense and Nonsense in Psychology.* New York: Penguin.

Fellows, O., and S. Milliken. 1972. *Buffon.* New York: Twayne Publishers, Inc.

Feyerabend, P. 1975. *Against Method: Outline of an Anarchistic Theory of Knowledge.* London: Verso.

———. 1978. *Science in a Free Society.* London: Verso.

Fichman, M. 1977. "Wallace: Zoogeography and the Problem of Land Bridges." *Journal of the History of Biology.* 10: 45–63.

———. 1981. *Alfred Russel Wallace.* Boston: Twayne Publishers.

Fogel, R. W., and G. R. Elton. 1983. *Which Road to the Past? Two Views of History.* New Haven: Yale University Press.

Foucault, M. 1972. *The Archeology of Knowledge and the Discourse on Language.* New York: Pantheon Books.

Frank, P. 1957. *Philosophy of Science: The Link Between Science and Philosophy.* Englewood Cliffs, NJ: Prentice-Hall.

Freud, S. 1957 (1910). *Leonardo da Vinci and a Memory of His Childhood.* In J. Strachey (ed. and trans.), *The Standard Edition of the Complete Psychological Works of Sigmund Freud* (Vol. 2). London: Hogarth Press.

Gardiner, B. G. 1995. "The Joint Essay of Darwin and Wallace." *The Linnean.* London: Linnean Society of London, 2/1: 13–24.

George, W. 1964. *Biologist Philosopher: A Study of the Life and Writings of Alfred Russel Wallace.* London: Abelard Schuman.

———. 1981. "Wallace and His Line." In T. C. Whitmore (ed.), *Wallace's Line and Plate Tectonics.* Oxford: Oxford University Press.

Ghiselin, M. T. 1969. *The Triumph of the Darwinian Method.* Berkeley: University of California Press.

Gillispie, C. C. 1960. *The Edge of Objectivity: An Essay in the History of Scientific Ideas.* Princeton: Princeton University Press.

Glick, T. F. 1988. *The Comparative Reception of Darwinism.* Chicago: University of Chicago Press.

Godfrey, L. R. 1985. *What Darwin Began. Modern Darwinian and Non-Darwinian Perspective on Evolution.* Boston: Allyn and Bacon.

Goerke, Heinz. 1973. *Linnaeus.* New York: Charles Scribner's Sons.

Goldberg, L. 1993. "The Structure of Phenotypic Personality Traits." *American Psychologist* 48, no. 1: 26–34.

Goldschmidt, R. 1940. *The Material Basis of Evolution.* New Haven: Yale University Press.

Gould, S. J. 1977. "Eternal Metaphors of Palaeontology." In A. Hallam (ed.), *Patterns of Evolution as Illustrated by the Fossil Record,* 1–26. New York: Elsevier.

———. 1980a. "Natural Selection and the Human Brain: Darwin vs. Wallace." In *The Panda's Thumb.* New York: W. W. Norton.

———. 1980b. "Is a New and General Theory of Evolution Emerging?" *Paleobiology* 6: 119–30.

———. 1982. *The Panda's Thumb.* New York: W. W. Norton.

———. 1985a. *The Flamingo's Smile: Reflection in Natural History.* New York: W. W. Norton.

———. 1985b. "The Paradox of the First Tier: An Agenda for Paleobiology. *Paleobiology* 11: 2–12.

———. 1987. *Time's Arrow, Time's Cycle.* Cambridge: Harvard University Press.

———. 1989a. *Wonderful Life: The Burgess Shale and the Nature of History.* New York: W. W. Norton.

———. 1989b. "The Horn of Triton." *Natural History* 12/89: 18–27.

———. 1989c. "An Asteroid to Die For." *Discoverer* 10, no. 10: 60–66.

Green, L. 1995. *Alfred Russel Wallace. His Life and Work.* Hertford and Ware Local Historical Society. Occasional paper No. 4.

Greene, J. C. 1959. *The Death of Adam.* Ames: Iowa State University Press.
———. 1975. "Reflections on the Progress of Darwin Studies." *Journal of the History of Biology* 8: 243–73.
———. 1982. *Science, Ideology, and World View. Essays in the History of Evolutionary Ideas.* Berkeley: University of California Press.
Gruber, H. E., and P. H. Barrett. 1981. *Darwin on Man: A Psychological Study of Scientific Creativity.* 2nd ed. Chicago: University of Chicago Press.
Guilford, J. P. 1959. *Personality.* New York: McGraw-Hill.
Gumerman, G. J., and M. Gell-Mann. 1994. *Understanding Complexity in the Prehistoric Southwest.* Reading, MA: Addison-Wesley.
Hacking, I. 1988. "Telepathy: Origins of Randomization in Experimental Design," *Isis* 79, no. 298: 427–51.
Hardison, R. 1988. *Upon the Shoulders of Giants.* 2nd ed. New York: University Press of America.
Harlan, D. 1989. "Intellectual History and the Return of Literature." *American Historical Review* 94, no. 3: 581–609.
Hawkins, J. A., and M. Gell-Mann. 1994. *The Evolution of Human Languages.* Reading, MA: Addison-Wesley.
Henderson, G. 1958. *Alfred Russel Wallace: His Role and Influence in Nineteenth Century Evolutionary Thought.* Doctoral Dissertation, University of Pennsylvania.
Henslow, G. 1871. *Genesis and Geology: A Plea for the Doctrine of Evolution.* London: Robert Hardwicke.
———. 1873. *The Theory of Evolution of Living Things and the Application of the Principles of Evolution to Religion, Considered as Illustrative of the "Wisdom and Beneficence of the Almighty."* London: Macmillan.
Herschel, J.F.W. 1830. *Preliminary Discourse on the Study of Natural Philosophy.* London: Longmans, Rees, Orme, Brown and Green.
Hill, C. 1980. *The Century of Revolution 1602–1714.* New York: W. W. Norton.
Hilton, I. 1967. "Differences in the Behavior of Mothers Toward First and Later Born Children." *Journal of Personality and Social Psychology* 7: 282–90.
Himmelfarb, G. 1959. *Darwin and the Darwinian Revolution.* Garden City, NY: Doubleday; London: Chatto and Windus.
———. 1980. "Review of *A Delicate Arrangement: The Strange Case of Charles Darwin and Alfred Russel Wallace.*" *The New York Times Book Review,* July 6.
———. 1986. *Marriage and Morals Among the Victorians.* New York: Knopf.
Hitching, F. 1982. *The Neck of the Giraffe.* New York: New American Library.
Hogben, L. T. 1918. *Alfred Russel Wallace: The Story of a Great Discoverer.* London: Society for Promoting Christian Knowledge.
Holton, G. 1988. *Thematic Origins of Scientific Thought.* Cambridge: Harvard University Press.
Hook, S. 1943. *The Hero in History: A Study in Limitation and Possibility.* Boston: Beacon Press.
Hooker, W. J. 1854. "Notices of Books—review of Wallace's *Palm Trees of the Amazon.*" *Hooker's Journal of Botany* 6: 61–62.
Hooykaas, R. 1970. "Historiography of Science, Its Aim and Methods." *Organon* 7: 37–49.
Horner, J. 1988. *Digging Dinosaurs.* New York: Harper and Row.
Houghton, W. E. 1957. *The Victorian Frame of Mind.* Chicago: University of Chicago Press.
Hull, D. L. 1973. *Darwin and His Critics. The Reception of Darwin's Theory of Evolution by the Scientific Community.* Chicago: University of Chicago Press.
———. 1984. Review of John Langdon Brooks' *Just Before the Origin* and Harry Clements's *Alfred Russel Wallace* in *Nature,* v. 308, 26 April.

Humboldt, A. von. 1818. Personal Narrative of Travels to the Equinoctial Regions of the New Continent During the Years 1799 to 1804. London: Longman, Hurst, Rees, Orme & Browne.

Hume, D. 1953. *Dialogues Concerning Natural Religion.* New York: Hafner.

Huxley, L. 1900. *Life and Letters of Thomas Henry Huxley,* 2 vols.

Huxley, T. H. 1896. *Darwiniana.* New York: Appleton.

Hyder, J. 1913. *The Case for Land Nationalisation.* London: Simpkin, Marshall, Hamilton, Kent & Col., Ltd.

Inkster, I. 1976. "The Social Context of an Educational Movement: A Revionist Approach to the English Mechanics' Institutes, 1820–1850." *Oxford Review of Education* 2: 277–307.

Irvine, W. 1955. *Apes, Angels, and Victorians: The Story of Darwin, Huxley, and Evolution.* New York: McGraw-Hill.

Jacob, M. C. 1976. *The Newtonians and the English Revolution: 1689–1720.* Ithaca, NY: Cornell University Press.

Jordan, K. 1905. "Der Gegensatz zwischen geographischer und nichtgeographischer Variation." *Z. wiss. Zool.* 83: 151–210.

Kammeyer, K. 1967. *Social Forces* 46: 71–80.

Keppel, G. 1973. *Design and Analysis: A Researcher's Handbook.* New York: Prentice-Hall.

Kerlinger, F. N., and E. J. Pedhazur. 1973. *Scientific Method in Psychology.* New York: Holt, Reinhart, Winston.

Kidwell, J. S. 1981. "Number of Siblings, Sibling Spacing, Sex, and Birth Order: Their Effects on Perceived Parent-Adolescent Relationships." *J. Marriage and Family,* May, 330–35.

Kinsey, A. C., W. B. Pomeroy, and C. E. Martin. 1948. *Sexual Behavior in the Human Male.* Philadelphia: W. B. Saunders.

Kloppenberg, J. T. 1989. "Objectivity and Historicism: A Century of American Historical Writing." *American Historical Review* 94, no. 4: 1011–30.

Knapp, S. 1999. *Footsteps in the Forest: Alfred Russel Wallace in the Amazon.* London: Natural History Museum.

Koch, H. L. 1956. "Attitudes of Young Children Toward Their Peers as Related to Certain Characteristics of Their Siblings." *Psychological Monographs* 70, no. 19.

Kohn, D. (ed.). 1985. *The Darwinian Heritage.* Princeton: Princeton University Press.

Kottler, M. J. 1974. "Alfred Russel Wallace, the Origin of Man, and Spiritualism." *ISIS* 65: 145–92.

———. 1985. "Charles Darwin and Alfred Russel Wallace: Two Decades of Debate over Natural Selection." In D. Kohn (ed.), *The Darwinian Heritage.* Princeton: Princeton University Press.

Kousser, J. M. 1980. "Quantitative Social-Scientific History." In M. Kammen (ed.), *The Past Before Us.* Ithaca, NY: Cornell University Press.

Kragh, H. 1987. *An Introduction to the Historiography of Science.* Cambridge: Cambridge University Press.

Krieger, L. 1977. *Ranke: The Meaning of History.* Chicago: University of Chicago Press.

Kuhn, T. S. 1962. *The Structure of Scientific Revolutions.* Chicago: University of Chicago Press.

———. 1977. *The Essential Tension: Selected Studies in Scientific Tradition and Change.* Chicago: University of Chicago Press.

Lakatos, I., and A. Musgrave (eds.). 1970. *Criticism and the Growth of Knowledge.* Cambridge: Cambridge University Press.

Laurent, J. 1984. "Science, Society and Politics in Late Nineteenth-Century England: A Further Look at Mechanics' Institutes." *Social Studies of Science* 14: 585–619.

Litchfield, H. (née Darwin). 1915. *Emma Darwin: A Century of Family Letters, 1792–1896.* 2 vols. London: Murray.
Loewenberg, B. J. 1959. *Darwin, Wallace and the Theory of Natural Selection: Including the Linnean Society Papers.* Cambridge, Mass.: Arlington Books.
———. 1965. "Darwin and Darwin Studies." *History of Science* 4:15–54.
Losee, J. 1987. *Philosophy of Science and Historical Enquiry.* Oxford: Clarendon Press.
Lovejoy, A. O. 1942. *The Great Chain of Being.* Cambridge: Harvard University Press.
Lyell, C. 1830–1833. *Principles of Geology, Being an Attempt to Explain the Former Changes of the Earth's Surface, by Reference to Causes Now in Operation.* 3 vols. London: John Murray.
———. 1863. *The Geological Evidences of the Antiquity of Man, with Remarks on Theories of the Origin of Species by Variation.* London: John Murray.
Lyell, K. M. (ed.). 1881. *The Life, Letters, and Journals of Sir Charles Lyell.* 2 vols. London: John Murray.
McKinney, H. L. 1966. "Alfred Russel Wallace and the Discovery of Natural Selection." *Journal of the History of Medicine and Allied Sciences* 21: 333–57.
———. 1972. "Wallace's Earliest Observations on Evolution: 28 December 1845." *ISIS* 60: 370–73.
———. 1972. *Wallace and Natural Selection.* New Haven: Yale University Press.
Malinchak, M. 1987. *Spiritualism and the Philosophy of Alfred Russel Wallace.* Drew University (Doctoral Dissertation).
Malthus, T. R. 1826. *An Essay on the Principle of Population, or, a View of its Past and Present Effects on Human Happiness; With an Inquiry into our Prospects Respecting the Future Removal or Mitigation of the Evils Which it Occasions.* 6th ed. 2 vols. London: John Murray.
Manier, E. 1980. "History, Philosophy and Sociology of Biology: A Family Romance." *Studies in the History and Philosophy of Science* 11: 1–24.
Marchant, J. 1916. *Alfred Russel Wallace, Letters and Reminiscences.* New York: Arno Press (1975).
Markus, H. 1981. "Sibling Personalities: The Luck of the Draw." *Psychology Today* 15, no. 6: 36–37.
Matthew, P. 1831. *On Naval Timber and Aboriculture.* London: Longmans.
Mayr, E. 1942. *Systematics and the Origin of Species.* New York: Columbia University Press.
———. 1954. "Change of Genetic Environment and Evolution." In J. Huxley, A. C. Hardy, and E. B. Ford (eds.), *Evolution as a Process.* London: Allen and Unwin.
———. 1957. "Species Concepts and Definitions," in *The Species Problem.* Washington DC: Amer. Assoc. Adv. Sci. Publ. no. 50.
———. 1963. *Animal Species and Evolution.* Cambridge: Harvard University Press.
———. 1970. *Populations, Species and Evolution.* Cambridge: Harvard University Press.
———. 1982. *The Growth of Biological Thought.* Cambridge: Harvard University Press.
———. 1988. *Toward a New Philosophy of Biology.* Cambridge: Harvard University Press.
Medawar, P. 1984. *Pluto's Republic: Incorporating the Art of the Soluble and Induction and Intuition in Scientific Thought.* Oxford: Oxford University Press.
Merton, R. K. 1973. *The Sociology of Science: Theoretical and Empirical Investigations.* Chicago: University of Chicago Press.
Merz, J. T. 1896–1914. *A History of European Thought in the Nineteenth Century.* 4 vols. Edinburgh: W. Blackwood and Sons.
Mill, J. S. 1843. *A System of Logic, Ratiocinative and Inductive, Being a Connected View of the principles of Evidence, and the methods of Scientific Investigation.* London: Longmans, Green.

Millhauser, M. 1959. "In the Air." In *Darwin: A Norton Critical Edition.* New York: W. W. Norton, 1970.

Milner, R. 1990. "Darwin for the Prosecution, Wallace for the Defense." *The North Country Naturalist.* Vol. 2, 19–50.

———. 1994. *Charles Darwin: Evolution of a Naturalist.* New York: Facts on File.

———. 1996. "Charles Darwin and Associates, Ghostbusters." *Scientific American,* October, 96–101.

Minkoff, E. 1983. *Evolutionary Biology.* New York: Addison-Wesley.

Mivart, St. George. 1871a. *On the Genesis of Species.* London: Macmillan.

———. 1871b. "Darwin's Descent of Man." *Quarterly Review* 131: 47–90.

———. 1873. *Man and Apes: An Exposition of Structural Resemblances and Differences Bearing Upon Questions of Affinity and Origin.* London: Macmillan.

———. 1876. *Contemporary Evolution.* London: Macmillan.

———. 1882. *Nature and Thought: An Introduction to a Natural Philosophy.* London: Kegan Paul, Trench.

———. 1889. *The Origin of Human Reason.* London: Kegan Paul, Trench.

Montagu, M.F.A. 1952. *Darwin: Competition and Cooperation.* New York: Henry Schuman.

Montaigne, M. 1952. *The Essays of Michel Eyquem de Montaigne.* Charles Cotton, Trans., Great Books of the Western World, v. 25. Chicago: University of Chicago Press.

Moore, J. 1979. *The Post-Darwinian Controversies.* London: Cambridge University Press.

Morgan, E. 1978. "From Alfon Nedd to Rio Negro. The Formative Years of Alfred Russel Wallace." *Transactions of the Neath Antiquarian Society,* 69–78.

Morton, S. G. 1839. *Crania Americana, or, a Comparative View of the Skulls of Various Aboriginal Nations of North and South America.* Philadelphia: John Pennington.

Myrdal, G. 1944. *An American Dilemma: The Negro Problem and Modern Democracy.* 2 vols. New York: Harper and Brothers.

Neumann, J. V., and Oskar Morgenstern. 1947. *Theory of Games and Economic Behavior.* Princeton: Princeton University Press.

Nickerson, R. 1998. "Confirmation Bias; A Ubiquitous Phenomenon in Many Guises." *Review of General Psychology* 2, no. 2: 175–220.

Nisbet, R. E. 1968. "Birth Order and Participation in Dangerous Sports." *Journal of Personality and Social Psychology* 8: 351–53.

Olson, R. 1971. *Science as Metaphor.* Berkeley: University of California Press.

———. 1982. *Science Deified and Science Defied: The Historical Significance of Science in Western Culture.* Vol. 1. Berkeley: University of California Press.

———. 1986. "On the Nature of God's Existence, Wisdom and Power: The Interplay Between Organic and Mechanistic Imagery in Anglican Natural Theology—1640–1740." In F. Furwick (ed.), *Approaches to Organic Form.* D. Reidel Publishing Co.

———. 1987. "The Mechanical Philosophy and Anglican Theology." In J. G. Burke (ed.), *Science and Culture in the Western Tradition.* Scottsdale, AZ: Gorsuch Scarisbrick.

———. 1991. *Science Deified and Science Defied: The Historical Significance of Science in Western Culture.* Vol. 2. Berkeley: University of California Press.

Oppenheim, J. 1985. *The Other World. Spiritualism and Psychical Research in England 1850–1914.* Cambridge: Cambridge University Press.

Osborn, H. F. 1913. "Alfred Russel Wallace, 1823–1913." *Popular Science Monthly,* 83.

———. 1929. *From the Greeks to Darwin.* New York: Charles Scribner's Sons.

Owen, R. 1813 (1963). *A New View of Society and Other Writings.* London: Everyman's Library.

Oxford English Dictionary. 2nd ed., 1989. Oxford: Clarendon Press.

Paley, W. 1802. *Natural Theology.* In *The Works of William Paley.* Edinburgh: Thomas Nelson and Peter Brown.

Palmer, R. D. 1966. "Birth Order and Identification." *Journal of Consulting Psychology* 30: 129–35.
Pantin, C.F.A. 1959a. "Alfred Russel Wallace, F.R.S. and his Essays of 1858 and 1855." *Royal Society Notes and Records* 14.1.67.
———. 1959b. "Alfred Russel Wallace." *Proceedings of the Linnean Society of London.* 170: 219–26.
———. 1960. "Alfred Russel Wallace: His Pre-Darwinian Essay of 1855." *Proceedings of the Linnean Society of London,* 171/2: 139–53.
Popper, K. 1968. *Conjectures and Refutations: The Growth of Scientific Knowledge.* Cambridge: Cambridge University Press.
Poulton, E. B. 1903. "What Is a species?" *Proc. Ent. Soc. London,* pp. lxxvi–cxvi.
———. 1923. "Alfred Russel Wallace. 1823–1913." *Proceedings of the Royal Society of London.* Series B. 95: 1–35.
Prance, G. T. 1999. "Alfred Russel Wallace." *The Linnean,* 15/1: 18–36.
Randall, J. H. 1926. *The Making of the Modern Mind.* Boston: Houghton Mifflin Co.
Reader, W. J. 1964. *Life in Victorian England.* New York: G. P. Putnam's Sons.
Richards, R. J. 1987. *Darwin and the Emergence of Evolutionary Theories of Mind and Behavior.* Chicago: University of Chicago Press.
Ritterbush, Philip. 1964. *Overtures to Biology.* New Haven: Yale University Press.
Rogers, J. 1992. "Darwin, Darwinism, and the Darwinian Culture." *Skeptic* 1/3: 86–89.
Romanes, G. J. 1889. "Mr. Wallace on Darwinism." *Contemporary Review* 56.
Romanes, E. 1896. *Life and Letters of George J. Romanes.* London: Longman, Green.
Rudwick, M. 1992. *Scenes from Deep Time: Early Pictorial Representations of the Prehistoric World.* Chicago: University of Chicago Press.
Runyan, W. M. (Ed.) 1988. *Psychology and Historical Interpretation.* Oxford: Oxford University Press.
Ruse, M. 1970. "The Darwin Industry: A Critical Evaluation." *History of Science* 7: 43–58.
Russett, C. E. 1976. *Darwin in America. The Intellectual Response 1865–1912.* San Francisco: W. H. Freeman and Co.
Sarton, G. 1960. *A History of Science: Ancient Science Through the Golden Age of Greece.* Cambridge: Harvard University Press.
Scarpelli, G. 1985. "Nothing in Nature That Is Not Useful: The Anti-Vaccination Crusade and the Idea of Harmonia Naturae in Alfred Russel Wallace." Paper presented at the University College London, History of Medicine Unit, Department of Anatomy and Embryology, November 6.
Schachter, F. F., E. Shore, S. Feldman-Rotman, R. E. Marquis, and S. Campbell. 1976. "Sibling Deidentification." *Developmental Psychology* 12: 418–27.
Schachter, F. F. 1982. "Sibling Deidentification and Split-Parent Identification: A Family Tetrad." In *Sibling Relationships: Their Nature and Significance Across the Lifespan* (eds. M. Lamb and B. Sutton-Smith). Hillsdale, NJ: Lawrence Erlbaum Associates.
Schuessler, R. 1989. "Exit Threats and Cooperation Under Anonymity." *Journal of Conflict Resolution* 33: 728–49.
Schwartz, J. 1984a. "Darwin, Wallace and the *Descent of Man." Journal of the History of Biology* 17, no. 2: 271–89.
———. 1984b. "Alfred Russel Wallace and 'Leonainie': A Hoax that Would Not Die." *Victorian Periodicals Review* 17: 3–15.
Shapin, S., and B. Barnes. 1977. "Science, Nature and Control: Interpreting Mechanics' Institutes." *Social Studies of Science* 7: 31–74.
Shapin, S. 1994. *A Social History of Truth.* Chicago: University of Chicago Press.
Sheets-Pyenson, S. 1985. "Popular Science Periodical in Paris and London: The Emergence of Low Scientific Culture, 1820–1875." *Annals of Science* 42: 549–72.

Shermer, M. B. 1988. *The Historical Matrix Model: A Theory of Historical Contingency.* Paper presented at the annual meeting of Interface 88, Atlanta.
———. 1990. "Darwin, Freud, and the Myth of the Hero in Science." *Knowledge: Creation, Diffusion, Utilization* 11, no. 3: 280–301.
———. 1991. "Science Defended, Science Defined." *Science, Technology, and Human Values* 16, no. 4: 517–39.
———. 1992. "The Mismeasure of History: Darwin, Gould, and the Nature of Change." *Skeptic* 1/3: 18–37.
———. 1993. "The Chaos of History: On a Chaotic Model That Represents the Role of Contingency and Necessity in Historical Sequences." *Nonlinear Science Today* 2, no. 4:1–13.
———. 1994. "Proving the Holocaust: The Refutation of Revisionism and the Restoration of History." *Skeptic* 2/4: 32–57.
———. 1995. "Exorcising LaPlace's Demon: Chaos and Antichaos, History and Metahistory." *History and Theory* 34, no. 1:59–83.
———. 1996. "Gould's Dangerous Idea: Contingency, Necessity, and the Nature of History." *Skeptic* 4/1: 91–95.
———. 1996b. "Rebel with a Cause: An Interview with Frank Sulloway." *Skeptic* 4/4, 68–73.
———. 1997. "The Crooked Timber of History." *Complexity* 2, no. 6: 23–29.
———. 1997b. *Why People Believe Weird Things: Pseudoscience, Superstitions, and Other Confusions of Our Time.* New York: W. H. Freeman.
———. 1998. "The Lost World of Jack Horner: An Interview with the World's Most Famous Dinosaur Digger." *Skeptic* 6/4, 72–80.
———. 2000. "The Grand Old Man of Evolution. An Interview with Evolutionary Biologist Ernst Mayr." *Skeptic* 8/1: 76–82.
———. 2001a. "Colorful Pebbles and Darwin's Dictum." *Scientific American* April, 38.
———. 2001b. In *The Borderlands of Science: Where Sense Meets Nonsense.* New York: Oxford University Press.
———. 2001c. "Contingencies and Counterfactuals: What Might Have Been and What Had to Be." *Skeptic* 8/3: 78–85.
———. In Press. "This View of History: Stephen Jay Gould as Historian of Science and Scientific Historian." *Science and Society.*
Smith, C. H. 1991. *Alfred Russel Wallace: An Anthology of His Shorter Writings.* Oxford: Oxford University Press.
———. 1999. "Alfred Russel Wallace on Evolution: A Change of Mind?" Paper presented at the Symposium on the History of Medicine and Science at the University of Southern Mississippi, Hattiesburg, February 26. Reprinted on http://www.wku.edu/~smithch/
Smith, R. 1972. "Alfred Russel Wallace: Philosophy of Nature and Man. *British Journal of the History of Science* 6: 177–99.
Somit, A., and S. A. Peterson (eds.). 1992. *The Dynamics of Evolution. The Punctuated Equilibrium Debate in the Natural and Social Science.* Ithaca: Cornell University Press.
Spencer, H. 1842. "The Proper Sphere of Government." In *Nonconformist,* June 15.
———. 1862. *First Principles.* London: Williams & Norgate.
———. 1893. *The Principles of Ethics.* 2 vols. Indianapolis: Liberty Classics.
———. 1904. *Autobiography.* 2 vols. New York: D. Appleton.
Spruce, R. 1855. Letter to Sir William Hooker. Archives, Royal Botanic Gardens, Kew.
Stannard, D. E. 1980. *Shrinking History: On Freud and the Failure of Psychohistory.* New York: Oxford University Press.
Stannard, J. 1978. "Natural History." In D. Lindberg (ed.), *Science in the Middle Ages.* Chicago: University of Chicago Press.
Steneck, N. 1976. *Science and Creation in the Middle Ages.* Notre Dame: University of Notre Dame Press.

Stern, F. (ed.). 1973. *The Varieties of History: From Voltaire to the Present.* New York: Vintage Books.

Stevens, M. D., and G. W. Roderick. 1972. "British Artisan Scientific and Technical Education." *Annals of Science* 29.

Stocking, G. W. 1987. *Victorian Anthropology.* New York: Free Press.

Stone, L. 1981. *The Past and the Present.* Boston: Routledge & Kegan Paul.

Stresemann, E. 1919. "Uber die europaischen Barmlaufer." *Verh. Orn. Ges. Bayern* 14: 39–74.

Sulloway, F. 1979. "Geographic Isolation in Darwin's Thinking: the Vicissitudes of a Crucial Idea." *Studies in the History of Biology* 3: 23–65.

———. 1982a. "Darwin and His Finches: The Evolution of a Legend." *Journal of the History of Biology* 15: 1–53.

———. 1982b. "Darwin's Conversion: The Beagle Voyage and its Aftermath." *Journal of the History of Biology* 15: 325–96.

———. 1987. "The Metaphor and the Rock: A Review of *Time's Arrow, Time's Cycle: Myth and Metaphor in the Discovery of Geological Time* by Stephen Jay Gould." *New York Review of Books,* May 2, 37–40.

———. 1990. "Orthodoxy and Innovation in Science: The Influence of Birth Order in a Multivariate Context." Preprint courtesy of author.

———. 1996. *Born to Rebel: Birth Order, Family Dynamics, and Creative Lives.* New York: Pantheon.

Sutton-Smith, B., and B. G. Rosenberg. 1970. *The Sibling.* New York: Holt, Rinehart and Winston.

Thomas, K. 1971. *Religion and the Decline of Magic.* New York: Charles Scribner's Sons.

Thurston, L. L. 1947. *Multiple Factor Analysis.* Chicago: University of Chicago Press.

Tielhard de Chardin, P. 1955. *The Phenomenon of Man.* London: Collins.

Tilly, C. 1978. *From Mobilization to Revolution.* Reading, MA: Addison-Wesley.

Tillyard, E.M.W. 1944. *The Elizabethan World Picture.* New York: Macmillan.

Trivers, R. L. 1971. "The Evolution of Reciprocal Altruism." *Quarterly Review of Biology* 46: 35–57.

Tuan, Y. 1963. "Latitude and Alfred Russel Wallace." *Journal of Geography* 62: 258–61.

Turner, F. M. 1974. *Between Science and Religion: The Reaction to Scientific Naturalism in Late Victorian England.* New Haven: Yale University Press.

Turner, J. S., and D. B. Helms. 1987. *Lifespan Development.* New York: Holt, Rinehart and Winston.

Voltaire. 1759 (1985). *Candide.* In B. R. Redman (ed.), *The Portable Voltaire.* New York: Penguin.

Vorzimmer, P. 1970. *The Years of Controversy:* The Origin of Species *and Its Critics.* Philadelphia: Temple University Press.

Weaver, J. H. 1987. *The World of Physics, Vol. II: The Einstein Universe and the Bohr Atom.* New York: Simon and Schuster.

Whewell, W. 1837. *History of the Inductive Sciences.* London: Parker.

———. 1840. *The Philosophy of the Inductive Sciences.* London: J. W. Parker.

Wiggins, J. S., K. E. Renner, G. L. Clore, and R. J. Rose. 1976. *Principles of Personality.* Reading, MA: Addison-Wesley.

Williams-Ellis, A. 1966. *Darwin's Moon: A Biography of Alfred Russel Wallace.* London: Blackie.

Wilson, J. G. 2000. *The Forgotten Naturalist: In Search of Alfred Russel Wallace.* Kew, Victoria (Australia): Arcadia (Australian Scholarly Publishing Pty Ltd.).

Wilson, L. G. (ed.) 1970. *Sir Charles Lyell's Scientific Journals on the Species Question.* New Haven: Yale University Press.

Winter, A. 1998. *Mesmerized: Powers of Mind in Victorian Britain.* Chicago: University of Chicago Press.

Woolgar, S. 1988. *Science: The Very Idea.* London: Tavistock Publishers.
Wrobel, A. 1987. *Pseudoscience and Society in 19th-Century America.* Lexington: University of Kentucky Press.
Young, R. M. 1970. *Mind, Brain and Adaptation in the Nineteenth Century: Cerebral Localization and its Biological Context from Gall to Ferrier.* Oxford: Clarendon Press.
———. 1985. *Darwin's Metaphor: Nature's Place in Victorian Culture.* Cambridge: Cambridge University Press.

INDEX

* Note – Page numbers in *italics* refer to photographs and illustrations.